Practice of Bayesian Probability Theory in Geotechnical Engineering

Wan-Huan Zhou · Zhen-Yu Yin ·
Ka-Veng Yuen

Practice of Bayesian Probability Theory in Geotechnical Engineering

同濟大學出版社
TONGJI UNIVERSITY PRESS

Springer

Wan-Huan Zhou
State Key Laboratory of Internet of Things for Smart City, Department of Civil and Environmental Engineering
University of Macau
Macau, China

Zhen-Yu Yin
Department of Civil and Environmental Engineering
The Hong Kong Polytechnic University
Hong Kong, China

Ka-Veng Yuen
State Key Laboratory of Internet of Things for Smart City, Department of Civil and Environmental Engineering
University of Macau
Macau, China

ISBN 978-981-15-9104-4 ISBN 978-981-15-9105-1 (eBook)
https://doi.org/10.1007/978-981-15-9105-1

Jointly published with Tongji University Press
The print edition is not for sale in China (Mainland). Customers from China (Mainland) please order the print book from: Tongji University Press.

This Springer imprint is published by the registered company Springer Nature Singapore Pte Ltd.
The registered company address is: 152 Beach Road, #21-01/04 Gateway East, Singapore 189721, Singapore

Dedicated to our families, teachers and students.

Foreword

The development of numerical simulations in geomechanics has promoted the construction of advanced constitutive models for soils. However, the limitations of available methods for model class selection and parameter identification make it difficult to efficiently apply these advanced models in practice. The selection of constitutive model affects simulation results, leading to different engineering decisions, which raises the problem of safety, economy, and risk in geotechnical engineering. Therefore, the influence of the choice of constitutive model on decision-making and its related consequences should be fully considered. Currently, the existing studies focus on how to develop more accurate constitutive modeling approaches, but there is still no systematic research on the evaluation of their applicability to engineering problems. The selection of a given constitutive model is often determined based on user preference and past experience, both of which tend to be subjective. Neglect in the selection of constitutive models has become one of the most important sources of risk for geotechnical engineering constructions. Therefore, constitutive model class selection and parameter identification methods based on optimization and probabilistic theory have important theoretical and practical significance for accurately predicting soil deformation, controlling engineering cost and preventing engineering disasters. In this regard, a clear identification of the model parameters has become central to the successful practice of soil mechanics and to the development of artificial intelligence in geotechnics.

I have known Dr. Wan-Huan Zhou since the years of her Ph.D. study at the Hong Kong Polytechnic University. Her research project focused on soil nail reinforcement mechanisms for landslide prevention. Her supervisor was Prof. Jianhua Yin, who is well known for his expertise in constitutive modeling and advanced physical experiments in geotechnical engineering. As one of her Ph.D. examiners, I was impressed by her enthusiasm in research and knowledge in the physical and numerical modeling of soil nail pullout resistance. After receiving her Ph.D. degree, Dr. Zhou joined the University of Macau as a young scholar in 2009. Since then, she has collaborated with Prof. Kelvin Yuen on the research of Bayesian methods in geotechnical engineering.

I came to know Dr. Zhen-Yu Yin during the Winter of 2011. At that time, he was at HKUST as a Visiting Scholar from Shanghai Jiao Tong University. We began academic discussions and collaboration from then on. His rigorous research approach, deep understanding of soil behavior and solid analytical skills have left a deep impression on me. He has led a dynamic research team and has published many well-cited papers. He also had 5 years of experience as an engineer before returning to academia, which prompted him to develop engineering application methods.

I have known Prof. Kelvin Yuen for more than 20 years since he was a Master's student at HKUST. Although his major research is on the development of Bayesian system identification for structural systems, many of his methods have been applied to the geotechnical engineering field, including those introduced in this book. I am sure that this book will raise a lot of attention because uncertainty quantification is very important in geotechnical engineering.

This book consists of a summary and a refinement of results obtained by the authors on the topic of model class selection and parameter identification. It gathers the authors' original results in optimization algorithms, model class selection methods, parameter identification processes, constitutive modeling approaches, and engineering applications. The development of the optimization theory in geotechnics is clearly presented based upon rigorous analyses. The illustrative examples allow the reader to quickly understand the concept of optimization-based parameter identification and its application to various practical cases. The authors have deliberately provided relevant source codes for optimization algorithms and constitutive models so that the reader can make use of this work for his or her own applications.

In summary, this book provides an informative exposition of the current advances in the formulation of numerical simulations in geomechanics and will benefit both scientists and practitioners. Students will also find in this book a stimulating introduction to the mechanics of geomaterials.

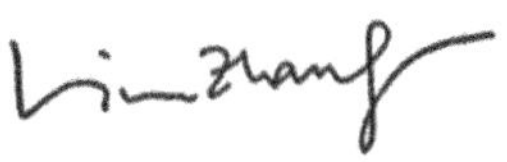

Limin Zhang
Ph.D., FASCE, FHKIE
Chair Professor of Geotechnical Engineering
Department of Civil and Environmental Engineering
The Hong Kong University of Science and Technology
Clear Water Bay, Kowloon, Hong Kong

Preface

The soil as the research object of soil mechanics has very complex property, and how to correctly understand or describe the soil physical or mechanical behaviors under different cases is crucial to the development of the soil mechanics. In geotechnical engineering, the laboratory and field tests are important ways to study and determine a certain soil physical property or parameters. Some soil properties can be measured directly, e.g., water content, density, specific gravity, etc. But some soil properties can only be estimated through their relationships with other soil properties, such as void ratio, porosity, and degree of saturation. There are still some cases in which the soil properties must be estimated through the complex or empirical formulae. In these situations, some unknown parameters in the model are required to be determined to describe the soil property. Back analysis is usually conducted based on the relevant observed responses, e.g., the determination of the compression index and the swelling index involved in the deformation predictions of one-dimensional compression problem. Therefore, it is worthwhile to evaluate the performance of models on the estimation of the real soil behavior.

It is well known that the soil has the complicated variability and uncertainty, and its state is affected by many factors, e.g., stress level and history, rainfall conditions, and physical and chemical changes. Therefore, the soil properties change in space and time even for the homogeneous soil, and this is considered to be the inherent variability of soil. The quantification of inherent variability is necessary for the appropriate evaluation of soil properties. To quantify the uncertainties of soil properties has great engineering significance. Moreover, due to the uncertainty, complexity, and time-consuming in the measurement of the soil properties, indirect methods are developed for predicting soil properties. Based on a large amount of experiments and engineering practices, the researchers have proposed the empirical or numerical formulae to estimate certain soil properties or mechanics index with appropriate assumptions, e.g., numerous empirical formulae for predicting SWCC and various soil constitutive models for describing the time-dependent behavior of soft soil or critical state behavior of granular soils, etc. These models usually contain unknown parameters, which are generally determined using the regression analysis based on test data. Many of empirical models may not reflect the mechanism of soil properties due to the complex mechanism of soil behavior, and the unknown parameters in the model may have no physical meanings.

Notably, most of empirical models for describing soil properties are obtained from data fitting, and this will introduce the uncertainty to the model performance. Even for the theoretical models, the uncertainty is also inevitable due to the idealizations and simplifications in the theories. Three primary sources of uncertainty, i.e., inherent soil variability, measurement error, and transformation uncertainty, contribute to the uncertainty in the model predictions. These components should be combined consistently by using a second-moment probabilistic approach. Besides the evaluation of model uncertainty, the efficiency of models for reflecting the actual soil properties is another valuable research issue. The investigation on the model efficiency can provide engineers with a straightforward suggestion for the engineering application of models.

Hence, research works in this book have been conducted to investigate different geotechnical problems with the use of Bayesian probabilistic approach. In the study of soil-water retention property, the effect of initial void ratio on the SWCC of same textured soils was investigated and new models were introduced in Chaps. 2 and 3. The involved predictive model was determined by using the Bayesian linear model class selection. Chapters 3 and 4 are dedicated to evaluating the model uncertainty for the creep behavior of soft soil and the effect of load duration while Chaps. 6 and 7 focus on shear behavior of granular soils. The Bayesian probabilistic method was adopted to determine all model parameters and evaluate the uncertainty of model parameters and model predictions. In Chap. 8, Bayesian linear model class selection is employed again for the estimation of pullout shear strength of grouted soil nails. Chapter 9 is dedicated to the study of the unconfined compression strength of natural fiber reinforced soil. The Bayesian nonparametric general regression method was introduced to select the most significant features from a set of experimental data. In Chap. 10, we introduce an efficient model updating method under the Bayesian framework for the deformation prediction of braced excavations.

This book begins with the most basic theoretical knowledge and combines geotechnical experiments to realize how Bayesian probabilistic method can be applied in practice, which can help readers quickly and accurately understand the constitutive simulation of soil characteristics and methods to evaluate the associated uncertainty in the predictions. In addition, this book provides source codes and a free software platform from the authors' work that can be directly used by readers for analysis and training. We also hope to enrich the developed application methods in the future by sharing with readers the latest scientific research results in the geotechnical field and promoting the application of these methods to achieve the purpose of geotechnical engineering innovation.

Macao, Hong Kong
May 2020

Wan-Huan Zhou

Zhen-Yu Yin

Ka-Veng Yuen

Acknowledgments

In the process of writing this book, we received much sincere guidance and help from experts and colleagues. Here, I would like to express my sincere gratitude to Prof. Pierre-Yves Hicher from Ecole Centrale de Nantes in France, Prof. Jian-Hua Yin from the Hong Kong Polytechnic University, Prof. Limin Zhang from Hong Kong University of Science and Technology, Prof. Shuilong Shen from Shantou University, and Prof. Hongwei Huang from Tongji University for their careful guidance and help during the book-writing process. And also we thank Dr. Zhao Linshuang, Dr. Tan Fang, Dr. Qi Xiaohui, and Ms. Tian Chen currently or formerly from the University of Macau; Dr. Yinfu Jin, Dr. Jie Yang, Dr. Pei Wang, and Dr. Hanlin Wang from the Hong Kong Polytechnic University for their participation in the research projects and hard work in the process of organizing and reviewing the book. We are also grateful to Tongji University Press and Springer for performing detailed work on the publication of this book.

Some of the results and publications in this book have been funded by the Science and Technology Development Fund, Macau SAR (File no. 0193/2017/A3, 0125/2014/A3, and SKL-IOTSC-2018-2020) and the University of Macau (MYRG2018-00173-FST). We would like to express our heartfelt thanks to National Natural Science Foundation of China (52022001, 51508585, 41372285, 51579179) and the Research Grants Council (RGC) of Hong Kong Special Administrative Region Government (HKSARG) of China (Grant No.: 15209119, 15217220, R5037-18F) for their funding.

About This Book

This book systematically introduces the application of Bayesian probabilistic approach in soil mechanics and geotechnical engineering. Four typical problems are analyzed using Bayesian probabilistic approach, i.e., to model the effect of initial void ratio on the soil-water characteristic curve (SWCC) of unsaturated soil, to select the optimal model for the prediction of the creep behavior of soft soil under one-dimensional straining, and to identify model parameters of soils and to select constitutive model of soils considering critical state concept. This book selects the simple and easy-to-understand Bayesian probabilistic algorithm, so that readers can master the Bayesian method to analyze and solve the problem in a short time. In addition, this book provides MATLAB codes for various algorithms and source codes for constitutive models so that readers can directly analyze and practice.

This book can be used as a postgraduate textbook for civil engineering, hydraulic engineering, transportation, railway, engineering geology, and other majors in colleges and universities, and as an elective course for senior undergraduates. It can also be used as a reference for relevant professional scientific researchers and engineers.

Contents

About the Authors

Wan-Huan Zhou is currently Head and Associate Professor of the Department of Civil and Environmental Engineering and Interim Director of Centre for Regional Oceans at the University of Macau. She received her B.Eng. degree in 2002 and M.Sc. degree in 2005 from Zhejiang University in China and Ph.D. in Geotechnical Engineering from The Hong Kong Polytechnic University in 2008. After graduation, she worked as a Lecturer at The Hong Kong Polytechnic University and later joined the University of Macau in September 2009. She has acted as Principal Investigator on 15 research projects, including grants from the National Natural Science Foundation (NSFC) and the Macao Science and Technology Development Fund (FDCT). She has published more than 40 SCI articles as the first or correspondent author in geotechnical journals, such as Journal of Geotechnical and Geoenvironmental Engineering, ASCE, and Canadian Geotechnical Journal. She won the 2011 best paper award of the International Association for Computer Methods and Advances in Geomechanics. In 2020, she was awarded the NSFC (National Natural Science Foundation of China)—the Excellent Young Scientist Fund (Hong Kong and Macau).

Zhen-Yu Yin is Associate Professor of Geotechnical Engineering at The Hong Kong Polytechnic University since 2018. He received his B.Sc. in Civil Engineering from Zhejiang University in 1997, followed by a 5-year engineering consultancy at the Zhejiang Jiahua Architecture Design Institute. Then, he obtained his M.Sc. and Ph.D. in Geotechnical Engineering at Ecole Centrale de Nantes (France) in 2003 and 2006, respectively. Then he has been working as Postdoctoral Researcher at Helsinki University of Technology (Finland), University of Strathclyde (Glasgow, UK), Ecole Centrale de Nantes, and University of Massachusetts (Umass-Amherst, USA). In 2010, he jointed Shanghai Jiao Tong University as Special Researcher and received "Professor of Exceptional Rank of Shanghai Dong-Fang Scholar". In 2013, he jointed Ecole Centrale de Nantes as Associate Professor before moving to Hong Kong. He has published over 170 articles in peer-reviewed international journals with an H index of 36 according to

Web of Science. Since 2012 he is an Associate Committee member of the granular materials committee of American Society of Civil Engineers.

Ka-Veng Yuen is the Dean of Graduate School and Distinguished Professor of Civil and Environmental Engineering (CEE) at the University of Macau. He received his Ph.D. in civil engineering from the California Institute of Technology (Caltech) after a study period of 2 years and 7 months. His major awards include Young Investigator Award, International Chinese Association on Computational Mechanics; QIAN Weichang programme scholar, K.C. Wong Education Foundation; Phi-Tau-Phi Honorary member; and the Housner's Fellowship. His paper "Model selection using response measurements: Bayesian probabilistic approach" (coauthored with J. L. Beck) is one of the top 10 most cited papers among all papers published in the Journal of Engineering Mechanics (established since 1983) of the American Society of Civil Engineers. Furthermore, his single-author book "Bayesian methods for structural dynamics and civil engineering" (published by the top publisher John Wiley & Sons in 2010) has attracted great attention of Bayesian methods for structural dynamics. It has become a classical reading in this research field. His H index is 33 according to Web of Science.

Abbreviations

a	Constant of fines content effect in silty sand (SIMSAND+fr)
a	Target inclination of yield surface related to volumetric strain (ASCM)
A_{d}	Constant of magnitude of stress-dilatancy (0.5 ~ 1.5)
b	Constant controlling the amount of grain breakage (SIMSAND+Br)
b	Target inclination of yield surface related to deviatoric plastic strain
C_{aei}	Intrinsic secondary compression index (remolded clay)
$\mathbf{D}$	Stiffness matrix of material
E	Young's modulus
e, e_0	Void ratio and initial void ratio
E_0	Referential Young's modulus (dimensionless)
e_{c0}	Initial critical state void ratio (SIMSAND); virgin initial critical state void ratio before breakage
e_{cuf}	Fractal initial critical state void ratio due to breakage
e_{d}	General shear strain
E_{h}, E_{v}	Horizontal and vertical Young's modulus
$e_{\mathrm{hc,c0}}$	Initial critical state void ratio of pure fine soils ($f_{\mathrm{c}} = 0\%$)
$e_{\mathrm{hf,c0}}$	Initial critical state void ratio of pure coarse soils ($f_{\mathrm{c}} = 100\%$)
$e_{\max}$	Maximum void ratio
E_{u}	Undrained Young's modulus
f_{th}	Threshold fines content from coarse to fine grain skeleton (20 ~ 35%)
f	Fines content
G	Shear modulus
G_0	Referencial shear modulus
G_{vh}	Shear modulus

I_1, I_2, I_3	The first, second, and third invariants of the stress tensor
$I_1^{'}$, $I_2^{'}$, $I_3^{'}$	The first, second, and third invariants of the strain tensor
J_1, J_2, J_3	The first, second, and third invariants of the deviatoric stress tensor
$J_1^{'}$, $J_2^{'}$, $J_3^{'}$	The first, second, and third invariants of the deviatoric strain tensor
K	Bulk modulus
K_0	The coefficient of earth pressure at rest
k_p	Plastic modulus-related constant in SIMSAND; plastic modulus-related parameter in ASCM
K_w	Bulk modulus of water
M	Constraint modulus in elasticity; slope of critical state line in p'-q plane
m	Constant of fines content effect in sandy silt
M_c	Slope of critical state line in triaxial compression in p'-q plane
n	Porosity of soil; elastic constant controlling nonlinear stiffness
n_d	Phase transformation angle-related constant (≈ 1)
n_p	Peak friction angle-related constant (≈ 1)
p	Mean effective stress
p_{at}	Atmosphere pressure
p_{b0}	Initial bonding adhesive stress
p_{c0}	Initial size of yield surface; initial size of yield surface of grain breakage (SIMSAND+Br)
p_{excess}	Excess pore pressure
p_{steady}	Steady pore pressure
q	Deviatoric stress
R_d	Ratio of mean diameter of sand to silt D_{50}/d_{50}
R_α	Stress relaxation coefficient
s_{ij}	Deviatoric stress tensor
u_x, u_y, u_z	Displacements
α_{k0}	Initial inclination of yield surface
β	Rate-dependency coefficient
χ^0	Initial bonding ratio
δ_{ij}	Kronecker symbol
ε_1, ε_2, ε_3	Principle strains
ε_a, ε_r	Axial strain and radial strain
ε_{ij}	Strain tensor
ε_m	Mean strain
ε_v	Volumetric strain
γ_{xy}, γ_{yx}, γ_{yz}, γ_{zy}, γ_{zx}, γ_{xz}	Engineering shear strains

φ	Friction angle
κ	Swelling index of the isotropic compression test (in *e*-ln*p'* plane)
κ_i	Intrinsic swelling index (of remolded soil, in *e*-ln*p'* plane)
λ	Lame constant in elasticity; compression index (in *e*-ln*p'* plane); constant controlling the nonlinearity of CSL in SIMSAND
λ'	Compression index under the plane of log*e*-log*p'*
λ_i	Intrinsic compression index (of remolded soil, in *e*-ln*p'* plane)
ν_u	Undrained Poisson's ratio
ν_{vh}'	Horizontal Poisson's ratio
ν_{vv}'	Vertical Poisson's ratio
θ	Lode angle
ρ	Constant controlling the movement of CSL
σ_a, σ_r	Axial stress and radial stress
σ_{ij}	Stress tensor
$\sigma_m(p)$	Mean stress
σ_n, σ_h	Vertical and horizontal stresses
σ_{p0}	Preconsolidation pressure
σ_w	Pore water pressure
σ_x, σ_y, σ_z	Normal stresses
σ_1, σ_2, σ_3	First, second, and third principle stresses
τ	Reference time (Oedometer test τ = 24h) (ANICREEP)
τ_{xy}, τ_{yx}, τ_{yz}, τ_{zy}, τ_{zx}, τ_{xz}	Shear stresses
υ	Poisson's ratio
ω	Absolute rotation rate of the yield surface
ω_d	Rotation rate of the yield surface related to the deviatoric plastic strain
ξ	Constant controlling the nonlinearity of CSL (SIMSAND); absolute rate of bond degradation
ξ_b	Degradation rate of the inter-particle cohesive bonding
ξ_d	Constant controlling the deviatoric strain-related bond degradation rate
ψ	Dilatancy angle

Chapter 1
Problem of Uncertainties in Geotechnical Engineering

1.1 Uncertainties in Geotechnical Engineering

Different types of uncertainties are involved in the geotechnical engineering, for example, uncertainty of concept, uncertainty of material classification, soil parameters, constitutive or empirical models, boundary conditions, other uncertainty caused by measurement error, and so on. All these uncertainties affect the evaluation of the soil or structure performance, and they are required to be considered and analyzed for reliable evaluation results. According to the type of uncertainty, different analysis methods have been utilized to evaluate the uncertainty, for example, stochastic theory, fuzzy mathematics, theory of gray systems, artificial neural networks, genetic algorithms, and so on. In some cases, these methods are coupled to solve the uncertain problems. Among these methods, the probabilistic method in stochastic theory has been widely applied to many aspects of geotechnical engineering, for example, slope analysis [1–4], liquefaction analysis [5, 6], reliability analysis of reinforced embankment [7], seepage analysis [8], soil–water characteristic curve [9–11], reducing uncertainties in empirical correlations [12], and so on. In the following sections, the sources of uncertainty and their analysis methods for predicting the performance of soil or structure are reviewed.

1.1.1 Uncertainties in Soil Properties

Soils are porous and discontinuous media and they are formed by the weathering of rocks. Under the natural conditions, the state of soil is affected by various factors, for example, stress level and stress history, water seepage, physical and chemical changes. Even in the homogeneous soil, the soil property also varies with the site, and it possesses the spatial variability. Due to these characteristics of soil, there are inevitable uncertainties involved in the determined soil properties. The uncertainties in measured soil properties are associated with the soil sample disturbance, test

W. Zhou et al., *Practice of Bayesian Probability Theory in Geotechnical Engineering*,
https://doi.org/10.1007/978-981-15-9105-1_1

conditions, human factors and statistical uncertainties. The uncertainties of estimated soil properties by models are caused by the limited test data of soil samples, the imperfection of adopted estimation formula and the uncertainty of other relevant soil properties.

Tang [13] focused on the uncertainty characterization of soil properties in a homogeneous soil layer, and stated three main sources of uncertainties: inherent spatial variability, systematic uncertainties from the tests and estimation errors from insufficient samples. He also proposed a simplified method to incorporate these major sources of uncertainties in determining the statistic of the spatial average soil properties.

Phoon and Kulhawy [14, 15] also summarized three primary sources for the geotechnical uncertainty: inherent variability, measurement error and transformation uncertainty, and pointed out that the relative impact of these three sources to the uncertainty in the designed soil property lies on the site conditions, degree of equipment and procedural control, and precision of the correlation models. They mentioned that it is necessary to evaluate these three primary uncertainties separately and to develop a systematic method that can combine these uncertainties consistently. In their studies, a second-moment probabilistic approach was adopted to evaluate the uncertainties in the design soil properties by combining the appropriate uncertainties. The variability of designed soil properties can be evaluated by the corresponding coefficient of variation (COV). There have been many studies on the common soil properties, that is, effective stress friction angle, in situ horizontal stress coefficient, undrained shear strength and Young's modulus.

Christian et al. [16] presented how probabilistic descriptions of soil properties can be achieved from laboratory and field data and utilized in the stability analysis. The uncertainties of soil properties for the reliability analysis of slope stability were considered to have two major sources: data scatter and systematic error. The former contains inherent spatial variability and random test error. The latter contains systematic statistical error caused by sampling and bias in the measurement processes. They also analyzed the relative contributions of the uncertainties of different parameters to the reliability of embankment using the first-order second-moment approach.

Many other researchers have also identified and classified major components of uncertainties associated with the estimated soil properties by different approaches. Gui et al. [17] studied the effect of stochastic hydraulic conductivity on the slope stability of an embankment dam by conducting the combination analysis of random field simulation, seepage analysis and slope analysis. Baecher and Christian [18] discussed the nature and philosophy of uncertainty in geological and geotechnical engineering and addressed fundamentals and limits of probabilistic and statistical methods. For the calculation of hydrodynamic pressures caused by rainfall, Zhang et al. [19] discussed the errors from model representation and from variability of model parameters separately. They found that different design assumptions and models can obtain significant different factors of safety of the slope. Zhang et al. [20] conducted numerous field pullout tests to evaluate four factors influencing the pullout resistance of soil nails.

1.1.2 Uncertainties in Geotechnical Model

Besides the uncertainties in estimated soil properties, the influence of two types of model uncertainty on the performance of soil or structure should also be noted, which are associated with the test of geotechnical physical model and performance model itself. The physical model tests, for example, centrifuge tests and full-scale field tests, are usually conducted to study the performance of soil or structure. The performance factors such as the pore water pressures and displacements can be measured directly, which are influenced by the inherent variability of soil properties. Similar to the soil property tests, the test results of physical model also suffer from the systematic and statistical uncertainties. The measurements may be influenced by model scaling, preparation of soil sample, instrumentation technology, interpretation of test results, and so on [21].

In addition, the uncertainties of model input variables, which are measured or estimated for soil parameters at a specific site, cause the variability of the model outputs. The systematic errors associated with the performance model itself, for example, uncertainties in constitutive behavior, imperfect model expression, initial condition, boundary condition [19, 22], can also contribute to the uncertainties of model outputs. In order to evaluate the performance of soil or underground structure, the model uncertainties must be considered in the analysis.

1.1.3 Back Analysis Methods

In geotechnical engineering, many complicated mathematical models have been proposed to study various properties. For instance, with respect to the stress–strain relationship of soil, suitable constitutive models can describe the nonlinear, irreversible and time-dependent behavior of soil. In order to obtain good predictions for the deformation behavior of soil, the parameters contained in the models often play very important role. Among the model parameters, some parameters are related to the soil properties, and they can be determined through the survey methods in the laboratory or field site. It has been noted that the reliability and applicability of measured soil parameters are significantly influenced by the test conditions which have large difference with the engineering practice. In addition, the soil parameters may change with time and the project process, such as construction technology and construction parameters. If the influences of these factors on the soil parameters are not considered, the design results or simulations with these soil parameters may have large errors. For the other model parameters, they may have no physical meanings and cannot be directly determined through the measurements. Fortunately, the stress, displacement and strain in the engineering practice can be measured, and then it became possible to determine the soil properties or model parameters using back analysis method based on the measured data.

Since the late 1970s, back analysis has been paid much attention and also obtained fast development. At first, the information, such as soil physical properties, constitutive models, boundary conditions and measured data, have been processed as deterministic quantities, that is, deterministic methods for back analysis. This method is easy to be implemented without considering many influencing factors, and its results basically meet for engineering applications. However, the soil engineering practice is an uncertainty system, and there is uncertainty in many aspects, such as the soil physical properties, the constitutive models and boundary conditions. The measured properties, for example, displacement and stress, are all uncertain variables containing the measurement error. Therefore, the deterministic analysis is not realistic to study the problems in geotechnical engineering.

At present, several methods for probabilistic back analysis have been conducted in geotechnical engineering, for example, maximum likelihood method [23–25], and Bayesian method [25, 28]. The Bayesian method is considered to be more effective for updating the soil parameters and response predictions [22, 26–28], and it can also evaluate the uncertainty of soil parameters by the obtained posterior probability distributions. Bayesian updating method has been successfully used to analyze different geotechnical problems. Kay [29] evaluated the safety factor for single pile in sand using this method, and Honjo et al. [24] studied the embankment on soft clay. Yan et al. [30] analyzed the correlation between the soil compression index and their influencing factors and proposed the predictive formulae for the compression index. Chiu et al. [10, 11] presented a Bayesian approach for estimating the water retention curve. Cao and Wang [31] and Wang et al. [32] developed Bayesian approaches to do underground soil stratum identification, soil classification and soil properties based on the cone penetration tests. Wang et al. [33] conducted the identification for the soil strata in the London Clay Formation using the water content data. Cao and Wang [34] presented a Bayesian model comparison approach to obtain the most plausible correlation function for a specific site based on the project-specific test results and previous site information. There are also many studies on the multistage excavation analysis by utilizing the Bayesian approach [35–38].

1.2 Bayesian Probabilistic Approach

Given a specific model and the observations, the above-mentioned back analysis methods can be utilized to evaluate the influences of the associated parameter uncertainties on the model predictions. However, different constitutive models or empirical models, which have different mathematical forms or complexity, are always proposed to solve the same issue in geotechnical engineering. It is unknown how these models perform in the engineering practice and which one has better ability to fit with the observations and provide better predictions in a specific project. The Bayesian approach as a useful tool can be adopted to investigate the predictive performance of models. Bayesian approach can not only update the model parameters and characterize the uncertainties using their posterior PDFs but also find the most plausible model for predicting the responses by Bayesian model class selection. In this section, the Bayesian approach for parametric identification and model class selection is briefly outlined. One may refer to the details from Yuen [39].

1.2.1 Parametric Identification

Consider a linear or nonlinear relationship (model class C) to describe the quantity of concern y:

$$y = y(\mathbf{x}; \boldsymbol{\theta}_{\mathrm{m}}, C) \tag{1.1}$$

where $\boldsymbol{\theta}_{\mathrm{m}}$ is the uncertain parameters contained in the model C; $\mathbf{x}$ is the measured variable vector. Use Y to denote the measurement of y and it is assumed that the difference can be adequately modeled as Gaussian random variable:

$$Y = y + \varepsilon \tag{1.2}$$

where ε is a Gaussian random variable with zero mean and variance σ_ε^2 referred hereafter as prediction-error variance, and it is adopted to represent the measurement noise and modeling error. The uncertain parameter vector $\theta = \left[\theta_{\mathrm{m}}^T, \sigma_\varepsilon^2\right]^T$ includes the model parameter vector $\boldsymbol{\theta}_{\mathrm{m}}$ and the prediction-error variance σ_ε^2. Their uncertainties can be represented by using the posterior PDFs, and the probability expression of posterior PDF given the data D is written as:

$$p(\boldsymbol{\theta}|D, C) = c_0 p(\boldsymbol{\theta}|C) p(D|\boldsymbol{\theta}, C) \tag{1.3}$$

where $c_0 = 1/p(D|C)$ is a normalizing constant; $p(D|C)$ is the evidence of model class C; $p(\boldsymbol{\theta}|C)$ is the prior PDF of the uncertain parameters in $\boldsymbol{\theta}$, which is based on the previous knowledge or user's judgment; $p(D|\boldsymbol{\theta}, C)$ is the likelihood function expressing the level of data fitting. If the prediction errors in different measured data are statistically independent, the likelihood function can be computed by:

$$p(D|\boldsymbol{\theta}, C) = (2\pi\sigma_\varepsilon^2)^{-\frac{N}{2}} \exp[-\frac{N}{2\sigma_\varepsilon^2} J_g(\boldsymbol{\theta}_{\mathrm{m}}; D, C)] \tag{1.4}$$

where N is the number of measured data; $J_g(\boldsymbol{\theta}_{\mathrm{m}}; D, C)$ is the goodness-of-fit function, and is given by:

$$J_g(\boldsymbol{\theta}_{\mathrm{m}}; D, C) = \frac{1}{N} \sum_{n=1}^{N} [Y(n) - y(n)]^2 \tag{1.5}$$

The most probable model parameter vector $\boldsymbol{\theta}_{\mathrm{m}}^*$ can be computed by maximizing the posterior PDF in Eq. (1.3). If N is large enough or adopt the improper prior, this will be equivalent to minimizing the goodness-of-fit function $J_g(\boldsymbol{\theta}_{\mathrm{m}}; D, C)$:

$$\boldsymbol{\theta}_{\mathrm{m}}^* = \arg \min_{\boldsymbol{\theta}_{\mathrm{m}}} J_g(\boldsymbol{\theta}_{\mathrm{m}}; D, C) \tag{1.6}$$

If $J_g(\boldsymbol{\theta}_{\mathrm{m}}; D, C)$ is implicit, numerical optimization is required to achieve the optimal model parameters [39].

The most probable value of the prediction-error variance σ_ε^2 can also be obtained by maximizing the posterior PDF, and its closed form solution is available:

$$\sigma_\varepsilon^{2*} = \min_{\boldsymbol{\theta}_{\mathrm{m}}} J_g(\boldsymbol{\theta}_{\mathrm{m}}; D, C) = J_g(\boldsymbol{\theta}_{\mathrm{m}}^*; D, C) \tag{1.7}$$

According to Beck and Katafygiotis [40], the posterior PDF $p(\theta|D, C)$ given a large number of data D is approximately Gaussian distribution, and its mean value is the optimal parameter vector $\boldsymbol{\theta}_{\mathrm{m}}^*$ and its covariance matrix can be obtained by computing the inverse of the Hessian matrix of the objective function $J(\boldsymbol{\theta}) = -\ln[p(\boldsymbol{\theta}|C)p(D|\boldsymbol{\theta}, C)]$ at $\boldsymbol{\theta}_{\mathrm{m}}^*$:

$$\mathbf{H}^{(l,l')}\left(\boldsymbol{\theta}^*\right) = -\frac{\partial^2}{\partial\theta_l\theta_{l'}} \ln[p(\boldsymbol{\theta}|C)p(D|\boldsymbol{\theta}, C)]\Big|_{\boldsymbol{\theta}=\boldsymbol{\theta}^*} \tag{1.8}$$

The posterior PDF in Eq. (1.3) provides the probabilistic description for the model parameters, but its topology may be very complicated in general. Especially for the case with large number of uncertain parameters, the distribution is difficult to be visualized. However, the numerical simulation based on the approximation method is found to be effective for solving this difficulty, and three types of methods will be presented in Sect. 1.2.3.

1.2.2 Model Class Selection

The probability/plausibility of a model class conditional on a database D can be obtained by using the Bayes' theorem as follows [41]:

$$P\left(C_j|D\right) = \frac{p\left(D|C_j\right) \cdot P\left(C_j\right)}{p(D)}, j = 1, 2, \ldots, N_c \tag{1.9}$$

where N_C is the number of model classes to be considered; $p(D) = \sum_{j=1}^{N_c} p(D|C_j)P(C_j)$ is a normalizing constant and $P(C_j)$ expresses the user's judgment on the initial plausibility of the model classes with $\sum_{j=1}^{N_c} P(C_j) = 1$. In general, uniform prior plausibility can be assumed, that is, $P(C_j) = 1/N_c$. The evidence of the model class C_j $p(D|C_j)$ can be obtained using the theorem of total probability:

$$p(D|C_j) = \int_{\Theta} p(D|\boldsymbol{\theta}, C_j)p(\boldsymbol{\theta}|C_j)d\boldsymbol{\theta},\ j = 1, 2, \ldots, N_c \tag{1.10}$$

where the parameter vector $\boldsymbol{\theta}$ belongs to the parameter space $\boldsymbol{\Theta}_j \subset \Re^{N_j}$ and it is defined in each model class (C_j). In general, Eq. (1.10) involves a high-dimensional integral and it is computationally expensive for direct computation. In the globally identifiable cases, the posterior PDF given large volume of data can be approximated by a Gaussian distribution, so the evidence $p(D|C_j)$ can be approximated by using Laplace's method for asymptotic expansion [42]:

$$p(D|C_j) \approx p(D|\boldsymbol{\theta}^*, C_j)p(\boldsymbol{\theta}^*|C_j)(2\pi)^{\frac{N_j}{2}}\left|\mathbf{H}_j(\boldsymbol{\theta}^*)\right|^{-\frac{1}{2}},\ j = 1, 2, \ldots, N_c \tag{1.11}$$

where $\boldsymbol{N}_j$ is the number of uncertain parameters for model class $\boldsymbol{C}_j$, $\boldsymbol{\theta}^*$ is the optimal parameter vector that maximizes $p(\boldsymbol{\theta}|D, C_j)$. The maximum likelihood value $p(D|\boldsymbol{\theta}^*, C_j)$ is larger for the model class that can better fit the data. Thus, the model class with more uncertain parameters is more likely to have greater maximum likelihood value. The remaining terms in Eq. (1.11) are called Ockham factor [43]:

$$O_j = p(\boldsymbol{\theta}^*|C_j)(2\pi)^{\frac{N_j}{2}}\left|\mathbf{H}_j(\boldsymbol{\theta}^*)\right|^{-\frac{1}{2}} \tag{1.12}$$

The Ockham factor provides a measurement for the robustness of the model class and its value decreases exponentially with the number of uncertain parameters in the model class.

On the other hand, the exact expression for $p(D|C_j)$ can be obtained based on the Bayes' theorem:

$$p(D|C_j) = \frac{p(D|\boldsymbol{\theta}^*, C_j)p(\boldsymbol{\theta}^*|C_j)}{p(\boldsymbol{\theta}^*|D, C_j)} \tag{1.13}$$

where the ratio $p(\boldsymbol{\theta}^*|C_j)/p(\boldsymbol{\theta}^*|D, C_j)$ is the Ockham factor.

Given the same prior plausibility for all model class candidates, the model class with the largest value in Eq. (1.11) is regarded to have a good tradeoff between the data-fitting capability and the robustness to model error. Once again, the updated or posterior PDF may have complex topology. In this case, the asymptotic approximation is not applicable and numerical simulation is necessary for the evaluation of the evidence integral.

1.2.3 Numerical Simulation for the Representation of the Updated PDF

The asymptotic approximation is accurate to solve the case when the updated PDF of the model parameters is very peak, for example, the amount of data is large enough. However, the stochastic simulation is more useful to handle the general cases, and it can achieve the goal of drawing samples from the updated PDF $p(\boldsymbol{\theta}|D, C)$. At present, different types of algorithms for Markov chain Monte Carlo (MCMC) simulation have been developed to implement this goal.

1.2.3.1 Metropolis–Hastings Algorithm

The Metropolis algorithm [44] is the first sampling method for iterative simulations, and it can be conducted through simple steps called the proposal-and-decision steps. Starting with the initial point, the algorithm proceeds by transferring the current state to the new state generated from a symmetric proposal distribution with a probability value. This algorithm was later generalized by Hastings [45] and the asymmetric proposal distribution is allowed to be used in generating the new state. This generalized algorithm is called the Metropolis–Hastings (MH) algorithm. In the MH method, the simulated Markov chain samples tend to follow the target PDF as the sample number approaches to infinite. With respect to the problem involved in Bayesian approach, the posterior PDF in Eq. (1.3) is the target PDF, and it is expressed as the product of a constant and a function, that is, $p(\boldsymbol{\theta}|D, C) = c_0 q(\boldsymbol{\theta})$.

(1) Select an appropriate proposal PDF $T(\xi|\boldsymbol{\theta})$, which is a PDF for ξ that depends on $\boldsymbol{\theta}$;
(2) At stage $k = 1$, choose an arbitrary point to be the initial sample $\boldsymbol{\theta}_1$;
(3) At the stage k (k starts from 2), a new point $\boldsymbol{\theta}_c$ randomly generates from the proposal distribution $T(\boldsymbol{\theta}_c|\boldsymbol{\theta}_{k-1})$, and compute the ratio of densities r:

$$r = \frac{q(\boldsymbol{\theta}_c)T(\boldsymbol{\theta}_{k-1}|\boldsymbol{\theta}_c)}{q(\boldsymbol{\theta}_{k-1})T(\boldsymbol{\theta}_c|\boldsymbol{\theta}_{k-1})} \tag{1.14}$$

(4) Generate a random number u from a uniform distribution U (0, 1);
(5) Determine whether $\boldsymbol{\theta}_c$ is acceptable with the following acceptance rule: if $u \le r$, $\boldsymbol{\theta}_c$ is accepted and set $\boldsymbol{\theta}_k = \boldsymbol{\theta}_c$; otherwise, set $\boldsymbol{\theta}_k = \boldsymbol{\theta}_{k-1}$. Then go back to step (3);

(6) Repeat steps (3)–(5) until the target number of samples (N) is reached, and a group of Markov chain samples $[\boldsymbol{\theta}_k : k = 1, 2, \ldots, N]$ can be obtained. Under the assumption of ergodicity, the Markov chain will converge to the stationary state as the sample number reaches to infinite, that is, the sample tends to $p(\boldsymbol{\theta}|D, C)$. Before the Markov chain reaches the stationary state, the generated samples as burn-in samples should be discarded. However, when the sample size is limited, the ergodicity of Markov chain is still an issue.

In the MH algorithm, the proposal PDF plays an important role, and its determination has significant influence on the efficiency of algorithm. In particular, it is not easy to conduct MH algorithm efficiently for these two cases: (a) If the scale of proposal PDF is too large, most of the candidate samples will be rejected and the accepted ratio of sample will be very low. The Markov chain will contain many repeated samples, and the correlation among samples will be large. (b) If its scale is too small, the accepted ratio will be very high and the Markov chain needs longer time to run over important regions of the posterior PDF. The Markov chain samples will be highly dependent. In order to enhance the efficiency of MH algorithm, the guidelines of the proposal PDF, which control the accepted ratio in a certain range (20–40%), were suggested [46]. However, the MH algorithm still may not be efficient in many cases, for example, the posterior PDF is very peaked, high-dimensional or multimodal, and so on [47]. Thus, it is not feasible to simulate samples according to the updated PDF directly using the MH algorithm.

1.2.3.2 Adaptive MH Algorithm

The adaptive MH (AMH) algorithm was proposed by Beck and Au [22] to settle the difficulty in sampling according to the posterior PDF directly. The algorithm introduces a sequence of intermediate PDFs that converge to the posterior PDF for sample generation, and finally achieves the samples from the posterior PDF.

Let $\left\{p^{(1)}, p^{(2)}, \ldots, p^{(s_0)}\right\}$ denotes a sequence of PDFs, and $p^{(s)}$ is chosen as updated PDF from Bayes' theorem based on an increasing amount of data:

$$p^{(s)} = p\left(\boldsymbol{\theta}|D^{(s)}, C\right) \tag{1.15}$$

where $D^{(1)} \subset D^{(2)} \subset \ldots \subset D^{(s_0)} = D$. If the updated PDF with data D has the following form:

$$p(\boldsymbol{\theta}|D, C) = c \exp\left(-\frac{J(\boldsymbol{\theta})}{2\varphi^2}\right) \tag{1.16}$$

where $J(\boldsymbol{\theta})$ is the goodness-of-fit function, and φ is a measure of the size of the prediction error. Therefore, the sequence $\left\{p^{(s)}\right\}$ can be constructed as:

$$p^{(s)} = c^{(s)} \exp\left(-\frac{J(\boldsymbol{\theta})}{2\left(\varphi^{(s)}\right)^2}\right) \tag{1.17}$$

where $\left(\varphi^{(s)}\right)^2 = 2^{s_0 - s}\varphi^2$ with $2^{s_0} \approx \varphi^{-2}$ if the length scale of the prior PDF is of order one. The following content shows the main steps of AMH algorithm:

(1) Starting with the prior PDF as the proposal PDF and $p^{(1)}$ as the target PDF, simulate the samples $\left\{\boldsymbol{\theta}_1^{(1)}, \boldsymbol{\theta}_2^{(1)}, \ldots, \boldsymbol{\theta}_N^{(1)}\right\}$ by using the MH algorithm;
(2) At the sth level, simulate the samples using the procedure similar to level 1 except that the proposal PDF in MH algorithm is constructed by the kernel density. The kernel density or proposal PDF at the sth level can be constructed by using the samples generated at the $(s-1)$th level:

$$p^{(s-1)}(\boldsymbol{\theta}) = \sum\nolimits_{n-1}^{N} w_n g\left(\boldsymbol{\theta}; \boldsymbol{\theta}_n^{(s-1)}, \sum\nolimits_n\right) \tag{1.18}$$

where $g\left(\boldsymbol{\theta}; \boldsymbol{\theta}_n^{(s-1)}, \Sigma_n\right)$ is the multidimensional Gaussian PDF evaluated at $\boldsymbol{\theta}$ with mean $\boldsymbol{\theta}_n^{(s-1)}$ and covariance matrix Σ_n; w_n is the weight with $\sum\limits_{n=1}^{N} w_n = 1$.

(3) Repeat step (2) until the s_0th simulation level is finished, and obtain the samples from the target updated PDF $p(\boldsymbol{\theta}|D, C) = p^{(s_0)}$.

With the help of AMH algorithm, the samples can be simulated from the posterior PDF that is very peaked or multimodal. However, this algorithm is still inefficient for the high-dimensional problems due to the inefficiency of kernel density estimations for high-dimensional problems. Based on experience, problem may arise when the number of uncertainty parameters is more than 4.

1.2.3.3 Transitional Markov Chain Monte Carlo Method

The transitional Markov chain Monte Carlo (TMCMC) method was proposed by Ching and Chen [47] on the basis of AMH algorithm. The TMCMC algorithm uses the same idea of "intermediate PDFs" as proposed by Beck and Au [22], but it does not require kernel density estimation. An additional advantage is that the intermediate PDFs can be automatically selected. The sequence of intermediate PDFs will ultimately converge to the posterior PDF and is defined by:

$$f_s(\boldsymbol{\theta}) \propto f(\boldsymbol{\theta}|C) \cdot f(D|\boldsymbol{\theta}, C)^{p_s}\left(s = 0, 1, \ldots, N_s; 0 = p_0 < p_1 < \cdots < p_{N_s} = 1\right) \tag{1.19}$$

where the subscript s denotes the stage number and N_S is the total number. The PDF of the initial stage with $p_0 = 0$ is proportional to the prior PDF, while the PDF of the final stage with $p_{N_s} = 1$ is proportional to the posterior PDF. The used intermediate PDFs can enable a good means of transitioning and obtaining the samples between the adjacent intermediate PDFs. Besides, the TMCMC algorithm can also estimate the evidence without solving any integral problem. The said algorithm can be enacted by following these steps:

(i) At the initial stage ($s = 0$), draw the samples $\{\boldsymbol{\theta}_{0,k}: k = 1, 2, \ldots, N_0\}$ according to the prior PDF $f_0(\boldsymbol{\theta})$;

(ii) Choose p_{s+1} such that the COV of $\left\{f(D|\boldsymbol{\theta}_{s,k}, C)^{p_{s+1}-p_s} : k = 1, 2, \ldots, N_s\right\}$ is equal to a prescribed threshold. Then, compute the plausibility weights $w\left(\boldsymbol{\theta}_{s,k}\right) = f(D|\boldsymbol{\theta}_{s,k}, C)^{p_{s+1}-p_s}$ for $k = 1, 2, \ldots, N_s$ and $S_s = \sum_{k=1}^{N_s} w\left(\boldsymbol{\theta}_{s,k}\right)/N_s$. Here, $\boldsymbol{\theta}_{s,k}$ denotes the kth sample that belongs to level s and the factor will be used to compute the evidence $f(D|C)$;

(iii) Apply the MH algorithm to generate the samples $\left\{\boldsymbol{\theta}_{s+1,k} : k = 1, 2, \ldots, N_{s+1}\right\}$ from $f_{s+1}(\boldsymbol{\theta})$. The candidate sample $\boldsymbol{\theta}^c$ of the sth sample in the Markov chain is generated from $N\left(\boldsymbol{\theta}_{s,l}, \Sigma_s\right)$, in which $\boldsymbol{\theta}_{s,l}$ is one of the current samples $\left\{\boldsymbol{\theta}_{s,l} : l = 1, 2, \ldots, N_s\right\}$. The lth initial sample $\boldsymbol{\theta}_{s,l}$ is chosen with probability $w\left(\boldsymbol{\theta}_{s,l}\right)/\sum_{l=1}^{N_s} w\left(\boldsymbol{\theta}_{s,l}\right)$. Σ_s is the covariance matrix of the Gaussian proposal PDF centered at the current sample $\left\{\boldsymbol{\theta}_{s,l} : l = 1, 2, \ldots, N_s\right\}$ and can be estimated using the following equation:

$$\sum_s = \beta^2 \sum_{k=1}^{N_s} w(\boldsymbol{\theta}_{s,k}) \left\{\boldsymbol{\theta}_{s,k} - \left[\sum_{l=1}^{N_s} w(\boldsymbol{\theta}_{s,l})\boldsymbol{\theta}_{s,l} \Big/ \sum_{l=1}^{N_s} w(\boldsymbol{\theta}_{s,l})\right]\right\} \times \left\{\boldsymbol{\theta}_{s,k} - \left[\sum_{l=1}^{N_s} w(\boldsymbol{\theta}_{s,l})\boldsymbol{\theta}_{s,l} \Big/ \sum_{l=1}^{N_s} w(\boldsymbol{\theta}_{s,l})\right]\right\}^T \tag{1.20}$$

where β is the prescribed scaling factor. Set $\boldsymbol{\theta}_{s+l,\,k} = \boldsymbol{\theta}^c$ and $\boldsymbol{\theta}_{s,l} = \boldsymbol{\theta}^c$ with probability $f_{s+1}(\boldsymbol{\theta}^c)/f_{s+1}(\boldsymbol{\theta}_{s,l})$; otherwise, set $\boldsymbol{\theta}_{s+l,\,k} = \boldsymbol{\theta}_{s,l}$.

(iv) Repeat steps (ii) and (iii) until level N_S has been reached, that is, $p_{N_S} = 1$. At this stage, samples $\left\{\boldsymbol{\theta}_{N_S,k} : k = 1, 2, \ldots, N_{N_S}\right\}$ are asymptotically distributed as $f(\boldsymbol{\theta}|D, C)$, and $S = \prod_{s=0}^{N_S-1} S_s$ is an asymptotically unbiased estimation of the evidence $f(D|C)$.

This procedure will be repeated for all model classes $\left\{C^{(m)} : m = 1, \ldots, N_C\right\}$ to obtain their evidence $S^{(m)}$. Thus, the model class selection can be done by performing the following equation:

$$P(C^{(m)}|D) = \frac{f(D|C^{(m)}) \cdot P(C^{(m)})}{\sum_{m=1}^{N_C} f(D|C^{(m)}) \cdot P(C^{(m)})} \approx \frac{S^{(m)} \cdot P(C^{(m)})}{\sum_{m=1}^{N_C} S^{(m)} \cdot P(C^{(m)})} \quad (1.21)$$

where N_C is the total number of model class candidates and $P(C^{(m)})$ is the prior PDF of the model class candidate $C^{(m)}$ with $\sum_{m=1}^{N_C} P(C^{(m)}) = 1$.

The TMCMC method possesses all advantages of AMH algorithm or other conventional methods, and it is more efficient to simulate samples for the high-dimensional problems. Moreover, the TMCMC method can compute the evidence for model class selection.

1.3 Soil Water Retention Property of Unsaturated Soil

1.3.1 Soil Suction

According to the level of the water contained in the soils, the soils can be divided into two broad categories: saturated soils and unsaturated soils. The unsaturated soil is commonly defined as having three phases: solid, water and air. However, the air–water interface or contractile skin must be considered as an independent phase when considering the stress state of unsaturated soil [48]. Since the 1930s, the geotechnical scholars began to study the issue of unsaturated soils. The suction is an important factor for studying the properties of unsaturated soils. It exhibits the intensity of interaction between soil–water and soil particles and the curvature state of the air–water interface in unsaturated soil [48].

The total suction consists of two components: matric suction (matric or capillary component of free energy) and osmotic suction (osmotic or solute component of free energy). Aitchison [49] gave the total suction as follows:

$$\phi = (u_a - u_w) + \pi \quad (1.22)$$

where ϕ denotes total suction; $(u_a - u_w)$ denotes matric suction, in which u_a and u_w represent pore-air and pore-water pressure, respectively; and π denotes the osmotic suction.

As shown in Fig. 1.1, water rises in a capillary tube immersed in water, similar to the pores of the soil with small radius acting as capillary tubes that cause the soil water to rise above the water table. The pressure of capillary water is negative relative to the air pressure, which is generally atmospheric (i.e., $u_a = 0$). Therefore, the capillary pressure, that is, matric suction, can be expressed as follows:

$$u_a - u_w = \frac{2T_s \cos\alpha}{r} \quad (1.23)$$

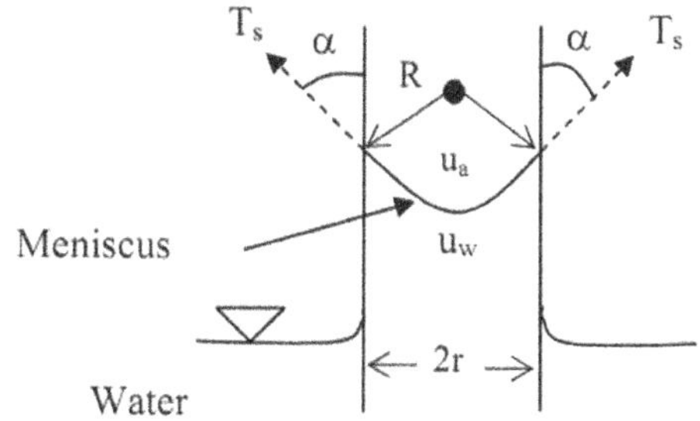

Fig. 1.1 Illustration of capillary tube immersed in water

where T_s is the surface tension of water; α is the contact angle; r is the radius of the capillary tube, that is, the radius of the pore in the soil. It can be seen from Eq. (1.23) that the matric suction increases as the pore radius decreases.

Some researchers have studied the effect of suction on the shear strength, permeability and compressibility of soils. Most experimental evidences show the shear strength increases nonlinearly with soil suction, and the brittleness and dilatancy of an unsaturated soil also increase [50–53]. Fredlund and Rahardjo [54] stated that the permeability of an unsaturated soil depends on degree of saturation, and it can be estimated using the relationship between degree of saturation and suction. Alonso et al. [55] pointed that the compressibility of an unsaturated soil generally increases with the decrease of suction. Wheeler and Sivakumar [56] and Chiu and Ng [57] found that the compressibility under unsaturated conditions is larger than that under saturated conditions through the experiments. Estabragh et al. [58] found that the compressibility of a compacted silty soil does not change monotonically with suction, and it reaches maximum at certain suction. On the other hand, the osmotic suction is insensitive to the changes of soil–water content [54].

1.3.2 Soil–Water Characteristic Curve (SWCC)

The soil–water characteristic curve (SWCC) describes a relationship between the suction and the water content in the soil. The suction refers to the matric suction, and the water content has three forms: gravimetric water content, volumetric water content and degree of saturation. The SWCC has been demonstrated to be very useful for the estimation of various unsaturated soil property functions [59, 60], such as shear strength function, permeability function and thermal property function. Given a reliable SWCC, the estimation of unsaturated soil property can avoid the cumbersome laboratory work for measuring the unsaturated soil properties. Fredlund et al. [48] presented a literature review on the early developments of the SWCC in soil physics.

A SWCC starts from saturation state at zero suction and ends at completely dry state approximately at 1,000,000 kPa suction. The entire SWCC can be divided into three stages, which are the boundary effect stage, the transition stage (i.e., primary and secondary) and the residual stage [61]. Figure 1.2 shows the typical SWCC curve with three stages, and describes its characteristics including the hysteresis between the desorption (or drying) curve and adsorption (or wetting) curve, air-entry value (AEV), residual water content and residual air content. Several factors contribute to the SWCC hysteresis, such as (a) contact angle variation during the wetting and drying process, (b) nonuniformity of pore size in soil (ink bottle effect), (c) entrapped air in soil and (d) thixotropic regain or aging due to dry and wetting history. It should be noted that the AEV is the matric suction at which air first enters the largest pores of a soil during the drying process [62].

Figure 1.3 presents three typical SWCCs for different types of soils, and it can be seen that the SWCCs gradually move up and to the right as the plasticity of soil increases. This is because that the water in the soil with more fine particles is harder

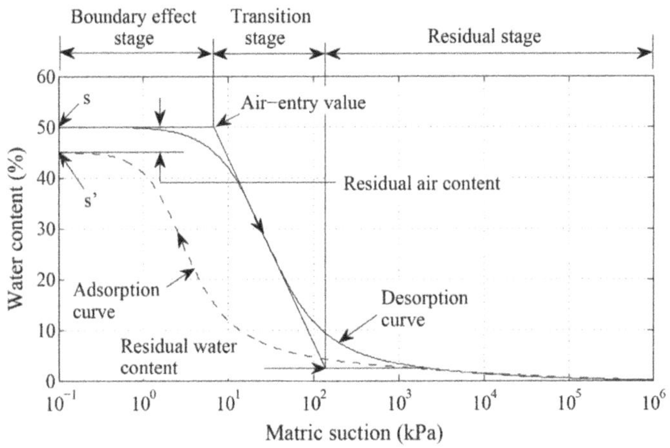

Fig. 1.2 Typical soil-water characteristic curve (SWCC) for a silty soil (Fredlund and Xing [63])

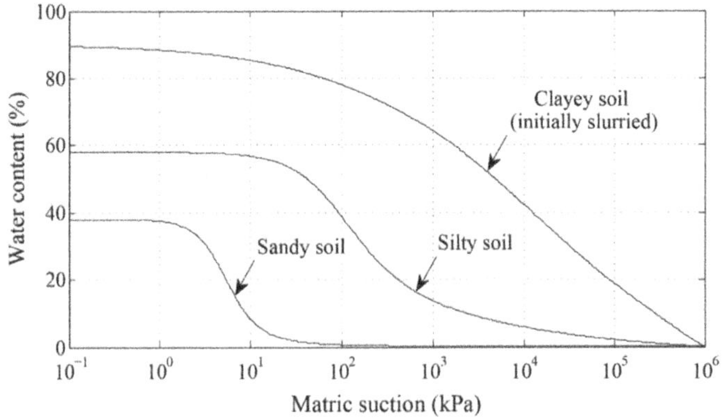

Fig. 1.3 Soil-water characteristic curves for a sandy, silty, and clayey soil (Fredlund and Xing [63])

to flow out. However, the SWCC is not only influenced by the soil texture but also by many other factors.

1.3.3 Influencing Factors on SWCC

Several experimental techniques or apparatuses have been developed to measure the SWCC, for example, filter paper method [64], different types of pressure plate apparatuses [54, 65–68] and modified direct shear apparatus [69]. These methods or apparatuses have their own features, and they may be applicable to different cases, for example, measurable range of suction, drying or wetting path.

Based on a large number of experimental results, researchers have found that a number of factors have significant influence on SWCC, such as the type of soil and mineralogy, grain-size distribution, initial void ratio, plastic limit, liquid limit, compaction energy, stress state and stress history, initial water content, clay fraction and temperature. Marinho and Chandler [70] indicated that there was a distinct relationship between the slope of SWCC and the liquid limit. Vanapalli et al. [71] studied the effect of soil structure and stress history on the SWCC based on the tests and found that the beginning of the SWCC was affected greatly. Kawai et al. [72] investigated the dependency of SWCC on initial void ratio. Uchaipichat and Khalili [73] conducted experimental work on the thermo-hydro mechanical behavior of an unsaturated silt. Iyer et al. [74] demonstrated that the initial water content only influenced the initial stage of the SWCC, and the specimen thickness had no significant influence on the shape of SWCC. Wang et al. [33] studied the effects of sample dimensions and shapes on the testing duration of SWCC. Elkady et al. [75] found that the SWCC of sand-natural expansive clay mixtures was mainly dependent on the clay content and compaction state.

Among these influencing factors, the effect of void ratio or soil compaction degree is the most significant in geotechnical engineering. Due to the limitations in measuring SWCC, for example, long-term cycle, high cost and discrete data points, the calculation model, which can simulate the effect of void ratio on SWCC, should be established to avoid the tedious experimental work. Considering that the shape and location of SWCC in the space depend on the parameters in SWCC model, some researchers have attempted to construct the relations of model parameters with the void ratio [76, 77]. There are some researchers trying to introduce the void ratio based on the existing empirical SWCC models [78–80].

1.3.4 Estimation Methods for SWCC

As mentioned above, the influencing factors of SWCC have been studied by laboratory data, but it is still difficult to determine the model for SWCC by theoretical analysis using the soil physical properties. The most commonly used models are usually obtained based on the accumulation of numerous experiments by the empirical analysis, and they are required to have good capacity to fit laboratory data. The four common methods for predicting SWCC are summarized below.

1.3.4.1 Empirical Equations

Numerous empirical equations have been proposed to fit the laboratory data, and they contain two or three fitting parameters. Table 1.1 lists several commonly used equations for SWCC. Leong and Rahardjo [81] reviewed and evaluated five popular SWCC equations and found that the Fredlund and Xing [82] equation possesses the best fitting ability. These models can provide the good fitting results with the measured data using the least-squares regression method [82]. They can deduce the whole SWCC using the limited test data, and the estimated results are suitable to be used in the computer model. Zhang and Chen [83] later extended the Fredlund and Xing [63] model and the van Genuchten [84] model to describe bimodal and multimodal SWCCs.

From the view of these empirical equations, some researchers have also investigated the relationships between the associated fitting parameters and the soil basis properties and established the estimation equations. Zapata [86] analyzed the three fitting parameters (a, k, m) in Fredlund and Xing [63] equation using the grain size diameter D_{60} and the product of the percent passing and the plasticity index for different soil types, that is, granular nonplastic materials and plastic materials, respectively. Chin et al. [87] based on regression analysis and one-point SWCC measurement proposed a simplified technique to estimate the SWCC, and the relationships of four fitting parameters (a, k, m, φ_r) in Fredlund and Xing equation [63] with basic soil properties were formulated for both coarse- and fine-grained soils, respectively. Torres [88] also obtained the correlation of the fitting parameter a with the grain-size diameter D_{10} for granular materials, and the other two fitting parameters k and m were found to be related to a and the relevant relation equations were also built.

Table 1.1 Equations for soil-water characteristic curve

References	Equation	Unknown coefficient
Gardner [85]	$\theta_w = \theta_r + \frac{\theta_s - \theta_r}{1 + a\varphi^b}$, where θ_s is saturated volumetric water content	θ_r, a, b
Brooks and Corey [62]	$\theta_w = \theta_r + (\theta_s - \theta_r)\left(\frac{a}{\varphi}\right)^b$	θ_r, a, b
Van Genuchten [84]	$\theta_w = \theta_r + \frac{\theta_s - \theta_r}{\left(1 + (ah)^k\right)^m}$, where h is the capillary head (cm)	θ_r, a, k, m
Fredlund and Xing [82]	$\theta_w = \frac{\theta_s}{\{\ln[\exp(1) + (\varphi/a)^k]\}^m}$	a, k, m
Fredlund and Xing [82]	$\theta_w = \frac{\theta_s}{\{\ln[\exp(1) + (\varphi/a)^k]\}^m}[1 - \frac{\ln(1 + \varphi/\varphi_r)}{\ln(1 + 10^6/\varphi_r)}]$	a, k, m, φ_r

The major drawback of the empirical methods is that the empirical equations and relevant fitting parameters have no clear physical meaning, and different equations may have their own applicable situations.

1.3.4.2 Pedotransfer Functions

The pedotransfer function can estimate the SWCC using the available data about the soil basic physical properties. There are two main types for pedotransfer function: point estimation model and parameter estimation model.

The point estimation model based on the soil property indexes can predict the water content or permeability coefficient at certain suction. Gupta and Larson [89] and Rawls et al. [90], respectively, proposed multiple linear regression equations to estimate the water contents at certain suctions. The merit of point estimation model is that its estimations about the specific water content have high accuracy and the factors related to the estimated water content can be seen clearly. The parameter estimation model assumes the SWCC can be expressed using the empirical equation, and then the relationships between the fitting parameters and the soil physical properties are built. Some studies [91–95] have presented the regression relations between the fitting parameters in SWCC model and the soil basic properties, for example, particle-size distribution, unit weight, organic matter content, density, gravel content, and salinity. The parameter estimation model is more direct, and the estimated fitting parameters can be used in the empirical equation to obtain the relevant SWCC.

1.3.4.3 Physico-empirical Models

The physico-empirical models are built to estimate the water content based on the particle-size distribution and the characteristic of soil-pore structure. Based on the observation of similarity between the shape of water retention curve and particle-size distribution, Arya and Paris [96] first proposed a physico-empirical model to estimate the SWCC using the particle-size distribution and unit weight. The model adopted three assumptions of soil particles and pore structure to deduce the estimated equation of the water content, and an important empirical parameter α was introduced in the model. In order to improve this model, Arya et al. [97] proposed the method of back analysis to determine α. Zhuang et al. [98] derived an analytical model using nonsimilar media method to estimate SWCC based on the measured soil physical properties, for example, particle-size distribution, unit weight and bulk density.

This model does not require measuring SWCC in advance; it can make use of most of the available data. However, the physico-empirical models cannot obtain the close predictions for soils where aggregation, cracking and root effects may be pronounced. Otherwise, the model works reasonably well [96].

1.3.4.4 Fractal Models

The fractal theory was first introduced to estimate the SWCC by Tyler and Wheatcraft [99], and then they derived the other fractal model of SWCC based on the Sierpinski carpet [100]. This fractal model considered only the fractal characteristics of the pore space but not the mass. At present, many models have been proposed for the fractal characteristic of SWCC [101–103]. They have been used to describe the water retention behavior and the variation of hydraulic conductivity in soils with different textural structures [104–107].

The disadvantage of the fractal models is that the fractal scaling considers only the effects of tortuosity of pore lengths, but not the effects of other factors on the water retention property, such as organic matter content, packing density, the chemical characteristic of particle surface and fluid property.

1.4 Creep Behavior of Soft Soil

1.4.1 Mechanism of Creep Behavior for Soft Soil

The behavior of soft soil is viscous, which results in time effects and strain rate effects. The creep behavior refers to the soil deformation with time under a constant effective stress. In the early days, it is often referred to as "secondary consolidation". Research has shown the existence of creep deformation in soft soils, and it is more obvious for the soils with lower permeability.

Creep behavior of soft soil has been considered as a challenging topic for engineers and researchers for several decades. An appropriate model is required for predicting the long-term settlement of soils. Le et al. [108] reviewed the five causes for soil creep: (a) the breakdown of interparticle bonds [109]; (b) jumping of molecule bonds [110]; (c) sliding among particles [111]; (d) water flows in a double-pore system [112] and (e) structural viscosity [113, 114].

Moreover, the mechanism of creep deformation can also be explained from the views of macro and micro deformation of soil, and this explanation is directly related to the study of settlement prediction [115]. In the macroscopic view, creep deformation is a result of soil structure rearrangement to reach a new equilibrium under the action of external force. In the microscopic view, creep deformation is considered as the deformation of microstructure caused by the drainage of adsorbed water or its structural viscosity. It is also suggested that the creep deformation will cease when there is no free water in the soil.

1.4.2 Time-Dependent Model for Creep Analysis

The time-dependent stress–strain behavior of soft soils has been investigated through the laboratory studies by the scholars [114–121]. In order to describe the viscous nature of soils, the strain rates were utilized in the models [118, 122]. A lot of work has been done by some researchers to model the time-dependent behavior under one-dimensional straining in oedometer tests [114, 118, 123]. The models for describing the time-dependent behavior under both triaxial stress states and general stress states have also been developed [124–126].

Most of the time-dependent constitutive models are based on Perzyna's overstress theory [128, 129]. These models can be classified into two categories: conventional overstress models and extended overstress models. The conventional overstress models assume that only elastic strains occur when the stress state is inside of the static yield surface [130–133]. The extended overstress models assume that the viscoplastic strains occur even though the stress state is located within the static yield surface [125, 127, 134, 135]. The hypothesis of the conventional overstress model has been proved to be in conflict with the experimental results, and the associated viscosity parameters cannot be determined easily through the low loading-rate tests. In contrast, the parameters of soil viscosity involved in the extended overstress models can be determined straightforward based on the constant strain-rate tests or conventional oedometer tests.

At present, some elastic viscoplastic models have been developed to incorporate the anisotropy and destructuration in the description of the stress–strain–time behavior of natural soft clays. Rocchi [136] used the concepts of initial natural yield locus and intrinsic reference yield locus and the overstress theory to consider the generation of viscoplastic strain and proposed a viscoplastic model that can incorporate the strain-rate dependence and destruction process. Kimoto and Oka [135] developed a rate-dependent model to consider the destructuration and inherent anisotropy, and the stress-induced anisotropy has not been considered. Based on the isotropic creep model by Vermeer and Neher [127], Leoni et al. [137] proposed an anisotropic model. Yin et al. [138–140141, 142] have conducted extensive work to model the strain-rate-dependency behavior of natural soft soil. The final version of their models can describe the initial anisotropy, induced anisotropy, destructuration and time-dependence simultaneously [143]. Yao et al. [144] proposed a modified unified-hardening model to describe the deformation of overconsolidated clays and discussed its potential of taking into account anisotropy and structural effects.

1.4.3 Bjerrum's Time Line Conceptual Model

Buisman [145] first modeled the effect of time on the compression of clay by introducing the term of secondary compression. Taylor and Merchant [109] later reported that one-dimensional compression of clay should be described using a family of curves, called "time lines", and each curve corresponds to a specific loading duration in a standard oedometer test. One implication of time lines is that the magnitude of preconsolidation pressure is different for each line. Bjerrum [114] has same observations and suggested to use the parallel lines to model the delayed compression in a $e - \log \sigma_2'$ diagram. The parallel lines represent a series of equilibrium relationships after different durations of sustained loading.

Bjerrum's time line model is illustrated in Fig. 1.4 for "young" and "aged" normally consolidation (NC) clays. Young NC clays denote the deposit sediments that reach equilibrium under their own weight without experiencing the delayed compression, whereas aged NC clays have undergone substantially delayed compression at constant loading. The compression of undisturbed samples of young and aged NC clays subjected to the uniaxial consolidation is presented in two bold curves in the figure. The upper curve shows the compression behavior of the young NC clay, and its preconsolidation pressure $\sigma_{z,pc}'$ is equal to $\sigma_{z,0}'$, that is the present vertical effective stress. Under this effective stress for 10,000 years, the young NC clay will develop the delayed compression. The compression of aged NC clay follows the lower curve, and its apparent preconsolidation pressure increases to $\sigma_{z,1}'$, which is caused by aging rather than by previous overloading. This implies that the reduction of void ratio caused by the delayed compression will lead to a more stable clay structure and then a larger preconsolidation pressure. It can be seen that the Bjerrum's time line model provides a better understanding of the apparent preconsolidation pressures that resulted from aging.

It should be noted that Bjerrum only vaguely mentioned the concept of "equivalent time" but did not distinguish the equivalent time from test time. He also did not give

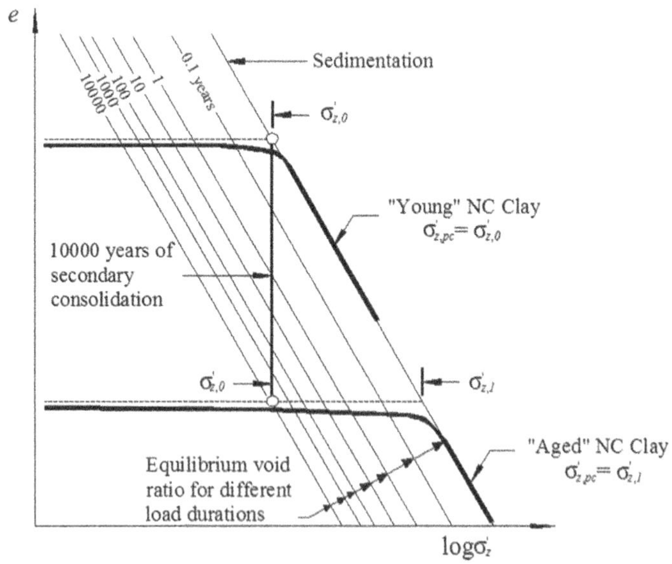

Fig. 1.4 Bjerrum's time-line model [114]

the time-dependent stress–strain model. Many researchers later developed Bjerrum's conceptual model into quantitative mathematical equations [118,146–150]. Yin and Graham [122, 123, 151] developed a series of models to describe 1-D time-dependent behavior of clay based on Bjerrum's conceptual model. Their models allow the treatment of strain-rate effects and relaxation, as well the more customary load increments with creep have been incorporated into the consolidation equation and verified by many consolidation data.

1.4.4 Yin and Graham's Elastic Viscoplastic (EVP) Model

1.4.4.1 1-D EVP Model by Yin and Graham

Yin and Graham [122, 123, 151] first clarified the concept of "equivalent time" and gave the physical explanation and mathematic definition of equivalent time. The other important concepts, such as "instant time line" and "reference time line", were also defined in their researches, as shown in the illustrations in Fig. 1.5.

"Instant time line" or "κ-line" is used to define the instantaneous vertical strain, which is pure elastic compression. It defines the instant elastic response of soil skeleton due to the variation of effective stress, but not the elastic-plastic response as assumed by Bjerrum [114]. It should be noted that the instant time line in Yin and Graham [122] has the same definition with that by Bjerrum [114], that is, it corresponds to the elastic-plastic deformations. "Reference time line" or "λ-line" has zero equivalent time ($t_e = 0$), and it is used as a reference to count the equivalent time. Equivalent time t_e has positive value ($0 < t_e < \infty$) below the reference time line, otherwise it is negative ($-t_0 < t_e < 0$) where t_0 is a material parameter. The

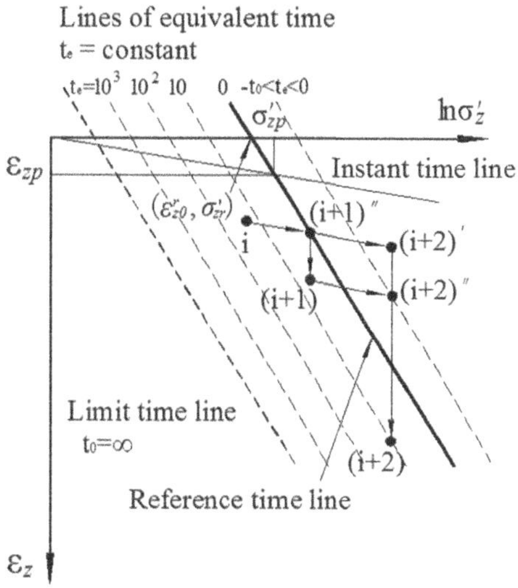

Fig. 1.5 Illustration of instant time line, reference time line, and equivalent-time lines for 1-D straining (after Yin and Graham [126])

vertical strain along the reference time line is the elastic-plastic compression. "Limit time line" is defined as a time line that has an infinite equivalent time value and a zero creep rate. Below the limit time line, the soil is considered to be elastic-plastic and time-independent.

The usage of "equivalent time" allows modeling the creep strains in both the overconsolidated and normally consolidated ranges using a constant creep parameter. Furthermore, it implies that the state at any point in Fig. 1.5 is independent of stress path and loading history. This also implies that the creep rate at point (i + 1) in the figure are the same for any loading path which finally reaches to point (i + 1). For example, the creep rate of soil experienced a loading and unloading path $\left[\mathrm{i} \rightarrow (\mathrm{i}+2)' \rightarrow (\mathrm{i}+2)'' \rightarrow (\mathrm{i}+1)\right]$ is equal to that in the loading and creep path $\left[\mathrm{i} \rightarrow (\mathrm{i}+1)'' \rightarrow (\mathrm{i}+1)\right]$ (see Fig. 1.5).

1.4.4.2 Development of EVP Models

The 1-D EVP model proposed by Yin and Graham [122, 123, 151] has been verified through numerous consolidation test data [151–154]. Yin et al. [134] developed a three-dimensional EVP model by incorporating a nonlinear function for the creep strain and a new loading surface. This 3-D model is more general than the original EVP model by Yin and Graham [126] and it can be utilized to model the time-dependent stress–strain behavior of both overconsolidated and normally clays. In order to consider the creep and swelling simultaneously, Yin and Tong [141] developed the original model while simulating the time-dependent behavior of saturated soils.

Kelln et al. [155] developed an improved EVP model, which incorporates a yield locus into the model based on the creep function, and it maintains good predictive capabilities to the time-dependent behavior under various test conditions. Kelln et al. [156, 157] also validated this model by analyzing the geotechnical structures in finite element program. Bodas Freitas et al. [158] proposed a new EVP model considering the isotach viscosity based on the overstress theory [129] and the EVP model by Yin et al. [126, 134, 159]. Zhou et al. [160] further analyzed anisotropy on the basis of the model by Yin [159]. Some researchers investigated different EVP models and adopted them to model the consolidation caused by vertical drains in soil subject to creep [152, 161].

These proposed models have their own advantages and disadvantages. In general, the complex model with more parameters can better capture the creep deformation of soft soil. Considering the practical application of engineering, a simple model with sufficient accuracy is preferred to be adopted. Therefore, given a specific soil sample, the most suitable model is required to be selected from the numerous existing models to analyze the deformation behavior.

1.4.4.3 Determination of Model Parameters

Regardless of the time-dependent model utilized in the analysis, the determination of model parameters is the first problem to be solved. At present, several studies demonstrated how to identify the parameters of an EVP model.

Yin and Graham [151] suggested determining parameters by gradually using the curve-fitting technique based on the experimental data at different consolidation stages. They pointed out that the loading increment that is located in the normally consolidated stage can give more information for the determination of creep parameters. The studies by Yin and Yin et al. [134, 159] presented the same way to determine the parameters. They suggested that the reference equivalent time (t_0) can be determined approximately referring to the time at which the excess pore water pressure has been dissipated completely, and then the other two parameters, that is, creep coefficient and creep strain limit, are obtained by fitting the measured data after the primary consolidation stage.

However, Mesri and Vardhanabhuti [162] pointed that it is not the best way to estimate the creep parameters just using the data after complete dissipation of the excess pore water pressure, and the determination of creep parameters should consider the consolidation data during the excess pore water pressures dissipation process. Le et al. [163] developed the trust-region reflective least-squares method to determine all model parameters simultaneously based on the test data from the whole consolidation stage at different loads, and the time parameter t_0 is assumed the unit minute. Ye et al. [164] used the simplex algorithm with random sampling method to optimize the creep and destructuration parameters of EVP model by Yin et al. [137, 139], and the other model parameters in EVP model were suggested to be determined in advance based on test data.

As can be noted, all the previous studies have adopted the deterministic method to determine the model parameters, and they only require the best fitting between the predictions and the observations. The deterministic method ignores the effect of uncertainty in test data and model itself on the model parameters. In Chaps. 4 and 5, the model parameters will be determined by using the Bayesian probabilistic method, and their uncertainties will also be evaluated at the same time. Moreover, the time parameter as an unknown model parameter with other creep parameters will be determined simultaneously using the whole consolidation data, and no parameter will be determined in advance. Although this study only focused on the parameter determination of 1-D elastic viscoplastic models by Yin et al. [139, 141], the adopted Bayesian probabilistic method can be utilized to analyze any time-dependent models.

1.5 Critical State Behavior of Granular Soils

1.5.1 Shear Behavior of Granular Soils

The shear properties of soil refer to the contraction/dilatancy and yield strength properties or friction properties during shearing. The conventional testing method includes (1) shear tests for uneven deformation of soil samples, such as direct shear tests and simple shear tests, and (2) triaxial shear tests that lead to relatively uniform deformation of soil samples. According to different design requirements, methods such as slow shearing or fast shearing and drained or undrained test conditions are usually adopted, and different deformation and strength indexes are obtained. Triaxial shear tests are widely used due to the advantage of sample uniformity during shearing.

Based on the consolidation history, the current void ratio and the stress state, clay can be divided into normally consolidated and overconsolidated clay, while sandy soil is divided into loose and dense sand. The results of a large number of triaxial shear tests are shown in (Fig. 1.6):

(a) Normally consolidated and slightly overconsolidated clays and loose sand exhibit volumetric contraction during shearing; the void ratio decreases under drained conditions, and the mean effective stress decreases under undrained conditions.

(b) Highly overconsolidated clay and dense sand exhibit volumetric expansion during shearing, that is, dilative characteristics (the void ratio becomes larger under drained conditions, and the mean effective stress becomes larger under undrained conditions), along with the peak stress ratio above the critical stress ratio.

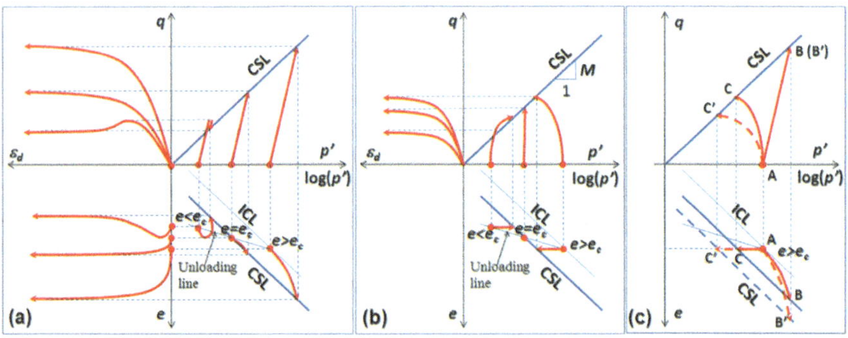

Fig. 1.6 Schematic diagram of the mechanical properties of different overconsolidated clays: **a** under drained conditions, **b** under undrained conditions, and **c** effects of the critical state line

(c) The stress ratios of clays with different overconsolidated ratios or sands with different densities all reach the same stress ratio line. This line is called the "critical state line", hereinafter referred to as the "stress ratio-related critical state line", which is related to the soil friction and determines the soil yield strength.
(d) The void ratios of a given clay at different overconsolidation ratios or a given sand at different densities also eventually reach the same "void ratio-log mean effective stress" line, which is also called the "critical state line", hereinafter referred to as the "void ratio-related critical state line".
(e) For clay, it is generally considered that the void ratio-related critical state line is a straight line parallel to the isotropic compression line; for sand, the void ratio-related critical state line is usually nonlinear. The relative position between the void ratio-related critical state line and the current state of the "void ratio-mean effective stress" determines the contractive or dilative behavior of the soil and the maximum volumetric strain during shearing.

Based on this phenomenon and the critical state concept, many researchers have developed various constitutive models under the framework of the elastoplasticity theory. These models can be divided into two categories according to their consideration of the void ratio-related critical state line:

(a) Implicit consideration of the critical state line: The position of the void ratio-related critical state line is controlled by the intersection of the yield surface and the stress ratio of the critical state line on the "p'–q" plane, such as in the Cam-Clay models. Notably, the multisurface model [165], the bounding surface model [166] and the sub-loading surface model [167] introduce the plastic strain within the yield surface so that the nonlinear stress–strain characteristics are improved in accordance with the yield strength characteristics. More details can be found in Chap. 6.

$$\varepsilon_{ij}^{p} = \Lambda \frac{\partial g}{\partial \sigma_{ij}}; \left(\text{such as Cam-Clay model: } g = f = \frac{q^2}{M^2 p'} + p' - p_c \right) \tag{1.24}$$

(b) Explicit consideration of the critical state line: The expression of the void ratio-related critical state line is directly introduced by establishing the relationship between the relative distance from the current "void ratio-stress" state and the critical state line, the peak friction angle and the dilatancy angle. The double-yield-surface model (Eq. 1.25) proposed by Yin et al. [168] simultaneously conforms to the nonlinear stress–strain and yield strength characteristics. More details can be found in Chap. 5.

$$\begin{cases} \text{Yield surface:} f_S = \dfrac{q}{p'} - \dfrac{M_p \varepsilon_d^p}{1/G_p + \varepsilon_d^p} \\ \text{Dilatancy/contraction:} \dfrac{d\varepsilon_v^p}{d\varepsilon_d^p} = D\left(M_{pt} - \dfrac{q}{p'}\right) \\ \text{Interlocking: } \tan\phi_p = \dfrac{e_c}{e}\tan\phi_\mu;\ \tan\phi_{pt} = \left(\dfrac{e_c}{e}\right)^{-1}\tan\phi_\mu \\ \text{Critical state line: } e_c = e_{cr0} - \lambda\ln\left(\dfrac{p'}{p_{cr0}}\right) \end{cases} \tag{1.25}$$

It should be pointed out that simple models, such as the Mohr–Coulomb and Duncan-Chang models, do not consider the void ratio-related critical state line and thus cannot describe contraction/dilation and peak/residual yield strength characteristics of soils associated with different overconsolidated ratios or densities. Therefore, this type of models is limited to volumetric deformation or geotechnical engineering with small change in pore water pressure (i.e., the strain is limited to before the occurrence of transformation between contraction and dilation and the occurrence of peak strength). Therefore, the rationality of the simulated results can only be guaranteed within a small strain range. However, it is more reasonable to use the model considering the void ratio-related critical state line.

The three-dimensional mechanical expression of the shear strength is one of the important properties of soils. The key to the problem lies in the influence of the intermediate principal stress on the shear strength. True triaxial tests, which allow control of the stresses in three independent directions, are usually used to perform loading along different stress paths on the π plane to obtain the strength characteristics under different intermediate principal stress conditions. There are three main theoretical modeling methods accounting for the effects of the intermediate principal stress on soil strength: (1) the lode angle modification method or $g(\theta)$ method, (2) the transformed stress tensor method and (3) the micromechanical analytical method. More details can be found in Chap. 4.

Notably, the stress path within geotechnical structures is generally complicated. The strength criterion of the commonly used Mohr–Coulomb model is generally conservative. On the other hand, the Duncan–Chang model implies the generalized von Mises strength criterion, which overestimates the material strength. Therefore, it is necessary to consider the influence of the 3D stress state based on the selected constitutive model. Furthermore, with the development of computer techniques, the development and application of micromechanical methods with physical insights into soils should receive increasing attention (see details in Chap. 9).

1.5.2 Critical State Line of Granular Soils

Several formulae for expressing the critical state concept have been proposed. The most typical formula is the linear formula in the e-log p' plane. The relationship has traditionally been written as follows:

$$e_c = e_{ref} - \lambda \ln\left(\frac{p'}{p_{ref}}\right) \tag{1.26}$$

where e_{ref} is a reference void ratio corresponding to a reference mean effective stress p_{ref}; and λ is the slope of the critical state line (CSL) in the e-log p' plane. Therefore, two parameters (e_{ref} and λ) are required for the definition of the CSL. The advantage of this formula is the simplicity of its form. However, experimental results have shown that the CSL is not always linear in the e-log p' plane, and mathematically, the critical void ratio e_c could become negative for high stress levels, which is meaningless. Note that a very high stress level still exists in geotechnical structures, such as at the pile tip during its installation.

More recently, another formula was proposed by Li and Wang [169], who assumed a nonlinear critical state line in the e-log p' plane, representing an extension of the linear formula with one additional parameter ξ:

$$e_c = e_{c0} - \lambda\left(\frac{p'}{p_{at}}\right)^{\xi} \tag{1.27}$$

where the additional parameter ξ controls the nonlinearity of the critical state line, giving a more flexible and accurate description according to experimental data, especially for very low to moderate stress levels. However, for high stress levels, the positiveness of the critical void ratio e_c could not be guaranteed, which could possibly cause numerical problems in some local elements in finite element modeling. To overcome this difficulty, Gudehus [170] suggested a third formula for the CSL, which is also a nonlinear formula but with an "s" form by considering an ultimate critical void ratio at very high stress levels. This formula can be expressed as follows:

$$e_c = e_{cu} + (e_{c0} - e_{cu}) \exp\left(-\left(\frac{p'}{p_{at} \cdot \lambda}\right)^{\xi}\right) \tag{1.28}$$

where e_{cu} is the critical void ratio when $p' \to \infty$. This expression eliminates the possibility of a negative value of the critical void ratio at high stress levels. However, two more parameters have to be determined.

Considering the advantage of the second formula, we adopted the expression of Eq. (1.27). The possible negative void ratio is a problem because the formula is based on a semi-logarithmic plot. If a double-logarithmic plot is used, then Eq. (1.27) can be revised as follows:

$$\log(e_c) = \log(e_{c0}) - \lambda\left(\frac{p'}{p_{at}}\right)^{\xi} \tag{1.29}$$

which can be re-written as follows:

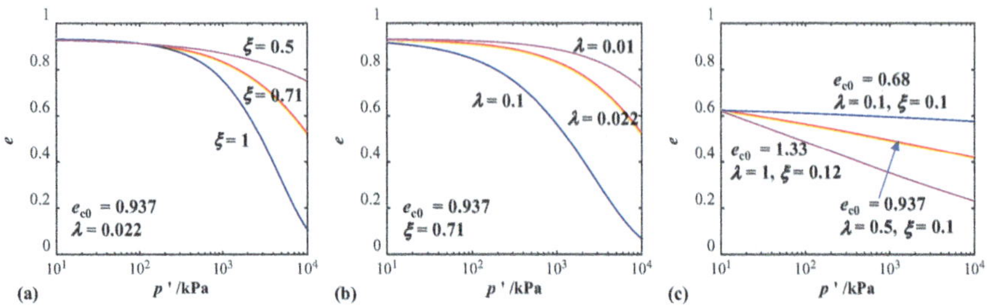

Fig. 1.7 Influence of material parameters on the linearity of the critical state line in a double logarithmic plot

$$e_c = e_{c0} \exp\left[-\lambda\left(\frac{p'}{p_{at}}\right)^{\xi}\right] \tag{1.30}$$

Given different values of the parameters e_{c0}, λ, and ξ, different curves can be obtained by using Eq. (1.30), as shown in Fig. 1.7. It can be concluded that ξ controls the slope of the CSL after the inflection point (Fig. 1.7a), λ controls the position of the inflection point (Fig. 1.7b), and the combination of the three parameters gives access to an almost constant slope of the CSL as in Eq. (1.26) (Fig. 1.7c). This enhanced formula has the advantage of giving a nonlinear CSL, and at the same time, this formula imposes the positiveness of the void ratio with three parameters as input.

With the critical state concept, the density state of the soil can be defined by the ratio e_c/e or the distance e_c-e, where e is the current void ratio and e_c is the critical void ratio at the current stress state p', which is obtained from the CSL equation. This implies $e_c/e < 1$ or $e_c-e < 0$ for loose packing, $e_c/e > 1$ or $e_c-e > 0$ for dense packing, and $e_c/e = 1$ or $e_c-e = 0$ when the initial state lies on the CSL in the e–p' plane. In agreement with the experimental results obtained by Biarez and Hicher [171], the density state effect on the stress–strain–strength behavior was implemented into the model by the following relation:

$$\phi_p = \arctan\left[\left(\frac{e_c}{e}\right)^{n_p} \tan\phi_c\right],\ \phi_{pt} = \arctan\left[\left(\frac{e}{e_c}\right)^{n_d} \tan\phi_c\right] \tag{1.31}$$

where n_p and n_d are parameters controlling the effect of particle interlocking; these parameters can be taken as 1 according to Biarez and Hicher [171].

The expression of the mobilized peak friction angle ϕ_p in Eq. (1.31) shows that in a loose structure, ϕ_p is initially smaller than the critical friction angle ϕ_c. When the stress ratio overcomes the phase transformation state, a dense structure provides a higher degree of interlocking. Thus, ϕ_p is greater than ϕ_c. When the loading stress reaches the peak stress, a dense structure will dilate, and the degree of interlocking will decrease. Consequently, the peak friction angle will be reduced, which results in a strain-softening phenomenon.

The mobilized phase transformation angle ϕ_{pt} in Eq. (1.31) implies that in a loose structure with $e > e_c$, ϕ_{pt} is larger than ϕ_{c}, which leads to a continuously contractive behavior; in a dense structure with $e < e_c$, ϕ_{pt} is smaller than ϕ_{c}, which imposes a dense structure that is initially contractive and then dilative during deviatoric loading.

For both loose and dense structures, when the stress state reaches the critical state line, the void ratio e approaches the critical void ratio e_c and volume changes will no longer occur. Thus, the constitutive equations guarantee that the stress and void ratio simultaneously reach the critical state in the p'–q–e space. Note that Eq. (1.31) gives a similar density effect as in the model suggested by Manzari and Dafalias [172] but in a different form.

Using the above mobilized peak friction angle ϕ_{p} and the mobilized phase transformation angle ϕ_{pt} and accounting for the Lode angle dependency, we can express the peak strength ratio M_{p} and the phase transformation stress ratio M_{pt} by the following relations:

$$M_p = \begin{cases} \frac{6\sin\phi_p}{3-\sin\phi_p} \text{ in compression} \\ \frac{6\sin\phi_p}{3+\sin\phi_p} \text{ in extension} \end{cases}, M_{pt} = \begin{cases} \frac{6\sin\phi_{pt}}{3-\sin\phi_{pt}} \text{ in compression} \\ \frac{6\sin\phi_{pt}}{3+\sin\phi_{pt}} \text{ in extension} \end{cases} \tag{1.32}$$

To enhance the model, Eqs. (1.30)–(1.32) were introduced in the model presented in the previous section and then used to simulate conventional drained and undrained triaxial tests on samples with different initial void ratios. As shown in Fig. 1.8, the enhanced model with the critical state concept can well reproduce the density effect on the mechanical behavior of granular materials. Note that Eq. (1.32) can also be applied to the basic model to enhance the extension strength following the strength criterion of Coulomb or Matsuoka-Nakai [173] without additional parameters.

Another option is to use the work of Been and Jefferies [174] directly:

$$M_p = M\exp\left[-n_p(e - e_c)\right], M_{pt} = M\exp[n_d(e - e_c)] \tag{1.33}$$

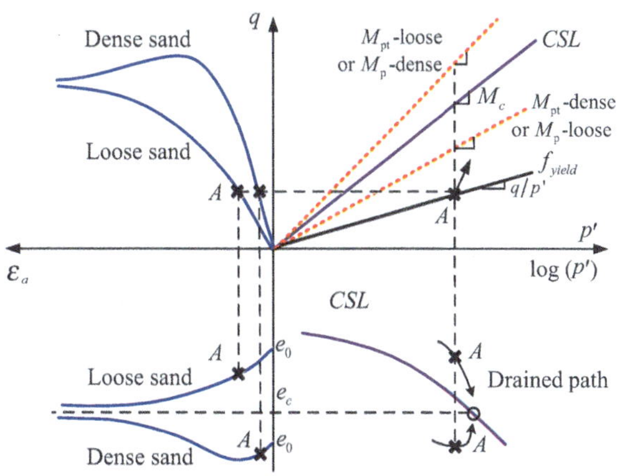

Fig. 1.8 Principle of the SIMSAND model

with

$$M = \begin{cases} \frac{6\sin\phi_c}{3-\sin\phi_c} \text{ in compression} \\ \frac{6\sin\phi_c}{3+\sin\phi_c} \text{ in extension} \end{cases} \tag{1.34}$$

Both the exponential form and the power form conform to the experimental results and can therefore be used to describe the characteristics of the soil. In summary, for loose (or normally consolidated) soil and dense (over-consolidated) soil, Fig. 1.8 shows the schematic diagram of the critical state-based nonlinear Mohr–Coulomb model, the so-called SIMSAND model.

1.6 Summary

This chapter reviewed the uncertainty in geotechnical engineering, and mainly discussed the uncertainties involved in the estimation of soil properties and geotechnical models. The influencing factors on the uncertainties and relevant studies were summarized. It was pointed out that the uncertainty of soil parameters should be considered and analyzed in estimating a certain soil property. Bayesian probabilistic approach as a useful tool was outlined from two application aspects, that is, parametric identification and model class selection. In view of the complex updated PDF, the chapter reviewed several available numerical simulation methods.

The chapter then reviewed previous studies on two problems of geotechnical engineering, that is, soil water retention property of unsaturated soil and creep behavior of soft soil. In the sections of soil water retention of unsaturated soil, the soil suction as an important factor for the development of the unsaturated soil mechanics was explained first, and its contribution on the soil shear strength, permeability and compressibility was discussed by reviewing the existing studies. The soil–water characteristic curve was then explained, and its influencing factors were summarized. The four commonly used methods for estimating SWCC were finally presented. The objective in the study of SWCC was mentioned, and a new model, which can consider the effect of initial void ratio on the SWCC of same textured soil sample, was required to be constructed and the relevant uncertainty analysis should also be conducted.

In the sections of creep behavior of soft soil, the mechanism of soil creep deformation was presented briefly, and the existing studies on the time-dependent models for describing the creep behavior were reviewed. As the basis of this study, the conceptual time line model proposed by Bjerrum [114] was illustrated, and the 1-D elastic viscoplastic model developed by Yin and Graham [122, 123, 149] based on the Bjerrum's model and the development of EVP models were reviewed. Several methods for determining the parameters of EVP model were summarized, and the objectives in the study of creep behavior were proposed, that is, to analyze the model parameters by using the Bayesian probabilistic method and to select the suitable model for the predictions of creep behavior of soft soil.

The shear behaviors of soil refer to the contraction/dilatancy, yield strength properties or friction properties and critical state during shearing. A constitutive model considering all these features has normally a number of parameters which poses difficulties in the engineering practice. Furthermore, a number of constitutive models have been developed by researchers during the last few decades. How to select a suitable model or models for a given problem is another issue for engineers in practice.

References

1. Tang WH, Yucemen MS, Ang AHS (1976) Probability-based short term design of soil slopes. Can Geotech J 13:201–215
2. Low BK, Tang WH (1997) Probabilistic slope analysis using Janbu's generalized procedure of slices. Comput Geotech 21(2):121–142
3. Cheung RWM, Tang WH (2000) Bayesian calibration of slope failure probability. In: GeoDenver conference 'Slope Stability 2000', Denver, USA, pp 72–85
4. Chong PC, Phoon KK, Tan TS (2000) Probabilistic analysis of unsaturated residual soil slopes. In: Application of statistics and probability proceedings of the ICASP 8 conference, Sydney, pp 375–382
5. Tang WH, Angulo M (1996) Bayesian liquefaction resistance analysis. In: Proceedings of the 1996 conference on uncertainty in the geologic environment, UNCERTAINTY'96, Part 2 (of 2), pp 1195–1206
6. Cetin KO, Der Kiureghian A, Seed RB (2002) Probabilistic models for the initiation of seismic soil liquefaction. Struct Saf 24(1):67–82
7. Low BK, Tang WH (1997) Reliability analysis of reinforced embankments on soft ground. Can Geotech J 34(5):672–685
8. Griffiths DV, Fenton GA (1998) Probabilistic analysis of exit gradients due to steady seepage. J Geotech Geoenviron Eng 124(9):789–797
9. Phoon KK, Santoso A, Quek ST (2010) Probabilistic analysis of soil-water characteristic curves. J Geotech Geoenviron Eng 136(3):445–455
10. Chiu CF, Yan WM, Yuen KV (2012) Estimation of water retention curve of granular soils from particle-size distribution-a Bayesian probabilistic approach. Can Geotech J 49(9):1024–1035
11. Chiu CF, Yan WM, Yuen KV (2012) Reliability analysis of soil-water characteristic curve and its application to slope stability analysis. Eng Geol 135-136:83–91
12. Zhang LM, Tang WH, Zhang LL, Zheng JG (2004) Reducing uncertainty of prediction from empirical correlations. J Geotech Geoenviron Eng 130(5):526–534
13. Tang WH (1984) Principles of probabilistic characterization of soil properties. In: Proceedings of symposium on probabilistic characterization of soil properties, ASCE National Convention, Atlanta, pp 74–89
14. Phoon KK, Kulhawy FH (1999) Characterization of geotechnical variability. Can Geotech J 36:612–624
15. Phoon KK, Kulhawy FH (1999) Evaluation of geotechnical property variability. Can Geotech J 36:625–639
16. Christian JT, Ladd CC, Baecher GB (1994) Reliability applied to slope stability analysis. J Geotech Eng 120(12):2180–2207
17. Gui S, Zhang R, Turner JR, Xue X (2000) Probabilistic slope stability analysis with stochastic soil hydraulic conductivity. J Geotech Geoenviron Eng 126(1):1–9
18. Baecher GB, Christian JT (2003) Reliability and statistics in geotechnical engineering. Wiley
19. Zhang LM, Zheng YR, Wang JL (2003) Errors in calculating hydrodynamic pressures for stability analysis of soil slopes subject to rainfall. In: Proceedings of ICASP9, San Francisco, USA, pp 1431–1438

20. Zhang LL, Zhang LM, Tang WH (2009) Uncertainties of field pullout resistance of soil nails. J Geotech Geoenviron Eng 135(7):966–972
21. Popescu R, Prevost JH (1995) Reliability assessment of centrifuge soil test results. Soil Dyn Earthq Eng 14:93–101
22. Beck JL, Au SK (2002) Bayesian updating of structural models and reliability using Markov chain Monte Carlo simulation. J Eng Mech 128(4):380–391
23. Finno RJ, Calvello M (2005) Supported excavations: the Observational Method and inverse modeling. J Geotech Geoenviron Eng 131(7):826–836
24. Honjo Y, Liu WT, Soumitra G (1994) Inverse analysis of an embankment on soft clay by extended Bayesian method. Int J Numer Anal Methods Geomech 18(10):709–734
25. Gioda G, Marier G (1980) Direct search solution of inverse problem in elastoplasticity: Identification of C, φ and E by pressure tunnel tests. Int J Numer Methods Eng 15:1823–1848
26. Tang WH (1971) A Bayesian evaluation of information for foundation engineering design. In: Statistics and probability in civil engineering, P. Lumb, ed., Hong Kong University Press, Hong Kong
27. Wang Y, Au SK, Cao Z (2010) Bayesian approach for probabilistic characterization of sand friction angles. Eng Geol 114:354–363
28. Ching J, Phoon KK, Chen YC (2010) Reducing shear strength uncertainties in clays by multivariate correlations. Can Geotech J 47(1):16–33
29. Kay JN (1976) Safety factor evaluation for single piles in sand. J Geotech Eng Div 102(10):1093–1108
30. Yan WM, Yuen KV, Yoon GL (2009) Bayesian probabilistic approach for the correlations of compression index for marine clays. J Geotech Geoenviron Eng 135(12):1932–1940
31. Cao Z, Wang Y (2013) Bayesian approach for probabilistic site characterization using cone penetration tests. J Geotech Geoenviron Eng 139(2):267–276
32. Wang Y, Huang K, Cao Z (2013) Probabilistic identification of underground soil stratification using cone penetration tests. Can Geotech J 50(7):766–776
33. Wang Y, Huang K, Cao Z (2014) Bayesian identification of soil strata in London clay. Géotechnique 64(3):239–246
34. Cao Z, Wang Y (2014) Bayesian model comparison and selection of spatial correlation functions for soil parameters. Struct Saf 49:10–17
35. Hsiao ECL, Schuster M, Juang CH, Kung TC (2008) Reliability analysis of excavation-induced ground settlement for building serviceability evaluation. J Geotech Geoenviron Eng 134(10):1448–1458
36. Juang CH, Luo Z, Atamturktur S, Huang H (2013) Bayesian updating of soil parameters for braced excavation using field observations. J Geotech Geoenviron Eng 139(3):395–406
37. Hsein JC, Luo Z, Atamturktur S, Huang HW (2013) Bayesian updating of soil parameters for braced excavations using field observations. J Geotech Geoenviron Eng 139(3):395–406
38. Wu SH, Ching JY, Ou CY (2014) Probabilistic observational method for estimating wall displacements in excavations. Can Geotech J 51:1111–1122
39. Yuen KV (2010a) Bayesian methods for structural dynamics and civil engineering. John Wiley & Sons (Asia) Pte Ltd, Singapore
40. Beck JL, Katafygiotis LS (1998) Updating models and their uncertainties. I: Bayesian statistic framework. J Eng Mech 124(4):455–461
41. Yuen KV (2010b) Recent developments of Bayesian model class selection and applications in civil engineering. Struct Saf 32(5):338–346
42. Papadimitriou C, Beck JL, Katafygiotis LS (1997) Asymptotic expansions for reliability and moments of uncertain systems. J Eng Mech 123(12):1219–1229
43. Gull SF (1988) Bayesian inductive inference and maximum entropy. In: Skilling J (ed) Maximum entropy and Bayesian methods. Kluwer Academic Publishers, Boston, MA, pp 53–74
44. Metropolis N, Rosenbluth AW, Rosenbluth MN, Teller AH, Teller E (1953) Equation of state calculations by fast computing machines. J Chem Phys 21(6):1087–1092

45. Hastings WK (1970) Monte Carlo sampling methods using Markov chains and their applications. Biometrika 57(1):97–109
46. Gelman A, Roberts GO, Gilks WR (1996) Efficient metropolis jumping rules. Bayesian Stat 5:599–608
47. Ching JY, Chen YC (2007) Transitional Markov Chain Monte Carlo method for Bayesian model updating, model class selection, and model averaging. J Eng Mech 133(7):816–832
48. Fredlund DG, Rahardjo H, Fredlund MD (2012) Unsaturated soil mechanics in engineering practice. Wiley, Hoboken, New Jersey
49. Aitchison GD (1964) Engineering concepts of moisture equilibria and moisture changes in soils. In: Statement of the review panel, in Moisture Equilibria and Moisture changes in soils beneath covered areas, a symposium in print, Butterworths, Sydney, pp 7–21
50. Fredlund DG, Rahardjo H, Gan JK (1987) Nonlinearity of strength envelope for unsaturated soils. In: Proceedings of the 6th international conference on expansive soils, vol 1. Balkema, Rotterdam, New Delhi, India, pp 49–54
51. Cui YJ, Delage P (1996) Yielding and plastic behaviour of an unsaturated compacted silt. Géotechnique 46(2):291–311
52. Ng CWW, Zhou RZB (2005) Effects of soil suction on dilatancy of an unsaturated soil. In: Proceedings of the 16th ICSMGE, vol 2. Osaka, Japan, pp 559–562
53. Hossain MA, Yin JH (2010) Shear strength and dilative characteristics of an unsaturated compacted completely decomposed granite soil. Can Geotech J 47(10):1112–1126
54. Fredlund DG, Rahardjo H (1993) Soil mechanics for unsaturated soils. Wiley, New York
55. Alonso EE, Gens A, Josa A (1990) A constitutive model for partially saturated soils. Géotechnique 40(3):405–430
56. Wheeler SJ, Sivakumar V (1995) An elasto-plastic critical state framework for unsaturated soil. Géotechnique 45(1):35–53
57. Chiu CF, Ng CWW (2003) A state-dependent elasto-plastic model for saturated and unsaturated soils. Géotechnique 53(9):809–829
58. Estabragh AR, Javadi AA, Boot JC (2004) Effect of compaction pressure on consolidation behaviour of unsaturated silty soil. Can Geotech J 41:540–550
59. Fredlund MD (1996) Design of a knowledge-based system for unsaturated soil properties. MSc Thesis, University of Saskatchewan, Saskatoon, Sask
60. Fredlund MD (1999). The role of unsaturated soil property functions in the practice of unsaturated soil mechanics. PhD dissertation, University of Saskatchewan, Saskatoon, Sask
61. Vanapalli SK (1994) Simple test procedures and their interpretation in evaluating the shear strength of an unsaturated soil. PhD Thesis, Department of Civil Engineering, University of Saskatchewan, Canada, pp 350
62. Brooks RH, Corey AT (1964) Hydraulic properties of porous media. Colorado State University Hydrology Paper, No. 3. Fort Collins, CO
63. Fredlund DG, Xing A (1994) Equations for the soil-water characteristic curve. Can Geotech J 31:521–532
64. Houston SL, Houston WR, Wagner AM (1994) Laboratory filter paper suction measurements. Geotech Test J 17(2):185–194
65. Richards LA (1928) The usefulness of capillary potential to soil moisture and plant investigators. J Agric Res 37:719–742
66. Reginato RJ, van Bavel CHM (1962) Pressure cell for soil cores. Soil Sci Soc Am Proc 26:1–3
67. Fredlund DG (2001) Ensuring a sound scientific basis for geotechnical engineering in developing and developed countries. In: Proceedings of the international conference on management of the land and water resources, Hanoi, pp 3–20
68. Fredlund DG, Stone J (2011) Development of the golder basic pressure plate cell. Internal report of Golder Associates, sponsored by Golder Innovation Research Fund, Saskatoon, SK
69. Gan JKM, Fredlund DG (1988) Multistage direct shear testing of unsaturated soils. ASTM Geotech Test J 11(2):132–138
70. Marinho FAM, Chandler RJ (1993) Aspects of the behavior of clays on drying. In: Unsaturated soils. Geotechnical special publication 39, ASCE, Reston, VA, pp 77–90

71. Vanapalli SK, Fredlund DG, Pufahl DE (1999) The influence of soil structure and stress history on the soil-water characteristics of a compacted till. Géotechnique 49(2):143–159
72. Kawai K, Karube D, Kato S (2000) The model of water retention curve considering effects of void ratio. In: Proceedings of the Asian conference on unsaturated soils, Balkema, Taylor and Francis Group, Rotterdam, Netherlands, pp 329–334
73. Uchaipichat A, Khalili N (2009) Experimental investigation of thermo-hydro mechanical behaviour of an unsaturated silt. Géotechnique 59(4):339–353
74. Iyer K, Jayanth S, Gurnani S, Singh DH (2013) Influence of initial water content and specimen thickness on the SWCC of fine-grained soils. Int J Geomech 13(6):894–899
75. Elkady TY, Al-Mahbashi A, Dafalla M (2015) Effect of compaction state on the soil water characteristic curves of sand-natural expansive clay mixtures. Eur J Environ Civil Eng. https://doi.org/10.1080/19648189.2015.1112844. (In Press)
76. Huang SY, Barbour SL, Fredlund DG (1998) Development and verification of a coefficient of permeability function for a deformable unsaturated soil. Can Geotech J 35(3):411–425
77. Wang TH, Lu J, Yue CK (2008) Soil-water characteristic curve for unsaturated loess considering temperature and density effect. Rock Soil Mech (In Chinese) 29(1):1–5
78. Gallipoli D, Wheeler SJ, Karstunen M (2003) Modelling the variation of degree of saturation in a deformable unsaturated soil. Géotechnique 53(1):105–112
79. Wei CF, Li H, Wang JL (2008) A constitutive model of unsaturated soils with consideration to the coupling of elastoplastic deformation and capillary hysteresis. In: 1st national symposium on geotechnical constitutive theory (In Chinese), Beijing, pp 259–266
80. Tarantino A (2009) A water retention model for deformable soils. Géotechnique 59(9):751–762
81. Leong EC, Rahardjo H (1997) Review of soil-water characteristic curve equations. J Geotech Geoenviron Eng 123(12):1106–1117
82. Fredlund DG, Xing A, Huang S (1994) Predicting the permeability function for unsaturated soils using the soil-water characteristic curve. Can Geotech J 31:533–546
83. Zhang LM, Chen Q (2005) Predicting bimodal soil-water characteristic curves. J Geotech Geoenviron Eng 131(5):666–670
84. van Genuchten MT (1980) A closed-form equation for predicting the hydraulic conductivity of unsaturated soils. Soil Sci Soc Am J 44:892–898
85. Gardner WR (1958) Some steady state solutions of the unsaturated moisture flow equation with application to evaporation from a water-table. Soil Sci J 85(4):228–232
86. Zapata CE (1999) Uncertainty in soil-water characteristic curve and impacts on unsaturated shear strength predictions. PhD dissertation, Arizona State University, Tempe, Ariz
87. Chin KB, Leong EC, Rahardjo H (2010) A simplified method to estimate the soil-water characteristic curve. Can Geotech J 47:1382–1400
88. Torres GH (2011) Estimating the soil-water characteristic curve using grain-size analysis and plasticity index. MSc Thesis, Arizona State University, Tempe, AZ
89. Gupta SC, Larson WE (1979) Estimating soil water retention characteristics from particle size distribution, organic matter percent, and bulk density. Water Resour Res 15(6):1633–1635
90. Rawls WJ, Brakensiek DL, Saxton KE (1982) Estimation of soil water properties. Trans Am Soc Agric Eng 25(5):1316–1320 & 1328
91. Cosby BJ, Hornberger GM, Clapp RB, Ginn TR (1984) A statistical exploration of the relationships of soil moisture characteristics to the physical properties of soils. Water Res 20:682–690
92. Saxton KE, Rawls WJ, Romberger JS, Papendick PI (1986) Estimating generalized soil water characteristics from texture. Soil Sci Soc Am J 50:1031–1036
93. Wösten JHM, van Genuchten MTh (1988) Using texture and other soil properties to predict the unsaturated soil hydraulic functions. Soil Sci Soc Am J 52:1762–1770
94. Vereecken JHM, Feyen J, Maes J, Darius P (1989) Estimating the soil moisture retention characteristic from texture, bulk density, and carbon content. Soil Sci 148:389–403
95. Saxton KE, Rawls WJ (2006) Soil water characteristic estimates by texture and organic matter for hydrologic solutions. Soil Sci Soc Am J 70(5):1569–1578

96. Arya LM, Paris JF (1981) A physicoempirical model to predict the soil moisture characteristics from particle-size distribution and bulk density data. Soil Sci Soc Am J 45(6):1023–1030
97. Arya LM, Leij FJ, van Genuchten MTh, Shouse PJ (1999) Scaling parameter to predict the soil water characteristic from particle-size distribution data. Soil Sci Soc Am J 63:510–519
98. Zhuang J, Jin Y, Miyazaki T (2001) Estimating water retention characteristic from soil particle size distribution using a non-similar media concept. Soil Sci 166:308–321
99. Tyler SW, Wheatcraft SW (1989) Application of fractal mathematics to soil water retention estimation. Soil Soc Am J 53(4):987–996
100. Tyler SW, Wheatcraft SW (1990) Fractal processes in soil water retention. Water Resour Res 26(5):1047–1054
101. Rieu M, Sposito G (1991) Fractal fragmentation, soil porosity, and soil water properties: I. Theory and II. Applications. Soil Soc Am J 55:1231–1244
102. Shepard JC (1993) Using a fractal moel to compute the hydraulic conductivity function. Soil Sci Soc Am J 57:300–306
103. Fallico C, Tarquis AM, De Bartoloa S, Veltria M (2010) Scaling analysis of water retention curves for unsaturated sandy loam soils by using fractal geometry. Eur J Soil Sci 61:425–436
104. Slawiński C, Sokolowska Z, Walczak R, Borówko M, Sokolowski S (2002) Fractal dimension of peat soils from adsorption and from water retention experiments. Colloids Surf A Physicochem Eng Asp 208(1–3):289–301
105. Hunt AG (2004) An explicit derivation of an exponential dependence of the hydraulic conductivity on relative saturation. Adv Water Resour 27(2):197–201
106. Xu YF, Dong P (2004) Fractal approach to hydraulic properties in unsaturated porous media. Chaos Solut Fractals 19:327–337
107. Perfect E (2005) Modeling the primary drainage curve of prefractal porous media. Vadose Zone J 4:959–966
108. Le TM, Fatahi B, Khabbaz H (2012) Viscous behaviour of soft clay and inducing factors. Geotech Geol Eng 30:1069–1083
109. Taylor DW, Merchant WA (1940) A theory of clay consolidation accounting for secondary compression. J Math Phys 19:167–185
110. Murayama S, Shibata T (1961) Rheological properties of clays. In: Proceedings of the 5th international conference on soil mechanics, vol 1, Paris, pp 269–273
111. Kuhn MR, Mitchell JK (1993) New perspectives on soil creep. J Geotech Eng 119(3):507–552
112. De Jong GJ, Verruijt A (1965) Primary and secondary consolidation of a spherical clay sample. In: Proceeding of the 6th *International Conference on Material Mechanism and Foundation Engineering*, Montreal, pp 254–258
113. Terzaghi K (1941) Undisturbed clay samples and undisturbed clays. J Boston Soc Civil Eng 29(3):211–231
114. Bjerrum L (1967) Seventh Rankine Lecture: Engineering geology of Norwegiannormally-consolidated marine clays as related to settlement of buildings. Géotechnique 17(2):81–118
115. Mesri G, Godlewski PM (1977) Time and stress-compressibility interrelationship. ASCE J Geotech Eng 103(5):417–430
116. Vaid YP, Campanella RG (1977) Time-dependent behavior of undisturbed clay. ASCE J Geotech Eng 103(7):693–709
117. Graham J, Crooks JHA, Bell AL (1983) Time effects on the stress-strain behavior of natural soft clays. Géotechnique 33(3):327–340
118. Leroueil S, Kabbaj M, Tavenas F, Bouchard R (1985) Stress-strain-strain rate relationship for the compressibility of sensitive natural clays. Géotechnique 35(2):159–180
119. Nash DFT, Sills GC, Davison LR (1992) One-dimensional consolidation testing of soft clay from Bothkennar. Géotechnique 42(2):241–256
120. Sheahan TC, Ladd CC, Germaine JT (1996) Rate-dependent undrained shear behavior of saturated clay. ASCE J Geotech Eng 122(2):99–108
121. Yin JH, Cheng CM (2006) Comparison of strain-rate dependent stress-strain behavior from K0-consolidated compression and extension tests on natural Hong Kong Marine deposits. Mar Georesour Geotechnol 24(2):119–147

122. Yin JH, Graham J (1989a) Viscous-elastic-plastic modelling of one-dimensional time-dependent behaviour of clays. Can Geotech J 26:199–209
123. Yin JH, Graham J (1989b). General elastic viscous plastic constitutive relationships for 1-D straining in clays. In: Proceedings of the 3rd international symposium numerical models geomechanics. Elsevier Applied Science, pp 108–117
124. Borja RI, Kavazanjian E (1985) A constitutive model for the stress-strain-time behavior of 'wet' clays. Géotechnique 35(3):283–298
125. Kutter BL, Sathialingam N (1992) Elastic-viscoplastic modelling of the rate-dependent behavior of clays. Géotechnique 42(3):427–441
126. Yin JH, Graham J (1999) Elastic viscoplastic modelling of the time-dependent stress-strain behaviour of soils. Can Geotech J 36:736–745
127. Vermeer PA, Neher HP (1999) A soft soil model that accounts for creep. In: Proceedings Plaxis symposium "Beyond 2000 in Computational Geotechnics", Amsterdam, pp 249–262
128. Perzyna P (1963) The constitutive equations for working hardening and rate sensitive plastic materials. Proc Vib Probl 4(3):281–290
129. Perzyna P (1966) Fundamental problems in viscoplasticity. In: Advances in applied mechanics, vol 9. Academic Press, New York, pp 244–368
130. Shahrour I, Meimon Y (1995) Calculation of marine foundations subjected to repeated loads by means of the homogenization method. Comput Geotech 17(1):93–106
131. Fodil A, Aloulou W, Hicher PY (1997) Viscoplastic behaviour of soft clay. Géotechnique 47(3):581–591
132. Hinchberger SD, Rowe RK (2005) Evaluation of the predictive ability of two elastic-viscoplastic constitutive models. Can Geotech J 42(6):341–371
133. Yin ZY, Hicher PY (2008) Identifying parameters controlling soil delayed behavior from laboratory and in situ pressuremeter testing. Int J Numer Anal Meth Geomech 32(12):1515–1535
134. Yin JH, Zhu JG, Graham J (2002) A new elastic viscoplastic model for time-dependent behaviour of normally and overconsolidated clays: theory and verification. Can Geotech J 39:157–173
135. Kimoto S, Oka F (2005) An elasto-viscoplastic model for clay considering destructuralization and consolidation analysis of unstable behavior. Soils Found 45(2):29–42
136. Rocchi G, Fontana M, Da Prat M (2003) Modelling of natural soft clay destruction processes using viscoplasticity theory. Géotechnique 53(8):729–745
137. Leoni M, Karstunen M, Vermeer PA (2008) Anisotropic creep model for soft soils. Géotechnique 58(3):215–226
138. Yin ZY, Karstunen M (2008) Influence of anisotropy, destructuration and viscosity on the behavior of an embankment on soft clay. In: Proceeding of the 12th international conference of the international association for computer methods and advances in geomechanics, Goa, India, CD-ROM
139. Yin ZY, Chang CS, Karstunen M, Hicher PY (2010) An anisotropic elastic viscoplastic model for soft clays. Int J Solids Struct 47(5):665–677
140. Yin ZY, Karstunen M, Hicher PY (2010) Evaluation of the influence of elasto-viscoplastic scaling functions on modelling time-dependent behaviour of natural clays. Soils Found 50(2):203–214
141. Yin JH, Tong F (2011) Constitutive modelling of the time-dependent stress-strain behaviour of saturated soils exhibiting both creep and swelling. Can Geotech J 48(12):1870–1885
142. Yin ZY, Karstunen M (2011) Modelling strain-rate-dependency of natural soft clays combined with anisotropy and destructuration. Acta Mech Solida Sin 24(3):216–230
143. Yin ZY, Karstunen M, Chang CS, Koskinen M, Lojander M (2011) Modeling time-dependent behavior of soft sensitive clay. J Geotech Geoenviron Eng 137(11):1103–1113
144. Yao Y, Gao Z, Zhao J, Wan Z (2012) Modified UH model: constitutive modelling of over-consolidated clays based on a parabolic Hvorslev envelope. J Geotech Geoenviron Eng 138(7):860–868

145. Buisman AS (1936) Results of long duration settlement tests. In: Proceedings of the 1st international conference on soil mechanics and foundation engineering, vol 1. Cambridge, Mass, pp 103–107
146. Garlanger JE (1972) The consolidation of soils exhibiting creep under constant effective stress. Géotechnique 22(1):71–78
147. Magnan JP (1984) Modélisation numériques du comportement des argiles molles naturelles. DSc thesis, Université Pierre et Marie Curie, Paris
148. Kabbaj M, Oka F, Leroueil S, Tavenas F (1986) Consolidation of natural clays and laboratory testing. In: Yong RN, Townsend FC (eds) Consolidation of soils: testing and evaluation, ASTM STP 892. American Society for Testing and Materials, Philadelphia, pp 378–404
149. Magnan JP, Lepidas I (1987) Influence of creep anisotropy and over consolidation on the results of finite element analyses of embankments on soft clays. In: Proceedings of international symposium on geotechnical engineering on soft soils, vol 1. Mexico City, pp 265–272
150. Imai G (1989) A unified theory of one-dimensional consolidation with creep. In: Proceedings of the 12th International Conference on SMFE, vol 1, pp 57–60
151. Yin JH, Graham J (1994) Equivalent times and one-dimensional elastic viscoplastic modelling of time-dependent stress-strain behaviour of clays. Can Geotech J 31:42–52
152. Nash DFT, Ryde SJ (2001) Modelling consolidation accelerated by vertical drains in soild subject to creep. Géotechnique 51(3):257–273
153. Matsuda H, Baek W, Baek J (2007a) Effect of creep on the settlement-time relation during primary consolidation of clay. In: Proceedings of international workshop on constitutive modelling, development, implementation, evaluation and application, Hong Kong, Jan12–13, pp 208–217
154. Matsuda H, Baek W, Baek J, Sato Y (2007b) Effect of creep on the settlement-time relation of clay during primary consolidation process. In: Proceedings of the 42th Japan national conference on geotechnical engineering, vol 135, pp 269–270
155. Kelln C, Sharma JS, Hughes D, Graham J (2008) An improved elastic-viscoplastic soil model. Can Geotech J 45(21):1356–1376
156. Kelln C, Sharma JS, Hughes D (2008) A finite element solution scheme for an elastic-viscoplastic soil model. Comput Geotech 35:524–536
157. Kelln C, Sharma JS, Hughes D, Graham J (2009) Finite element analysis of an embankment on a soft estuarine deposit using an elastic-viscoplastic soil model. Can Geotech J 46(12):357–368
158. Bodas Freitas TM, Potts DM, Zdravkovic L (2011) A time dependent constitutive model for soils with isotach viscosity. Comput Geotech 38(6):809–820
159. Yin JH (1999) Non-linear creep of soils in oedometer tests. Géotechnique 49(5):699–707
160. Zhou C, Yin JH, Zhu JG, Cheng CM (2005) Elastic anisotropic viscoplastic modeling of the strain-rate-dependent stress-strain behaviour of *K0*-consolidated natural marine clays in triaxial shear tests. Int J Geomech 5(3):218–232
161. Nash DFT, Brown MA (2015) Influence of destructuration of soft clay on time-dependent settlements: comparison of some elastic visco-plastic models. Int J Geomech. A4014004. https://doi.org/10.1061/(ASCE)GM.1943-5622.0000281
162. Mesri G, Vardhanabhuti B (2005) Secondary compression. J Geotech Geoenviron Eng 131:398–401
163. Le TM, Fatahi B, Khabbaz H (2015) Numerical optimisation to obtain elastic viscoplastic model parameters for soft clay. Int J Plast 65:1–21
164. Ye L, Jin YF, Shen SL, Sun PP, Zhou C (2016) An efficient parameter identification procedure for soft sensitive clays. J Zhejiang Univ Sci A (Appl Phys Eng) 17(1):76–88
165. Baudet B, Stallebrass S (2004) A constitutive model for structured clays. Géotechnique 54(4):269–278
166. Dafalias Y, Herrmann L (1982) Bounding surface formulation of soil plasticity. In: Soil mechanics-transient and cyclic loads, vol 10, pp 253–282
167. Asaoka A, Nakano M, Noda T (2000) Superloading yield surface concept for highly structured soil behavior. Soils Found 40(2):99–110

168. Yin Z-Y, Xu Q, Hicher P-Y (2013) A simple critical-state-based double-yield-surface model for clay behavior under complex loading. Acta Geotech 8(5):509–523
169. Li XS, Wang Y (1998) Linear representation of steady-state line for sand. J Geotechn Geoenviron Eng 124(12):1215–1217
170. Gudehus G (1997) Attractors, percolation thresholds and phase limits of granular soils. Proc Powder Grains 97
171. Biarez J, Hicher P-Y (1994) Elementary mechanics of soil behaviour: saturated remoulded soils. AA Balkema,
172. Manzari MT, Dafalias YF (1997) A critical state two-surface plasticity model for sands. Geotechnique 47(2):255–272
173. Matsuoka H, Nakai T (1985) Relationship among tresca, mises, mohr-coulomb and matsuoka-nakai failure criteria. Soils Found 25(4):123–128
174. Been K, Jefferies M (1985) A state parameter for sands. Geotechnique 35(2):99–112

Chapter 2
Estimation of SWCC and Permeability for Granular Soils

Abstract This chapter presents a new estimation model for soil–water characteristic curve (SWCC) of granular soils with the effect of initial density. The proposed model, namely the modified Fredlund and Xing (MFX) model, is based on the Fredlund and Xing equation without increasing the number of parameters. The uncertainty of measurement data is analyzed by using the Bayesian method and applied in the estimation of the relative permeability function (k_r). Different levels of confidence intervals of SWCC are obtained by using the Bayesian approach. It is found that the proposed method can better predict the k_r values with smaller uncertainty for the soil with different initial dry densities than the Fredlund and Xing equation. The new MFX model is readily adopted for engineers to quickly estimate the SWCC and relative permeability of granular soil under different initial dry densities.

2.1 Introduction

The measurement of soil parameters for unsaturated soils requires complicated laboratory tests and it is always time-consuming. To avoid cumbersome laboratory work, the soil–water characteristic curve (SWCC) is often used to estimate various unsaturated soil property functions, such as shear strength function, permeability function and thermal property function. At present, many methods or soil property indexes have been adopted to estimate SWCC. The statistical correlations between water potential at certain matric suction and soil texture have been developed [1, 2]. Some researchers [3, 4] established empirical relations between soil properties and fitting parameters of the SWCC equation. Some researchers used a physics-based conceptual model to estimate SWCC, such as the works by Arya and Paris [5] and Fredlund et al. [6] among the others. This approach involves conversion of the grain-size distribution into pore-size distribution of a soil. Similarly, Fredlund and Xing [7] proposed a SWCC equation based on the assumption that the shape of SWCC depends on the pore-size distribution of the soil. Leong and Rahardjo [8] reviewed and evaluated five popular SWCC equations and found that the Fredlund and Xing [7] equation gave

W. Zhou et al., *Practice of Bayesian Probability Theory in Geotechnical Engineering*,
https://doi.org/10.1007/978-981-15-9105-1_2

the best fit among the others. Zhang and Chen [9] extended the Fredlund and Xing [7] model and the van Genuchten [10] model to describe bimodal and multimodal SWCCs.

The influencing factors of SWCC have been studied by previous researchers. Zhou and Yu [11] found that the initial water content and stress state were important influencing factors. It is well accepted that the initial dry density of soil has a significant influence on the SWCC. Yang et al. [12] showed that the shape of SWCC was similar to the grain-size distribution of the soil and was affected by the dry density of soil. Similarly, Rajkai et al. [13] studied the estimation of water retention characteristic from bulk density and particle-size distribution parameters. Vanapalli et al. [14] revealed that the effect of initial water content and porosity on SWCC tends to diminish at high suction level. Gallipoli et al. [15] suggested an improved form of the van Genuchten's [10] SWCC model to account for the influence of void ratio. Tarantino [16] proposed a SWCC equation for deformable soils based on an empirical power function of the water ratio. Gallage and Uchimura [17] investigated the effect of dry density and grain-size distribution on SWCC's parameters of sandy soils. Sheng and Zhou [18] studied the effect of deformation on the soil–water retention properties and found the influence of initial void ratio on the air-entry value and the slope of SWCC. Therefore, the effect of initial dry density of soil should be sufficiently considered in estimating the SWCC of granular soils.

Generally speaking, the measured data are needed to obtain the optimal curve-fitting parameters in the SWCC models. However, it should be noted that the measured data of SWCC and the fitting parameters always exhibit high level of uncertainty due to the complicated unmodeled influencing factors. For the geotechnical uncertainties, Phoon and Kulhawy [19, 20] summarized its three primary sources: inherent variability, measurement error and transformation uncertainty. The uncertainty of empirical models can be considered by regarding the curve-fitting parameters as a random vector [21–23]. Some probabilistic measures have also been adopted to consider the uncertainties of empirical models or parameters in other civil or geotechnical engineering problems [24–35]. Recently, Zhai and Rahardjo [36] developed equations for the confidence limits of the best-fit SWCC to quantify the uncertainties in the associated fitting parameters.

For the same granular soil under different initial dry densities, their SWCCs are different. In this chapter, considering the effect of initial dry density, we propose a new estimation method based on Fredlund and Xing [7] equation. The data from the Unsaturated Soil Hydraulic Database (UNSODA, Leij et al. [37]) are used for fitting and verification of the proposed method. The estimation results by Fredlund and Xing [7] model and Tarantino [16] model will also be presented for comparison. It is found that the proposed method provides better predictions for the granular soils with different initial dry densities than the Fredlund and Xing model. Both the Tarantino's model and the present model consider the effect of the initial void ratio on the SWCC. From the comparison and statistical analysis, it is found that the estimations by the present method exhibit higher correlations with experimental data than the Tarantino's model for the granular soil samples. In addition, the uncertainty of the fitting parameters in the SWCC equations was analyzed by using the Bayesian method, and then the SWCC and relative permeability function (k_r) with different percentiles were predicted. The study shows that the estimated values of k_r by the proposed method are associated with smaller uncertainty than those by the Fredlund and Xing model for the soils with different initial dry densities.

2.2 Estimation of SWCCs with Different Initial Dry Densities

2.2.1 *Fredlund and Xing [7] Equation for SWCC*

Based on the assumption that the shape of the SWCC depends on the pore-size distribution of the soil, Fredlund and Xing [7] proposed a three-parameter equation for the SWCC, with flexibility to fit a wide variety of soils:

$$\theta = \frac{\theta_s}{\{\ln[\exp(1) + (\varphi/a)^k]\}^m}\left[1 - \frac{\ln(1 + \varphi/\varphi_r)}{\ln(1 + 10^6/\varphi_r)}\right] \tag{2.1}$$

where θ_s is the saturated volumetric water content; a is the fitting parameter related to the air-entry value for the soil; k is the fitting parameter related to the maximum slope of the curve; m is the fitting parameter related to the curvature of the curve; φ is the soil suction; φ_r represents the soil suction related to the residual volumetric water content. In order to reduce the complexity of the equation, it was suggested that the residual suction φ_r takes the value of 3000 kPa, regardless of the soil types [38]. In addition, the normalized volumetric water content, that is, $S = \theta/\theta_s$, is introduced and Eq. (2.1) can be re-written as:

$$S = \frac{1}{\{\ln[\exp(1) + (\varphi/a)^k]\}^m}\left[1 - \frac{\ln(1 + \varphi/\varphi_r)}{\ln(1 + 10^6/\varphi_r)}\right] \tag{2.2}$$

The fitting parameters, that is, a, k and m, can be obtained by nonlinear least-squares methods. Although the Fredlund and Xing equation is capable of describing the SWCC by fitting the measured data, it cannot distinguish the SWCCs for the same soil under different initial dry densities. It is because the same set of fitting parameters (a, k, m) is used in the Fredlund and Xing equation to describe their SWCCs. In other words, for the same soils with different initial dry densities, their SWCCs from Eq. (2.2) are identical. This induces larger level of prediction uncertainty in the estimation of unsaturated soil properties as the soil initial dry density has a significant influence on the SWCC, permeability and shear strength.

2.2.2 *Effect of Initial Dry Density and Proposed Estimation Method*

Soil is a porous medium with different particle sizes and complex texture. Granular soils at different compaction states are associated with different dry densities and porosities. The water in the porous media with different pore size exhibits different tension due to capillary effect. In the drying process, water is more difficult to flow out from soil with larger dry density, and hence the soil has different water retentions under different densities. Generally speaking, granular soil with larger dry density has larger air-entry value and larger residual water content. As a result, its SWCC will associate with smaller maximum slope [12, 17]. It should be noted that the contribution of the capillary mechanisms and associated pore structure to water retention is more important under relatively low suction. At high suction, the water retention is dominated by the surface adsorption mechanism. Although the clayey soil basically has this feature, the volume shrinkage of samples in the SWCC test is not negligible. For this reason, only granular soil is considered in this chapter.

From the above discussion about the effect of initial dry density on the SWCC of granular soil, a new estimation model is proposed. Sheng and Zhou [18] found that the modified suction can reflect the influence of porosity (n) on the hydraulic relationship between the water content and suction:

$$\varphi^* = n \cdot \varphi \tag{2.3}$$

In this study, the volume change during the desorption process is ignored. Inspired by the concept of modified suction, the initial porosity (n_0) is used in the estimation to consider the effect of initial dry density on the SWCC. After substantial investigations, the fitting parameters are proposed to be modified as follows:

$$a = a^*/n_0, \ k = k^* \cdot n_0, \ m = m^* \cdot n_0 \tag{2.4}$$

where n_0 is the measured initial porosity which is introduced to the set of fitting parameters (a^*, k^*, m^*) to consider the effect of initial dry density on soil–water characteristic curve. In particular, $a = a^*/n_0$ reflects the influence of the initial porosity on the air-entry value, and the air-entry value decreases as the initial porosity n_0 increases; $k = k^* \cdot n_0$ and $m = m^* \cdot n_0$ reflect the effect of initial porosity on the maximum slope and curvature of the SWCC. The water in the soil with larger n_0 is easier to flow out, so its SWCC is associated with larger maximum slope and curvature. Thus, the new estimation model is:

$$S = \frac{1}{\left\{\ln\left[\exp(1) + (\varphi \cdot n_0/a^*)^{k^* \cdot n_0}\right]\right\}^{m^* \cdot n_0}} \left[1 - \frac{\ln(1 + \varphi/\varphi_r)}{\ln(1 + 10^6/\varphi_r)}\right] \tag{2.5}$$

The effect of n_0 is further shown in Fig. 2.1. The three curves in this figure represent the SWCCs of the same soil ($a^* = 1, k^* = 15, m^* = 2$) with different initial porosities.

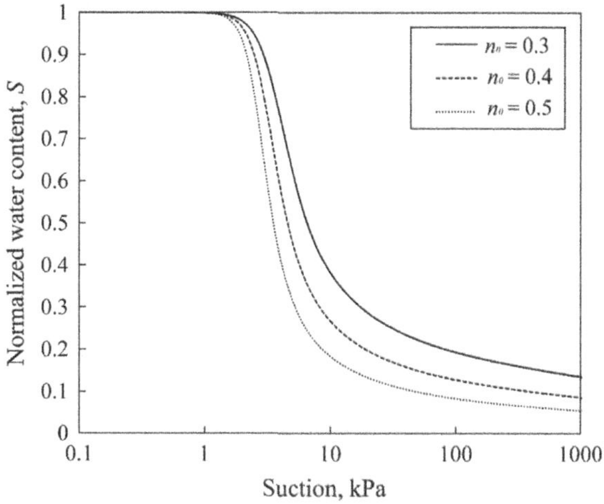

Fig. 2.1 The SWCCs for the same soils under different initial porosities ($a^* = 1$, $k^* = 15$, $m^* = 2$)

It is seen that the air-entry value, the maximum slope and the curvature of the SWCC are all shifted due to the change of the value of n_0. For the soil with a smaller value of n_0, its SWCC is always above that of the soil with a larger initial value of n_0. At the same suction level, the soil with a smaller value of n_0 is always associated with larger volumetric water content. Meanwhile, its maximum slope is smaller, implying that it is more difficult for the water to drain away from the soil with smaller value of n_0. Therefore, its corresponding residual water content is larger. Li and Zhang [40] and Zhang and Li [41] studied the soil microporosity structure and showed that the soils with different void ratios are evidently different in their inter-aggregate structures but almost the same in their intra-aggregate structures. Therefore, at low suction levels, the soil SWCCs at different void ratios would be distinctly different due to the different compressions of inter-aggregate pores. While at high suction, the soil water retention characteristics are mainly attributed to the intra-aggregate pores, which do not differ substantially for the soils of different initial densities. It can be seen from Fig. 2.1 that at high suction levels, the three curves tend to overlap gradually. In the next section, the proposed estimation method will be used to analyze the granular soils.

2.3 Verification and Discussion

The lab test data of SWCCs for three types of sands, namely Berlin coarse sand, Berlin medium sand and Wagram sand, with different initial porosities were collected from the UNSODA [37] database. According to the database manual [37], the tensiometry and gamma-ray attenuation were used to measure the suction and volumetric water content of Berlin coarse and Berlin medium sands, respectively. The pressure outflow method was used to determine the water retention of Wagram sand. The relevant soil properties and texture contents are summarized in Table 2.1. Figure 2.2 shows the process of data partitioning, curve-fitting and verification. The data from the same soil are divided into two groups randomly and repeatedly according to proportion of 2:1: the first group of two-thirds of the data points for obtaining the fitting parameters and the rest of the one-third for verification. In the fitting process, different sets of fitting parameters, that is, (a, k, m) for the Fredlund and Xing equation and (a^*, k^*, m^*) for the proposed estimation method, are obtained for different data sets.

It was found that the calculated fitting parameters could depend heavily on the partitioning of the two groups of data for fitting and verification due to the limited number of data points. Therefore, to assure the reliability of the data fitting, a simple criterion is imposed for data partitioning. According to the process of Fig. 2.2, 2250 sets of data are randomly selected from each soil type and 2250 sets of fitting parameters are computed by both Fredlund and Xing equation and the proposed method. Sorting the values of each fitting parameter, their values at 97.5 and 2.5 percentiles are obtained. Thus, the corresponding 95% confidence interval (CI) for each parameter can be constructed. 1000 sets of test data are randomly selected to analyze statistically if their fitting parameters, (a, k, m) and (a^*, k^*, m^*), are all located in their corresponding 95% CIs. Figure 2.3 shows an example of frequency distribution

Table 2.1 Soil properties and texture contents

Soil code	Soil type	n_0	θ_s	ρ_d (g/cm^3)	k_s (m/s)	Sand (%) 0.05–2 mm	Silt (%) 0.002–0.05 mm	Clay (%) <0.002 mm
1460	Berlin coarse sand	0.297	0.261	1.85	2.91e−5	98	0.8	1.2
1461	Berlin coarse sand	0.373	0.362	1.65	2.31e−4	99	0.7	0.3
1462	Berlin medium sand	0.430	0.315	1.50	1.16e−4	97	1.4	1.6
1463	Berlin medium sand	0.399	0.388	1.58	8.01e−5	97	1.9	1.1
1464	Berlin medium sand	0.365	0.357	1.67	2.31e−5	93	4.4	2.6
1140	Wagram sand	0.428[a]	0.385	1.56	2.05e−5	92	5	3
1141	Wagram sand	0.336[a]	0.302	1.70	4.47e−6	90	8	2
1142	Wagram sand	0.272[a]	0.245	1.80	1.55e−6	87	11	2

Note [a]The initial porosity is obtained through the saturated volumetric water content divided by 0.9 [39]

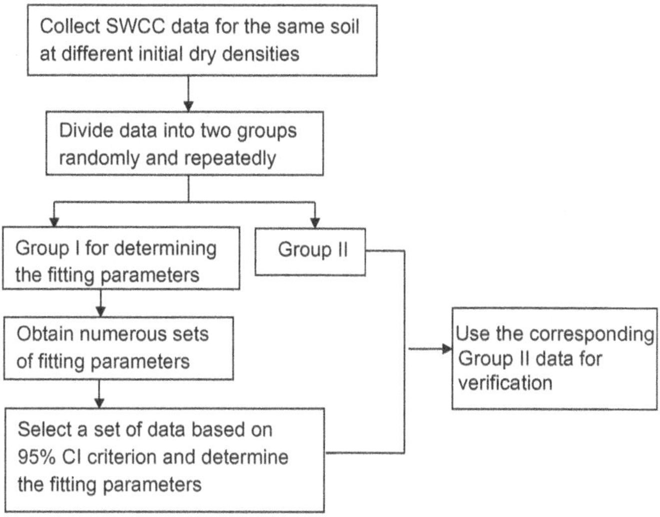

Fig. 2.2 Process for data partitioning, fitting and verification

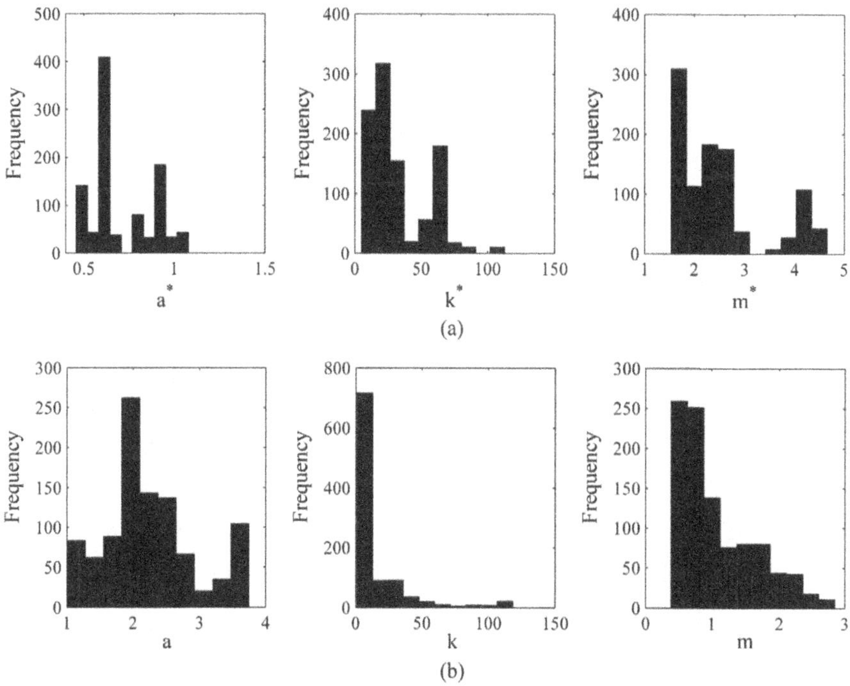

Fig. 2.3 Frequency distribution about fitting parameters of 1000 sets of random selected samples for Berlin coarse sand by **a** the proposed method and **b** Fredlund and Xing equation [7]

about the relevant fitting parameters by Fredlund and Xing equation and the proposed method. Finally, a random data set is used as Group I data, and then the rest of data points are categorized into Group II data for verification.

Figure 2.4 shows the fitting and verification results for the three types of sands with different initial porosities, where "FX" represents the Fredlund and Xing equation, that is, Eq. (2.2), and "MFX" represents the proposed model, that is, Eq. (2.5). The test results of the three types of sands in Fig. 2.4 show that the initial dry density (or initial porosity) of the soil has a significant influence on the soil water retention properties. Again, it is noted that the FX model provides only a single SWCC for the soil with the measured data. It cannot distinguish the SWCCs for soils with

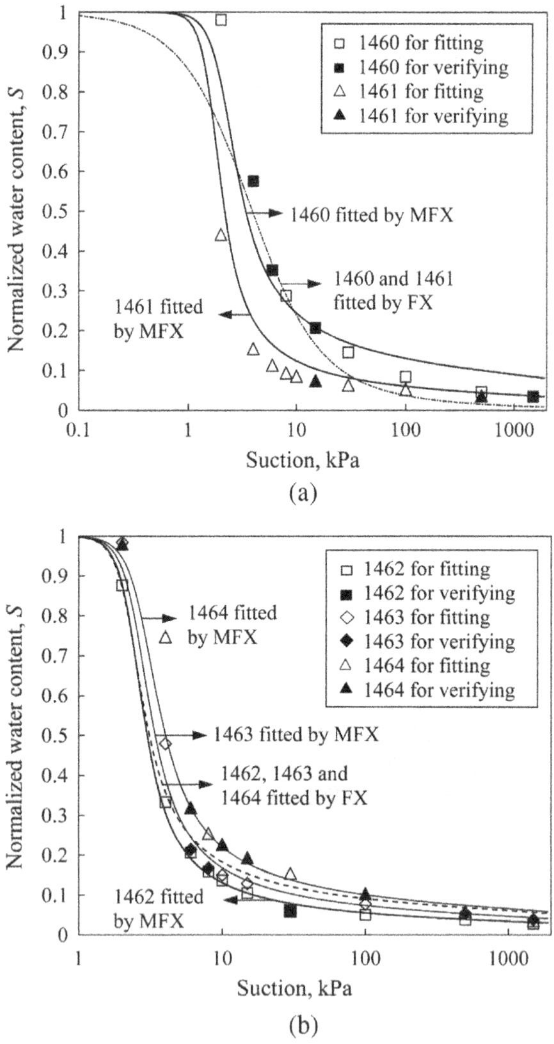

Fig. 2.4 Comparisons between the measured and the estimated SWCC by Fredlund and Xing [7] equation and the proposed method for **a** Berlin coarse sand, **b** Berlin medium sand, and **c** Wagram sand

Fig. 2.4 (continued)

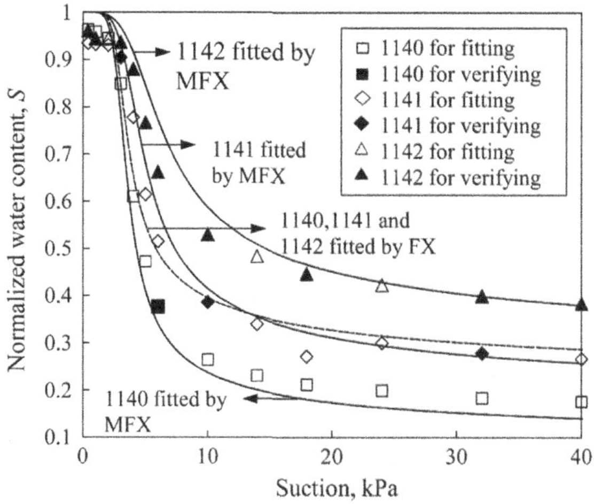

(c)

Table 2.2 Statistic index of the verification results from 1000 random selected samples for three types of soil

Soil type	FX		MFX	
	MAE	*RMSE*	*MAE*	*RMSE*
Berlin coarse sand	0.0731	0.1237	0.0377	0.0860
Berlin medium sand	0.0403	0.0707	0.0193	0.0316
Wagram sand	0.0698	0.0860	0.0337	0.0412

different initial porosities. With the consideration of initial porosity, the MFX can better estimate the SWCCs of the sands with different initial porosities than the FX model.

Table 2.2 presents the verification results of two statistical error indicators, that is, mean absolute error (*MAE*) and root mean square error (*RMSE*), of the 1000 sets of randomly selected test data. It can be seen from Table 2.2 that *MAE* and *RMSE* by the proposed MFX are always smaller than the corresponding values by FX. This confirms that the MFX model has improved the FX model for the prediction of the SWCCs.

Some researchers have proposed modifications to the van Genuchten's [10] model to account for the effect of the void ratio. In particular, Gallipoli et al. [15] suggested an improved form of the van Genuchten's SWCC model to incorporate volume changes. There are four fitting parameters in the model suggested by Gallipoli et al. [15]. The variation of the specific volume can be taken into account by using an elasto-plastic stress–strain model. In addition, Tarantino [16] modified the van Genuchten's [10] model and proposed a void-ratio-dependent SWCC model as written in the following form, for deformable soils:

$$S_r = \left\{ \frac{1}{1 + \left[(e/a)^{1/b} \varphi \right]^c} \right\}^{b/c} \tag{2.6}$$

where a, b and c are three fitting parameters. The Tarantino's model [16] has similar structure as the model proposed by Gallipoli et al. [15], but it has one less fitting parameter than the Gallipoli et al.'s equation since it assumes an empirical power function of the water ratio. The model has been validated by using experimental data for coarse-grained soils with significant finer fraction and fine-grained soils with low plasticity. In this chapter, the estimations by the Tarantino's model for the three granular soils (with very little fine fraction as shown in Table 2.1) at different initial void ratios are presented and compared with the present model. Both models, that is, Eqs. (2.5) and (2.6), contain three fitting parameters and they can account for the effect of initial void ratio on SWCC. It is noted that the void ratio e in Eq. (2.6) is assumed as a constant, which is equal to initial void ratio $e_0 = n_0/(1 - n_0)$, due to lack of experimental data about the volume changes with suction. Nevertheless, most granular soils undergo only a small amount of volume change under a wide range of applied soil suctions. Fredlund et al. [42] suggested that the volume change due to suction increase can be assumed as zero for coarse-grained soils. Zhang and Li [41] also pointed out that the soil with coarse-controlled structure has limited total volume change after drying. For this reason, it is assumed that the volume change of coarse-grained soils upon an increase in soil suction is zero. In the later section, the predictive equations for the estimation of permeability of unsaturated soil from a SWCC are based on this assumption.

Figure 2.5 shows the experimental data and estimations of normalized volumetric water content by the present model and the Tarantino's model [16] for three types of granular soils. It can be seen that Eq. (2.5) generally has better correlations with experimental data than Eq. (2.6), especially for low level of water contents. Table 2.3

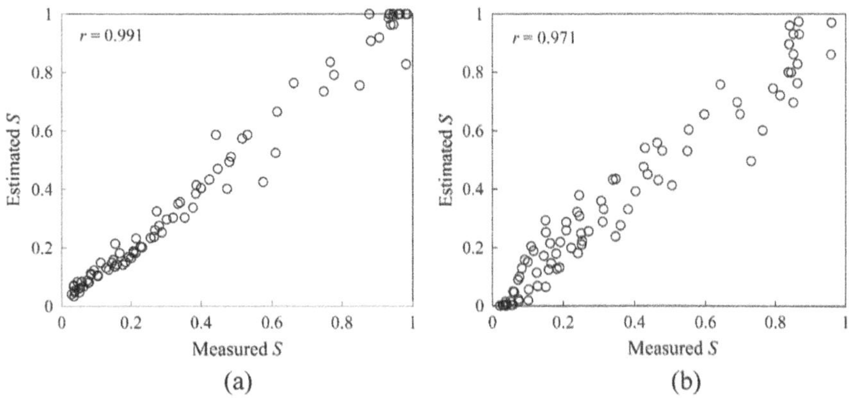

Fig. 2.5 Experimental data and estimations of normalized volumetric water content by **a** the present method and **b** Tarantino's model [16] for three types of granular soils

Table 2.3 Statistic index of results estimated by the present model and the Tarantino's model [16] for three types of granular soils

Soil type	Present model, Eq. (2.5)			Tarantino's model, Eq. (2.6)		
	MSE	*RMSE*	*r*	*MSE*	*RMSE*	*r*
Berlin coarse sand	0.046	0.067	0.9663	0.057	0.068	0.9472
Berlin medium sand	0.018	0.027	0.9962	0.056	0.073	0.9604
Wagram sand	0.038	0.046	0.9918	0.059	0.070	0.9635

shows the statistic results estimated by Eqs. (2.5) and (2.6). It is clear to see that the *MSE* and *RMSE* of the estimations by Eq. (2.5) are always smaller than that by Eq. (2.6). It is also demonstrated from the correlation coefficient *r* that Eq. (2.5) has better fitting power than Eq. (2.6) with the experimental data for the three granular soils.

2.4 Bayesian Approach and Confidence Interval of SWCC

In this section, the Bayesian approach is used to obtain the optimal values and the associated uncertainty of the three parameters from the MFX model and FX equation. Considering the uncertainty of the fitting parameters, Eq. (2.2) is analyzed using the Bayesian approach in this section. The equation, $S_{\mathrm{m}} = S + \varepsilon$, includes the model output error ε, which reflects the measurement noise and modeling error. This model output error is assumed to follow Gaussian distribution with zero mean and variance σ_ε^2. Therefore, the posterior probability density function (PDF) of the unknown parameter $\zeta = [a, k, m]^{\mathrm{T}}$ given the database D can be computed as follows:

$$p(\zeta|D) = c_0 p(\zeta)(2\pi)^{-N/2}\sigma_\varepsilon^{-N} \exp\left[-\frac{N}{2\sigma_\varepsilon^2} J_{\mathrm{g}}(\zeta|D)\right] \tag{2.7}$$

where N is the number of measured data points; c_0 is the normalizing constant; $p(\zeta)$ is the prior PDF of the model parameters in ζ representing the user's judgment; and $J_{\mathrm{g}}(\zeta|D)$ is the goodness-of-fit function indicating the level of data-fitting, and it has the following form:

$$J_g(\zeta|D) = \frac{1}{N}\sum_{n=1}^{N} [S_m(n) - S(\varphi;\zeta)]^2 \tag{2.8}$$

In this study, the unknown parameters include the three model parameters, (a, k, m) for FX model, (a^*, k^*, m^*) for MFX and the model output error variance σ_ε^2. The prior probability density function of fitting parameter is assumed to follow the uniform distribution, so the results will depend solely on the measurements. Firstly, the optimal values of the fitting parameters can be computed by minimizing the goodness-of-fit function, and then the optimal value of the model output error variance σ_ε^2 can be obtained by maximizing the posterior PDF, that is, Eq. (2.7). In addition, the covariance matrix Σ of unknown parameters is given by the inverse of the Hessian matrix which can be computed by the second derivatives of the function $-\ln p(\zeta|D)$. The measurements from two soils including Berlin coarse sand (soil codes 1460 and 1461) and Berlin medium sand (soil codes 1462 and 1464) are used, respectively, to obtain the optimal values of fitting parameters and the corresponding uncertainties are computed by using the Bayesian method. The relevant results, that is, $(\bar{a}, \bar{k}, \bar{m}, \sigma_a^2, \sigma_k^2, \sigma_m^2)$ and coefficient of determination R^2, are shown in Table 2.4. It should be noted that for MFX, the parameters (a, k, m) are computed by Eq. (2.4) after obtaining (a^*, k^*, m^*). The variances of the fitting parameters represent the uncertainty of the estimation. Except the variances for k, the variances for the parameters a and m by the proposed MFX equation are all smaller than those by the FX equation. Moreover, the R^2 obtained from MFX equation are obviously larger than that from FX equation, implying that the proposed method provides better estimation results than the Fredlund and Xing model.

From the optimal values and the corresponding covariance matrix of fitting parameters, different levels of confidence intervals (CIs) of SWCC corresponding to the particular soil can be obtained by the Markov chain Monte Carlo (MCMC) simulations. In this study, 10,000 samples of these fitting parameters were generated by MCMC. For each sample, SWCCs can be estimated by both FX and MFX together with the measurement noise. After sorting the obtained estimations, the water content and matric suction at different percentiles can be counted, and then the corresponding fitting parameters are obtained. Table 2.5 presents the fitting parameters for different percentiles of SWCC samples, where the quantitative difference between the parameters of the two methods can be clearly seen. These obtained fitting parameters will be used to estimate the relative permeability function (k_r) in the next section, so that the estimated k_r values will correspond to different percentile levels.

Table 2.4 Comparison of Bayesian analysis results between the proposed MFX and Fredlund and Xing equation (FX) [7]

	Soil code	$\bar{a}$	σ_a^2	$\bar{k}$	σ_k^2	$\bar{m}$	σ_m^2	R^2
MFX	1460	2.03	0.015	5.02	15.58	0.723	0.071	0.919
	1461	1.62	0.009	6.31	24.58	0.909	0.112	0.838
	1462	3.19	0.003	19.97	34.28	0.651	0.004	0.970
	1464	3.76	0.005	16.96	24.70	0.552	0.003	0.979
FX	1460	1.82	0.146	3.16	4.87	1.114	0.301	0.756
	1461							0.038
	1462	3.23	0.161	6.52	33.55	0.830	0.107	0.833
	1464							0.971

Table 2.5 Fitting parameters for different percentiles of SWCC samples

Soil code	Model parameters		Percentiles								
			2.5	5	10	25	50	75	90	95	97.5
1460	MFX	*a*	1.94	1.95	1.97	2.01	2.05	2.10	2.12	2.04	1.79
		k	6.15	5.87	5.52	4.92	4.16	3.34	2.61	2.06	2.00
		m	1.14	1.08	1.00	0.88	0.74	0.61	0.47	0.40	0.28
1461	MFX	*a*	1.49	1.54	1.56	1.59	1.60	1.54	1.42	1.24	0.98
		k	7.96	8.94	8.69	7.80	6.01	3.85	2.28	1.77	1.52
		m	1.63	1.42	1.28	1.07	0.88	0.71	0.52	0.40	0.28
1462	MFX	*a*	3.11	3.14	3.15	3.16	3.19	3.22	3.22	3.24	3.23
		k	23.25	23.52	23.09	21.42	19.49	17.33	14.12	12.26	10.06
		m	0.78	0.75	0.72	0.69	0.65	0.61	0.59	0.57	0.57
1464	MFX	*a*	3.68	3.70	3.72	3.74	3.76	3.78	3.79	3.79	3.80
		k	20.62	20.48	20.08	18.60	16.88	14.54	12.16	10.39	8.56
		m	0.65	0.63	0.61	0.58	0.55	0.52	0.50	1.80	0.49
1460/1461	FX	*a*	1.43	1.54	1.65	1.79	1.89	1.95	1.85	1.74	1.74
		k	3.45	3.86	3.59	3.39	3.00	2.58	2.37	2.74	2.96
		m	2.25	1.93	1.77	1.41	1.06	0.71	0.41	0.23	0.14
1462/1464	FX	*a*	2.49	2.60	2.73	2.99	3.27	3.28	3.15	2.96	2.82
		k	4.92	4.93	5.03	5.83	6.77	4.89	3.56	3.21	4.07
		m	1.91	1.76	1.55	1.16	0.79	0.59	0.39	0.27	0.15

2.5 Estimation of Relative Permeability Function k_r

The soil dry density has significant influence not only on the SWCC but also on other properties of unsaturated soil. In this section, the proposed SWCC model is applied to estimate the relative permeability function of unsaturated soil. The relative permeability function, k_r, is the ratio of the effective permeability to the saturated permeability for the unsaturated soil $k_r = k_w/k_s$. Several statistical formulae have been proposed to estimate the relative permeability function from SWCC equation by previous researchers. Fredlund et al. [42, 43] proposed using the FX equation to predict the coefficient of permeability for unsaturated soil. The formula has the following integration form:

$$k_r(\phi) = \Theta \frac{\int_{\phi}^{\phi_0} \frac{\theta(\nu)-\theta(\phi)}{\nu^2} \theta'(\nu) \mathrm{d}\nu}{\int_{\phi_s}^{\phi_0} \frac{\theta(\nu)-\theta_s}{\nu^2} \theta'(\nu) \mathrm{d}\nu} \tag{2.9}$$

where ν is a dummy variable associated with the suction; φ_0 is the suction corresponding to complete dryness, $\varphi_0 = 10^6$ kPa; φ_s is the suction corresponding to the initial saturated water content θ_s and it has a small finite value, $\varphi_s = 0.001$ kPa,

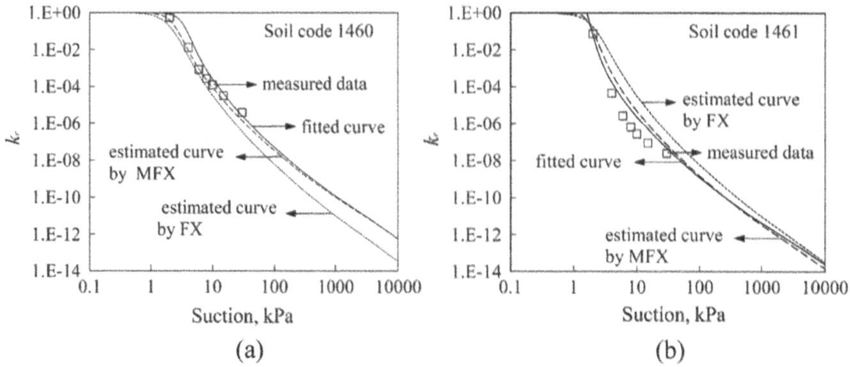

Fig. 2.6 Comparisons of k_r values between the measured and estimated results by Fredlund and Xing equation [7] and the proposed method for Berlin coarse sand: **a** soil code 1460 **b** soil code 1461

to avoid singularity in the integration; θ' is the derivative of the volumetric water content function; Θ is the correction factor used here which can take tortuosity into consideration and increases the flexibility of the permeability estimation equation.

By using Eq. (2.9), both MFX and FX can be used to estimate k_r values. Figure 2.6 presents the measured and estimated k_r values from MFX and FX for Berlin coarse sand, with the use of the fitting parameters shown in Table 2.4. The SWCC of each soil code was also computed by curve-fitting of Eq. (2.2), and then the SWCC was used in Eq. (2.9) to compute k_r, denoted as "fitted curve" in Fig. 2.6. Comparing the fitted curve with the measured k_r, it can be seen that the fitted curve does not perfectly agree with the measured data. This disagreement is suspected to be due to the error of measurements or the estimation model, that is, Eq. (2.9) in this case. In order to eliminate the effect of measurement error and model error, the fitted curve is used as a reference to compare with the estimated k_r by MFX and FX. It is found in Fig. 2.6 that the MFX can well predict the k_r for the soils with different initial porosities, while the estimated k_r by FX deviates substantially from the "fitted curve".

In order to clearly show the difference between the estimated k_r by MFX and FX, the "fitted curve" is used to compute the relative error of the k_r values. The relative error (RE) is defined as:

$$RE = \frac{k_{\mathrm{r,estimated}} - k_{\mathrm{r,\ fitted\ curve}}}{k_{\mathrm{r,\ fitted\ curve}}} \times 100\% \tag{2.10}$$

where $k_{\mathrm{r,\ estimated}}$ is the value estimated by the SWCC of MFX or FX and $k_{\mathrm{r,\ fitted\ curve}}$ is the value in the fitted curve. Figure 2.7 depicts the relative error of k_r values estimated by the SWCCs of MFX and FX for Berlin coarse sand. It is clearly seen that the RE value of k_r estimated by MFX is much smaller than that by FX for a wide range of suction.

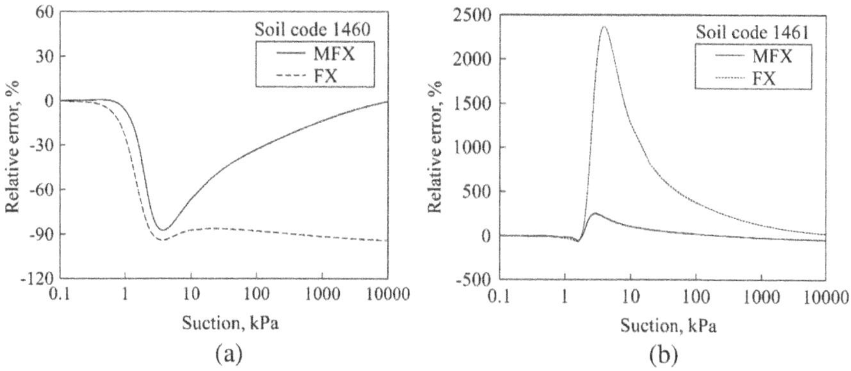

Fig. 2.7 Relative error of k_r values estimated by Fredlund and Xing equation [7] and the proposed method for Berlin coarse sand: **a** soil code 1460 **b** soil code 1461

The obtained optimal values and the variance of the fitting parameters can be used to estimate the SWCCs corresponding to different percentiles. Then, the corresponding k_r curves can be obtained by Eq. (2.9). Using the fitting parameters with different percentiles in Table 2.5, the estimated k_r at different percentiles were computed and compared with the fitted curves. Figure 2.8 presents the upper and lower boundaries of the predicted k_r from 50% CI SWCCs and the mean curve estimated by MFX and FX for soil codes 1460 and 1462. It can be seen that the fitted curve is mostly located in the predicted area of k_r from 50% CI SWCCs of FX and MFX, and closer to the mean curve estimated by MFX. It is also found that the estimations from the 50% CI SWCCs by MFX have smaller range than those by FX. In other words, the newly proposed model provides significantly smaller estimation uncertainty of the relative permeability function for the soils with different initial porosities than the Fredlund and Xing equation.

Figure 2.9 shows the influence of the initial porosity on the predicted k_r from 50% CI SWCCs by MFX and FX. Since the Fredlund and Xing model does not take into account the effect of initial porosity, the k_r curves estimated from the boundaries of 50% CI SWCCs by FX are identical for different values of n_0. On the other hand, it can be observed from Fig. 2.9 that the predicted k_r curves of $n_0 = 0.4$ by MFX are always below the curves of $n_0 = 0.3$ for the 25 and 75 percentiles. In other words, smaller k_r values are predicted by the proposed method for the soil with larger initial porosity.

2.6 Conclusions

A new estimation model for SWCC is introduced in this chapter to consider the influence of the initial dry density of granular soils. The proposed method is based on the Fredlund and Xing equation [7] without increasing the number of adjustable parameters. The test data from the literature were used for fitting and verification analysis.

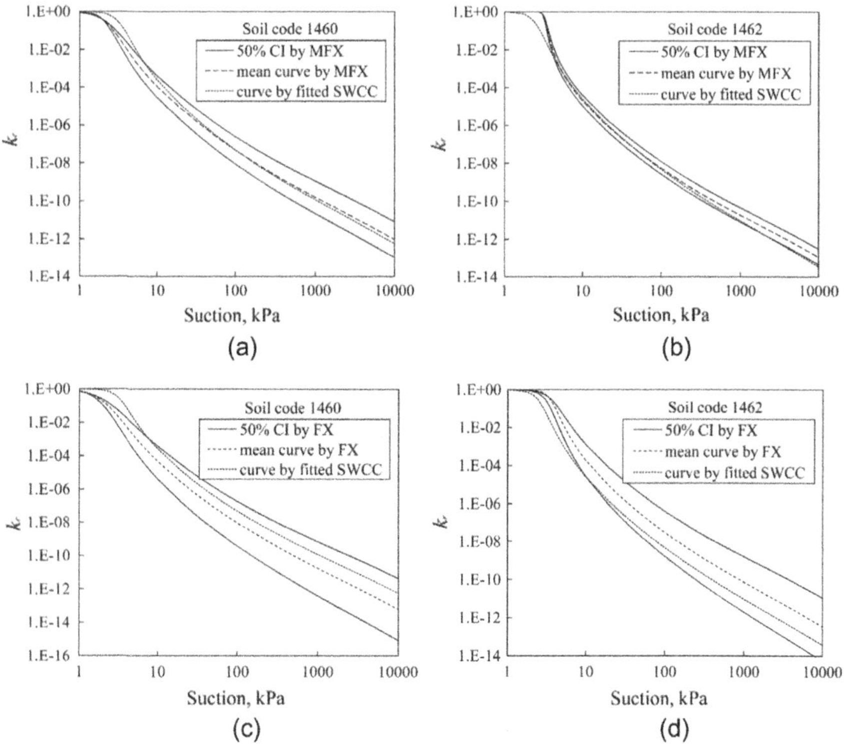

Fig. 2.8 Upper and lower boundaries of the predicted kr curves from 50% CI SWCCs and the mean curve estimated by the proposed method: **a** soil code 1460 and **b** soil code 1462; and the curves estimated by Fredlund and Xing equation [7]: **c** soil code 1460 and **d** soil code 1462

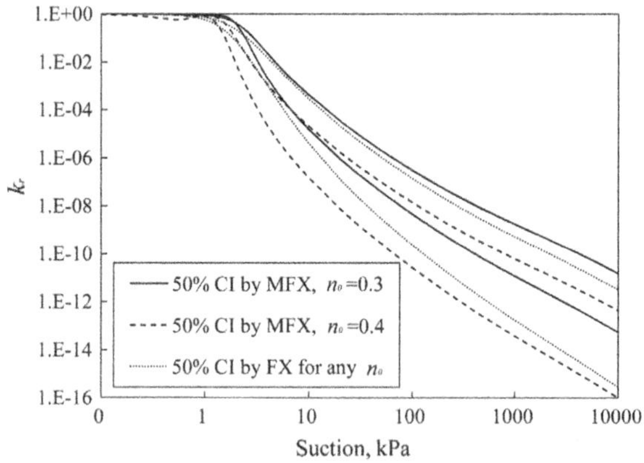

Fig. 2.9 Influence of initial porosity on predicted kr curves from 50% CI SWCCs by the proposed method and Fredlund and Xing equation [7]

It is found that with the consideration of the initial porosity, the proposed MFX method can better estimate the SWCCs of the soils with different initial dry densities than the Fredlund and Xing equation. From the comparison with the Tarantino's model [16], the statistical analyses show that the estimations by the proposed MFX model exhibit higher correlations with the experimental data for the granular soil samples. Bayesian approach is adopted to consider the uncertainty of the estimated SWCC. Furthermore, the proposed MFX method is applied to predict the k_r values for the soils with different initial porosities. It is found that the proposed method can better predict the k_r values with smaller uncertainty for the soils with different initial porosities than the Fredlund and Xing equation. The new method is readily used for engineers to quickly estimate the unsaturated properties (SWCC, relatively permeability, etc.) of the same granular soil under different initial dry densities.

It shall also be pointed out that the new proposed method captures the SWCC and permeability characteristics of coarse-grained soils under different initial dry densities in the drying process and the volume chance upon an increase in soil suction is assumed to be zero. Cautions shall be taken to apply this model to wetting process as the mechanisms for drying and wetting processes are different.

References

1. Gupta SC, Larson WE (1979) Estimating soil water retention characteristics from particle size distribution, organic matter percent, and bulk density. Water Resour Res 15(6):1633–1635
2. Saxton KE, Rawls WJ, Romberger JS, Papendick PI (1986) Estimating generalized soil water characteristics from texture. Soil Sci Soc Am J 50:1031–1036
3. Ahuja LR, Naney JW, Williams RD (1985) Estimating soil water characteristics from simpler properties or limited data. Soil Sci Soc Am J 49(5):1100–1105
4. Chin KB, Leong EC, Rahardjo H (2010) A simplified method to estimate the soil-water characteristic curve. Can Geotech J 47:1382–1400
5. Arya LM, Paris JF (1981) A physicoempirical model to predict the soil moisture characteristics from particle-size distribution and bulk density data. Soil Sci Soc Am J 45(6):1023–1030
6. Fredlund MD, Fredlund DG, Wilson GW (1997) Prediction of the soil-water characteristics curve from grain size distribution and volume-mass properties. In: Proceedings of the 3rd Brazilian symposium on unsaturated soils, Rio de Janeiro, Brazil, 22–25 Apr 1997. Freitas Bastos, Rio de Janeiro, Brazil, pp 13–23
7. Fredlund DG, Xing A (1994) Equations for the soil-water characteristic curve. Can Geotech J 31:521–532
8. Leong EC, Rahardjo H (1997) Review of soil-water characteristic curve equations. J Geotech Geoenviron Eng 123(12):1106–1117
9. Zhang LM, Chen Q (2005) Predicting bimodal soil-water characteristic curves. J Geotech Geoenviron Eng 131(5):666–670
10. van Genuchten MT (1980) A closed-form equation for predicting the hydraulic conductivity of unsaturated soils. Soil Sci Soc Am J 44:892–898
11. Zhou J, Yu JL (2005) Influence affecting the soil-water characteristic curve. J Zhejiang Univ Sci 6A(8):797–804
12. Yang H, Rahardjo H, Leong EC, Fredlund DG (2004) Factors affecting drying and wetting soil-water characteristic curves of sandy soils. Can Geotech J 41:908–920

13. Rajkai K, Kabos S, Genuchten V, Jansson PE (1996) Estimation of water-retention characteristics from the bulk density and particle-size distribution of Swedish soils. Soil Sci 161(12):832–845
14. Vanapalli SK, Fredlund DG, Pufahl DE (1999) The influence of soil structure and stress history on the soil-water characteristics of a compacted till. Geotechnique 49(2):143–159
15. Gallipoli D, Wheeler SJ, Karstunen M (2003) Modelling the variation of degree of saturation in a deformable unsaturated soil. Geotechnique 53(1):105–112
16. Tarantino A (2009) A water retention model for deformable soils. Geotechnique 59(9):751–762
17. Gallage CPK, Uchimura T (2010) Effects of dry density and grain size distribution on soil-water characteristic curves of sandy soils. Soil Found 50(1):161–172
18. Sheng D, Zhou AN (2011) Coupling hydraulic with mechanical models for unsaturated soils. Can Geotech J 48:826–840
19. Phoon KK, Kulhawy FH (1999) Characterization of geotechnical variability. Can Geotech J 36:612–624
20. Phoon KK, Kulhawy FH (1999) Evaluation of geotechnical property variability. Can Geotech J 36:625–639
21. Phoon KK, Santoso A, Quek ST (2010) Probabilistic analysis of soil-water characteristic curves. J Geotech Geoenviron Eng 136(3):445–455
22. Wang Y, Au SK, Kulhawy FH (2011) Expanded reliability-based design approach for drilled shafts. J Geotech Geoenviron Eng 137(2):140–149
23. Chiu CF, Yan WM, Yuen KV (2012) Reliability analysis of soil-water characteristic curve and its application to slope stability analysis. Eng Geol 135–136:83–91
24. Phoon KK, Liu SL, Chow YK (2003a) Estimation of model uncertainties for reliability-based design of cantilever walls in sand. In: Proceedings of the international workshop on limit state design in geotechnical engineering practice (LSD2003), Massachusetts Institute of Technology, Cambridge (CDROM)
25. Phoon KK, Kulhawy FH (2003b) Evaluation of model uncertainties for reliability-based foundation design. In: Proceedings of the ninth international conference on applications of statistics and probability in civil engineering, San Francisco, pp 1351–1356
26. Zhang LM, Tang WH, Zhang LL, Zheng JG (2004) Reducing uncertainty of prediction from empirical correlations. J Geotech Geoenviron Eng 130(5):526–534
27. Ching J, Chen YC, Phoon KK (2009) Reducing uncertainty in undrained shear strength. Geotechnical risk and safety. Taylor and Francis Group, London, ISBN 978-0415-49874-6
28. Yan WM, Yuen KV, Yoon GL (2009) Bayesian probabilistic approach for the correlations of compression index for marine clays. J Geotech Geoenviron Eng 135(12):1932–1940
29. Wang Y, Au SK, Cao ZJ (2010) Bayesian approach for probabilistic characterization of sand friction angles. Eng Geol 114:354–363
30. Yuen KV (2010) Recent developments of Bayesian model class selection and applications in civil engineering. Struct Saf 32(5):338–346
31. Yuen KV, Kuok SC (2011) Bayesian methods for updating dynamic models. Appl Mech Rev 64(1): 010802-1–010802-18
32. Yuen KV, Mu HQ (2011) Peak ground acceleration estimation by linear and nonlinear models with reduced order Monte Carlo simulation. Comput Aided Civil Infrastructural Eng 26(1):30–47
33. Ching J, Chen JR, Yeh JY, Phoon KK (2012) Updating uncertainties in friction angles of clean sands. J Geotech Geoenviron Eng 138(2):217–229
34. Juang CH, Ching J, Lou Z, Ku CS (2012) New models for probability of liquefaction using standard penetration tests based on an updated database of case histories. Eng Geol 133–134:85–93
35. Zhou WH, Yuen KV, Tan F (2013) Estimation of maximum pullout shear stress of grouted soil nails using Bayesian probabilistic approach. Int J Geomech 13(5):659–664
36. Zhai Q, Rahardjo H (2013) Quantification of uncertainties in soil-water characteristic curve associated with fitting parameters. Eng Geol 163:144–152

37. Leij FJ, Alves WJ, van Genuchten MTh, Williams JR (1996) The UNSODA unsaturated hydraulic database. USEPA, Cincinnati, Ohio. EPA/600/R-96/095
38. Sillers WS, Fredlund DG (2001) Statistical assessment of soil-water characteristic curve models for geotechnical engineering. Can Geotech J 38:1297–1313
39. Pachepsky YA, Timlin D, Ahuja LR (1999) Estimating saturated soil hydraulic conductivity using water retention data and neural networks. Soil Sci 164:552–560
40. Li X, Zhang LM (2009) Characterization of dual-structure pore-size distribution of soil. Can Geotech J 46:129–141
41. Zhang LM, Li X (2010) Microporosity structure of coarse granular soils. J Geotech Geoenviron Eng 136(10):1425–1436
42. Fredlund DG, Rahardjo H, Fredlund MD (2012) Unsaturated soil mechanics in engineering practice. Wiley, Hoboken, New Jersey
43. Fredlund DG, Xing A, Huang S (1994) Predicting the permeability function for unsaturated soils using the soil-water characteristic curve. Can Geotech J 31:533–546

Chapter 3
Modeling SWCC for Coarse-Grained and Fine-Grained Soil

Abstract Following Chap. 2, the work is extended to predict SWCC of a soil with different initial densities by using Bayesian method, for both coarse-grained and fine-grained soil. An adjustment parameter (β) is introduced to express the relationships between the matric suctions of two soil samples. The parameter β is a function of the initial void ratio, matric suction or volumetric water content. Bayesian model class selection is adopted to determine the optimal predictive models of β. The optimal models of β are validated by comparing the estimated matric suction and measured data. The comparisons show that the newly proposed method produces more accurate SWCCs than the other models. Furthermore, the influence of the model parameters of β on the predicted matric suction and SWCC is evaluated using Latin hypercube sampling.

3.1 Introduction

Over the past few decades, the unsaturated soil theory has been substantially developed. The measurement of unsaturated soil properties, such as unsaturated hydraulic conductivity and shear strength, requires complicated laboratory tests, which are generally costly and time-consuming. To avoid cumbersome laboratory work, the soil–water characteristic curve (SWCC) and saturated soil properties are commonly used to estimate various unsaturated soil properties.

As a key function used to evaluate the properties of unsaturated soils, the SWCC can be directly measured in the laboratory. The conventional methods include tensiometer, suction-controlled test, pressure plate method and centrifuge tests. However, these methods suffer from various drawbacks, such as long-term cycles, high costs and discrete data points. Numerous methods have been proposed to estimate the SWCC using easily measurable soil properties. These soil properties, including the particle-size distribution, bulk density and organic matter content, are commonly used to estimate the water contents at certain suction levels [1–4]. Generally speaking, regression analysis is conducted in these studies. The relationships between these soil properties and the fitting parameters in empirical or fractal

W. Zhou et al., *Practice of Bayesian Probability Theory in Geotechnical Engineering*,
https://doi.org/10.1007/978-981-15-9105-1_3

SWCC models have also been investigated [5–13]. Additionally, some researchers have proposed SWCC prediction models using only the particle-size distribution [14–16].

The particle-size distribution (PSD) is an important soil texture property that can be easily obtained in the laboratory. Previous studies have shown that the PSD plays an essential role in controlling the shape of the SWCC. Based on the Fredlund and Xing equation [17], Fredlund et al. [15, 18] proposed two PSD fitting models, namely unimodal and bimodal models, to estimate the SWCC. Fredlund et al. [19] further noted that this method performed well for sands and silts. However, the method has low estimation accuracy for clays, tills and loams due to the complex fabrics of finer soils. Liu et al. [20] proposed mathematical functions for both bimodal and multimodal SWCCs, in which the involved parameters have certain physical significance. Nasta et al. [21] expanded upon the scaling technique of Kosugi and Hopmans [22] and the physico-empirical model of Arya et al. [14], correlating the water retention function with PSD. Recent studies have examined the effects of soil density and other factors on the water retention properties of the same soil. Some experimental results showed that initial dry densities greatly influence the SWCC in terms of degree of saturation [23]. Beckett and Augarde [24] incorporated porosity and particles of different sizes, shapes and separation distances in the prediction of the soil water retention properties. They also noted that determining multiple pore-size distributions over a range of applied suctions is crucial for obtaining good SWCC prediction. Li and Zhang [25] and Zhang and Li [26] studied the microstructures of several types of soil samples through measuring pore-size distributions, which directly determine the shape of SWCC, and investigated their influencing factors, for example, compaction conditions, water content and coarse fraction. They found that the pore volumes of soil larger than a certain pore size decrease significantly with the void ratio, and the pore-size distributions of saturated soil samples with different void ratios are different.

Therefore, the same-textured soil samples (i.e., with the same PSD) under different initial void ratios exhibit different water retention properties due to different pore-size distributions caused by different initial void ratios. In this chapter, a method is introduced to estimate the SWCCs of same-textured soils with different initial void ratios using a set of SWCC data and the Fredlund and Xing empirical equation [17]. SWCC measurements of coarse- and fine-grained soils are analyzed. Based on the physico-empirical hypotheses and weight–volume relationships, the relationships of the volumetric water contents and matric suctions between the same-textured soil samples with different initial void ratios are established. In the matric suction relationship, an adjustment parameter, which is a function of the initial void ratio, matric suction and volumetric water content, is investigated using the Bayesian method. Individual optimal models of this adjustment parameter are obtained for coarse- and fine-grained soils. The influence of the model parameters in the optimal models on the predicted matric suction and SWCC is analyzed using Latin hypercube sampling [27].

3.2 Establishment of Relationships

Arya and Paris [14] presented a SWCC model that combined physical hypotheses with an empirical presentation. Based on the shape similarity between the SWCC and the cumulative PSD curve, the Arya and Paris (AP) model was established using the PSD and dry unit weight of a soil. The model considers the water flow paths in the soil as a bundle of capillary tubes and assumes that the size of the soil particles is related to the corresponding pore diameters of the capillary tubes.

Given a set of SWCC data for a soil sample (1) with a known initial void ratio ($e^{(1)}$), the fitting SWCC can be obtained using the Fredlund and Xing empirical equation [17]:

$$\theta_{\mathrm{w}} = \frac{\theta_{\mathrm{s}}}{\left\{\ln\left[\exp(1) + (\varphi/a)^n\right]\right\}^m} \tag{3.1}$$

where θ_{s} is the saturated volumetric water content; a, n and m are three adjustable parameters; and φ is the matric suction of soil. The fitting results for the set of measured SWCC data can be written as ($\theta_{\mathrm{w}}^{(1)}$, $\varphi^{(1)}$). Using the dimensionless volumetric water content, $S = \theta_{\mathrm{w}}/\theta_{\mathrm{s}}$, Eq. (3.1) can be re-written as follows:

$$S = \frac{1}{\left\{\ln\left[\exp(1) + (\varphi/a)^n\right]\right\}^m} \tag{3.2}$$

The estimation of the SWCC ($\theta_{\mathrm{w}}^{(2)}$, $\varphi^{(2)}$) of a soil sample (2) that has the same PSD curve as the sample (1) but a different initial void ratio ($e^{(2)}$) is presented as follows.

3.2.1 *Relationship of Volumetric Water Contents*

The particle-size distribution curve is divided into n fractions. The solid particles in each fraction are assumed to be of uniform particle size. Assembling all the fractions, a natural structure sample is formed. The dry density of the sample is assumed equal to that of each assemblage of the n fractions.

The pore volume of each fraction per unit sample mass is given as follows:

$$V_{\mathrm{v}i} = m_{\mathrm{s}i} e/\rho_{\mathrm{s}} \quad i = 1, 2, \ldots, n \tag{3.3}$$

where V_{vi} is the soil pore volume of the ith fraction per unit sample mass; m_{si} is the soil particle mass of the ith fraction per unit sample mass; ρ_s is the soil particle mass density; and e is the initial void ratio. The first fraction of the assemblage has the smallest pore size.

It is assumed that soil water will first fill in the small pores in the soil at certain water contents. Thus, the volumetric water content of the soil sample when the pores of the ith fraction are filled with water can be calculated by the accumulated pore volume:

$$\theta_{wi} = \sum_{j=1}^{j=i} V_{vj} \Big/ V \quad i = 1, 2, \ldots, n \tag{3.4}$$

where V is the sample volume per unit sample mass.

Based on Eqs. (3.3) and (3.4), the volumetric water contents of two soil samples (1) and (2) satisfy the following relationship:

$$\frac{\theta_{wi}^{(2)}}{\theta_{wi}^{(1)}} = \frac{e^{(2)} V^{(1)} \sum_{j=1}^{j=i} m_{sj}^{(2)}}{e^{(1)} V^{(2)} \sum_{j=1}^{j=i} m_{sj}^{(1)}} \tag{3.5}$$

where $\theta_{wi}^{(1)}$ is the volumetric water content when the ith soil pore is filled with water in soil sample (1) with initial void ratio $e^{(1)}$, and $\theta_{wi}^{(2)}$ is that of soil sample (2) with initial void ratio $e^{(2)}$.

Based on the PSD curve, the particle mass in each fraction is the same in both soil samples, that is, $m_{sj}^{(2)} = m_{sj}^{(1)}$, $j = 1, 2, \ldots, i$. Considering the phase relationships $\rho_s V_s = \rho_d V$ and $\rho_d = G_s \rho_w / (1 + e)$, the following equation can be obtained:

$$\frac{V^{(2)}}{V^{(1)}} = \frac{1 + e^{(2)}}{1 + e^{(1)}} \tag{3.6}$$

Thus, the ratio of the volumetric water content can be obtained by substituting Eqs. (3.6) into (3.5):

$$\frac{\theta_{wi}^{(2)}}{\theta_{wi}^{(1)}} = \frac{e^{(2)} \left(1 + e^{(1)}\right)}{e^{(1)} \left(1 + e^{(2)}\right)} \tag{3.7}$$

The volumetric water content of soil sample (2) can be estimated using the volumetric water content data of soil sample (1) and the initial void ratios.

3.2.2 Relationship of Suctions

The matric suction of the ith fraction is related to the pore radius r_i through the following relationship:

$$\varphi_i = 2T_s \cos\phi / r_i \tag{3.8}$$

where T_s is the surface tension of water and ϕ is the contact angle between water and the soil. Then, for a same-textured soil, the following equation can be established for each fraction of the two soil samples:

$$\frac{\varphi_i^{(2)}}{\varphi_i^{(1)}} = \frac{r_i^{(1)}}{r_i^{(2)}} \tag{3.9}$$

Since it is difficult to directly measure the pore radiusr_i in the laboratory, indirect estimation is usually adopted. In this study, the following assumptions are adopted from the AP model:

(a) The soil particles are spherical, and the equivalent particle size is R_t for the ith fraction;
(b) The pore is cylindrical, and its radius $r_i$depends on the equivalent particle size R_t of the ith fraction;
(c) The ith fraction is composed of n_i soil particles with equivalent particle size R_t and the corresponding pore length l_i.

According to these assumptions, the particle volume and pore volume of ith fraction can be calculated as follows:

$$V_{si} = 4n_i\pi R_i^3 / 3 = m_{si} / \rho_s \tag{3.10}$$

$$V_{vi} = \pi r_i^2 l_i = \left(m_{si} / \rho_s\right)e \tag{3.11}$$

where the pore length is assumed to be $l_i = 2n_i^\alpha R_i$ and the scaling parameter α is introduced to account for the fact that the pore length is larger than the particle size. Substituting Eqs. (3.10) into (3.11), the pore radius can be expressed as follows:

$$r_i = R_i\left(2en_i^{1-\alpha_i} / 3\right)^{1/2} \tag{3.12}$$

The scaling parameter α varies with the corresponding fraction, and it is denoted as α_i for the ith fraction. With sufficient data, some researchers have studied the dependence of α on matric suction [28] and volumetric water content [29]. For the ith fraction per unit soil mass, two soil samples should have the same equivalent particle size R_i and the same particle number n_i. Therefore, the relationship between their pore radii has the following form:

$$\frac{r_i^{(1)}}{r_i^{(2)}} = \sqrt{\frac{e^{(1)}}{e^{(2)}} n_i^{\alpha_i^{(2)} - \alpha_i^{(1)}}} \tag{3.13}$$

Substituting Eqs. (3.9) into (3.13), the matric suction relationship between both soil samples can be obtained as follows:

$$\left[\frac{\varphi_i^{(2)}}{\varphi_i^{(1)}}\right]^2 = \frac{e^{(1)}}{e^{(2)}} \cdot \beta_i \tag{3.14}$$

where $\beta_i = n_i^{\alpha_i^{(2)} - \alpha_i^{(1)}}$ is an adjustment parameter that reflects the relationship between two matric suctions caused by different initial void ratios. In this study, Eqs. (3.7) and (3.14) are used to estimate the SWCC of same-textured soil with a different initial void ratio. In SWCC, if the scaling parameter α is the same for two different soil samples, the adjustment parameter β is equal to 1. In the following analysis, β is considered as a function that is associated with the known SWCC of sample (1) and the initial void ratio.

Because the adjustment parameter β cannot be measured directly in the laboratory or by field tests, the "measured data" of β used to determine the optimal predictive model are obtained by the following four steps:

(a) The measured SWCC data of soil sample (1) are fitted using the Fredlund and Xing equation (Eq. (3.1)).
(b) The volumetric water content of soil sample (2) is computed from Eq. (3.7) using the fitted SWCC of soil sample (1) and two initial void ratios.
(c) Based on the estimated volumetric water content of soil sample (2), the corresponding matric suction of soil sample (2) is obtained by computing the inverse function of Eq. (3.1).
(d) The β values between the two soil samples in group i are estimated from Eq. (3.14), that is, $\boldsymbol{\beta}_i = \left[\beta_{1i}, \beta_{2i}, \ldots, \beta_{n_i i}\right]^{\mathrm{T}}$, where n_i is the number of selected β values in group i. Then, the β values in all groups of soil samples, that is, $\boldsymbol{\beta} = \left[\boldsymbol{\beta}_1, \boldsymbol{\beta}_2, \ldots, \boldsymbol{\beta}_i, \ldots, \boldsymbol{\beta}_{n_s}\right]^{\mathrm{T}}$, where n_s is the number of groups of soil samples, are taken as the "measurements" in Bayesian model class selection to determine the optimal predictive models of β for coarse- and fine-grained soils.

3.3 Determination of the Adjustment Parameter β

Based on the existing studies of the scaling parameter in the AP model [28, 29], the adjustment parameter β in this study is assumed related to the water content and matric suction of soil sample (1). In addition, the initial void ratios of the two soil samples are added to consider the effect of the initial void ratio on the relationship of matric suctions. For simplicity, a linear predictive model is built for β. Due to the large difference between the orders of magnitude of β over the whole SWCC, the natural logarithm of β is adopted as a model output. A full predictive model of $\ln\beta$ is established as a linear function of the possible forms of three important factors: $S^{(1)}$, $e^{(2)}/e^{(1)}$ and $\varphi^{(1)}$. This model is as follows:

$$\ln\beta = b_1 + b_2\cdot S^{(1)} + b_3\cdot\left(S^{(1)}\right)^2 + b_4\cdot\ln S^{(1)} + b_5\cdot\exp\left(S^{(1)}\right) + b_6\cdot e^{(2)}/e^{(1)} \\ + b_7\cdot\ln\varphi^{(1)} + b_8\cdot e^{(2)}/e^{(1)}\cdot\ln\varphi^{(1)} \tag{3.15}$$

where $e^{(2)}/e^{(1)}$ is the initial void ratios of the two soil samples with the same texture. $S^{(1)}$ is the dimensionless volumetric water content of sample (1) and $\varphi^{(1)}$ is the corresponding matric suction. $b_1, b_2, \ldots, b_8$ are the regression coefficients that must be estimated. Measured data of $\ln\beta$ are obtained for coarse- and fine-grained soils based on the method presented in the previous section.

3.3.1 Established of Predictive Model

As briefly introduced in Sect. 1.2, the Bayesian approach can not only update the model parameters and characterize the uncertainties using their posterior probability distribution functions (PDFs) but also find the most plausible model for predicting the responses using Bayesian model class selection. In this section, the Bayesian model class selection is applied to establish the predictive model for adjustment parameter β.

A linear predictive model class C can be written as follows:

$$\ln\beta(\mathbf{x};\mathbf{b},C) = \sum_{l=1}^{N_b} b_l x_l \tag{3.16}$$

where $x_1, x_2, \ldots,$ and x_{N_b} are the corresponding data, that is, variables such as $S^{(1)}$, $\left(S^{(1)}\right)^2$, $\ln S^{(1)}$, $\exp\left(S^{(1)}\right)$, $e^{(2)}/e^{(1)}$ and $\ln\varphi^{(1)}$. $b_1, b_2, \ldots,$ and b_{N_b} are uncertain coefficients required for identification. N_b is the number of uncertain coefficients in the predictive formula.

The measurement of is denoted by y and is modeled as follows:

$$y = \ln\beta(\mathbf{x};\mathbf{b},C) + \varepsilon \tag{3.17}$$

where ε is the predictive error. Additionally, it is a zero-mean normal random variable with prediction-error variance σ_ε^2. Therefore, all the uncertain parameters in $\boldsymbol{\theta} = \left[\mathbf{b}^{\mathrm{T}}, \sigma_\varepsilon^2\right]^{\mathrm{T}}$, and the number of components N_j is $N_b + 1$. Their uncertainties can be evaluated using the posterior PDFs, and the expression of the posterior PDF for data D is written as follows:

$$P(\boldsymbol{\theta}|D,C) = c_0 p(\boldsymbol{\theta}|C)p(D|\boldsymbol{\theta},C) \tag{3.18}$$

where $c_0 = 1/p(D|C)$ is a normalizing constant; $p(D|C)$ is the evidence of model class C; $p(\boldsymbol{\theta}|C)$ is the prior PDF of the uncertain parameters in $\boldsymbol{\theta}$, which is based on the previous knowledge or user's judgment; and $p(D|\boldsymbol{\theta}, C)$ is the likelihood function expressing the level of data-fitting. If the prediction errors in different measured data are statistically independent, the likelihood function can be computed as follows:

$$\mathrm{p}(D|\boldsymbol{\theta}, C) = \left(2\pi\sigma_\varepsilon^2\right)^{-\frac{N}{2}} \exp\left[-\frac{N}{2\sigma_\varepsilon^2} J_g(\mathbf{b}; D, C)\right] \tag{3.19}$$

where N is the number of measured data. $J_g(\mathbf{b}; D, C)$ is the goodness-of-fit function, which is shown in Eq. (3.20):

$$J_g(\mathbf{b}; D, C) = \frac{1}{N}\sum_{n=1}^{N}\left[y(n) - \sum_{l=1}^{N_b} b_l x_l(n)^2\right]^2 \tag{3.20}$$

With a uniform prior PDF of the coefficients, the optimal coefficient vector $\mathbf{b}^*$ can be obtained by minimizing $J_g(\mathbf{b}; D, C)$ when the range of the uniform distribution is sufficiently large. This can be performed by solving the linear algebraic equation $\partial J_g(\mathbf{b}; D, C)/\partial \mathbf{b} = 0$, and the result is as follows:

$$\mathbf{b}^* = \mathrm{A}^{-1}\begin{bmatrix} \frac{1}{N}\sum\limits_{n=1}^{N} x_1(n)y(n) \\ \frac{1}{N}\sum\limits_{n=1}^{N} x_2(n)y(n) \\ \cdots \\ \frac{1}{N}\sum\limits_{n=1}^{N} x_{N_b}(n)y(n) \end{bmatrix} \tag{3.21}$$

where $\mathbf{A}$ is an $N_b \times N_b$ symmetric matrix with (l, l') component $A^{(l,l')} = \frac{1}{N}\sum_{n=1}^{N} x_l(n)x_{l'}(n)$.

Furthermore, the updated fitting-error variance σ_ε^{2*} can be obtained by minimizing "goodness-of-fit" value:

$$\sigma_\varepsilon^{2*} = \min J_g(\mathbf{b}; D, C) = J_g(\mathbf{b}^*; D, C) \tag{3.22}$$

The uncertainty in the parameter estimations can be represented by the covariance matrix given by $\sum_{\boldsymbol{\theta}} = \mathbf{H}(\boldsymbol{\theta}^*)^{-1}$, where the Hessian matrix $\mathbf{H}(\boldsymbol{\theta}^*)$ is given as follows:

$$\mathbf{H}(\boldsymbol{\theta}^*) = N\begin{bmatrix} \left(\sigma_\varepsilon^{2*}\right)^{-1}\mathbf{A} & 0_{N_b\times 1} \\ 0_{1\times N_b} & \frac{1}{2}\left(\sigma_\varepsilon^{2*}\right)^{-2} \end{bmatrix} \tag{3.23}$$

3.3.2 Model Class Selection

A number of model class candidates can be generated by including different terms on the right side of Eq. (3.15). In all candidate model classes, the constant b_1 is retained as an intercept term. Therefore, there are 127 candidates in total. The most suitable predictive model class among the model class candidates can be obtained by Bayesian model class selection. The plausibility of a predictive model class C_j given the data D can be obtained by using Eq. (1.9). The evidence of the model class C_j can be approximated as follows:

$$p(D|C_j) \approx p(\boldsymbol{\theta}^*|C_j)\exp\left(-\frac{N}{2}\right)\frac{\pi^{\frac{N_j-N}{2}}\left(\sqrt{2}\sigma_\varepsilon^*\right)^{N_j+1-N}}{\sqrt{N^{N_j}|\mathbf{A}|}} \tag{3.24}$$

where $\boldsymbol{\theta}^*$ is the optimal parameter vector for a given model class C_j, that is, $\boldsymbol{\theta}^* = \left[\mathbf{b}^{*\mathrm{T}}, \sigma_\varepsilon^{2*}\right]^{\mathrm{T}}$. The Ockham factor $p(\boldsymbol{\theta}^*|C_j)/p(\boldsymbol{\theta}^*|D, C_j)$ provides a measurement of the robustness of the model class, and its value decreases exponentially with the number of uncertain parameters in the model class. Therefore, given the same plausibility for all model class candidates, the model class with the largest value given by Eq. (3.24) is regarded to have the best tradeoff between the data-fitting capability and the robustness to model error. A MATLAB code for the linear model class selection is presented in Appendix A.

3.4 Results

3.4.1 Optimal Model for β

The SWRC data resulting from the lab tests of 34 soils were collected from the Unsaturated Soil Hydraulic Database (UNSODA) [30] for the analysis of the optimal model of the adjustment parameter. The data are divided into two categories according to the Unified Soil Classification System (USCS) [31], that is, 10 coarse-grained soils and 24 fine-grained soils, as shown in Table 3.1. The PSDs of some fine-grained samples are missing. The soil samples in the same row have the same PSD but different initial void ratios. Samples with the same PSD are considered to have the same texture and grouped. Therefore, 6 and 12 groups of soil samples are used for the analysis of coarse- and fine-grained soils, respectively. Each group of soil samples has a series of measured β values. It should be noted that it is not possible to obtain the same PSDs, even for two soil samples with the same texture. Hence, soil samples in the same group can have slightly different PSDs.

Tables 3.2 and 3.3 show the top 10 and full model classes based on the model class selection of coarse- and fine-grained soils, respectively. These selected model classes

Table 3.1 Soil samples used for analysis [30]

Classification	Soil type	Soil samples			Soil samples		
		Soil code	Initial void ratio e_0	Saturated volumetric water content θ_s	Soil code	Initial void ratio e_0	Saturated volumetric water content θ_s
Coarse-grained soils	S	1460	0.423	0.261	1461	0.595	0.362
	S	1464	0.575	0.357	1462	0.754	0.430
	S	2210	0.499	0.300	2211	0.572	0.328
	S	2210	0.499	0.300	2213	0.560	0.323
	S	2210	0.499	0.300	2216	0.585	0.332
	S	4445	0.577	0.290	4441	0.689	0.300
Fine-grained soils	SL#	1340	0.736	0.424	1350	0.934	0.483
	SL#	1352	0.742	0.426	1342	0.957	0.489
	SC	1361	0.779	0.438	1362	1.155	0.536
	SC	2032*	1.160	0.500	2031*	1.336	0.515
	SC	2041*	1.058	0.484	2040*	1.653	0.549
	SC	2071*	1.203	0.502	2070*	1.660	0.561
	SC	2090*	1.404	0.535	2080*	1.577	0.557
	SL	2240	0.845	0.458	2242	1.193	0.544
	L	2531	0.808	0.401	2530	0.938	0.426
	SC	2540	0.458	0.270	2550	0.684	0.336
	SL#	2732	0.672	0.402	2731	0.894	0.471
	L	4592	1.028	0.507	4590	1.481	0.597

Note soil code "*" denotes a soil sample with no particle-size distribution data; "S", "SC", "SL", "SL#" and "L" denote sand, silty clay, sandy loam, silty loam and loam, respectively

Table 3.2 Model class selection results for coarse-grained soils

Number	Parameters	σ_ε^2	$\ln p(D\|\boldsymbol{\theta}^*, C_j)$	$\ln o_j$	$p(C_j\|D)$
1	1 3 6 8	1.469	−357.705	−26.398	0.571
2	1 3 7 8	1.483	−358.748	−27.203	0.090
3	1 5 6 8	1.489	−359.207	−26.792	0.086
4	1 3 6 7 8	1.463	−357.228	−28.797	0.084
5	1 2 5 6 8	1.454	−356.568	−30.344	0.034
6	1 2 3 6 8	1.461	−357.099	−30.513	0.017
7	1 2 6 8	1.533	−362.410	−25.403	0.014
8	1 5 6 7 8	1.482	−358.679	−29.186	1.326e−2
9	1 5 7 8	1.504	−360.312	−27.598	1.268e−2
10	1 3 4 6 8	1.469	−357.694	−30.470	9.828e−3
41	1 2 3 4 5 6 7 8	1.424	−354.222	−38.424	1.112e−4

Table 3.3 Model class selection results for fine-grained soils

Number	Parameters	σ_ε^2	$\ln p(D\|\boldsymbol{\theta}^*, C_j) \times 10^3$	$\ln o_j$	$p(C_j\|D)$
1	1 7 8	3.956	−0.950	−20.278	0.662
2	1 3 7 8	3.894	−0.946	−26.069	0.074
3	1 6 7 8	3.950	−0.950	−22.933	0.069
4	1 2 7 8	3.913	−0.948	−25.266	0.056
5	1 5 7 8	3.899	−0.947	−26.523	0.035
6	1 4 7 8	3.924	−0.948	−25.233	0.030
7	1 3 4 5 7 8	3.806	−0.941	−32.938	0.013
8	1 2 3 4 7 8	3.809	−0.942	−33.004	0.010
9	1 3 6 7 8	3.885	−0.946	−28.733	0.009
10	1 2 4 5 7 8	3.814	−0.942	−33.070	0.007
24	1 2 3 4 5 6 7 8	3.790	−0.940	−37.037	5.792e−4

have the relatively larger posterior plausibilities, and they are ranked according to the posterior plausibility $P(C_j|D)$. Figure 3.1 presents a histogram of the posterior plausibilities. The second column in Tables 3.2 and 3.3 shows the terms included in each model class, for example, "1 3 6 8" in the second column of Table 3.2 denotes a model class with free parameters b_1, b_3, b_6 and b_8 in Eq. (3.15). In addition to the constant term, all these model classes include the term $e^{(2)}/e^{(1)} \cdot \ln\varphi^{(1)}$, representing the multiplication effect of the ratio of initial void ratio and matric suction on the adjustment parameter for both types of soils. In addition, for fine-grained soils, the seventh term of $\ln\varphi^{(1)}$ is another particularly important factor, as it is present in the top 10 model classes.

The optimal values of the regression coefficients in the selected predictive model classes are shown in Tables 3.4 and 3.5. Their posterior variances are also shown in parentheses. Based on the analysis results shown in Tables 3.2, 3.3, 3.4 and 3.5, the most plausible model classes of $\ln\beta$ are as follows:

$$\ln\beta = 7.371 - 3.115 \cdot \left(S^{(1)}\right)^2 - 4.404 \cdot e^{(2)}/e^{(1)} - 0.896 \cdot e^{(2)}/e^{(1)} \cdot \ln\varphi^{(1)} \quad \text{for the coarse-grained soils} \tag{3.25}$$

$$\ln\beta = -2.219 + 3.508 \cdot \ln\varphi^{(1)} - 2.731 \cdot e^{(2)}/e^{(1)} \ln\varphi^{(1)} \quad \text{for the fine-grained soils} \tag{3.26}$$

Based on Eqs. (3.7), (3.14), (3.25) and (3.26), the associated estimations are shown in Figs. 3.2, 3.3 and 3.4. Three performance indices are computed, including the correlation coefficient (ρ; Eq. (3.27)), mean absolute percentage error (*MAPE*; Eq. (3.28)) and root mean square error (*RMSE*; Eq. (3.29)):

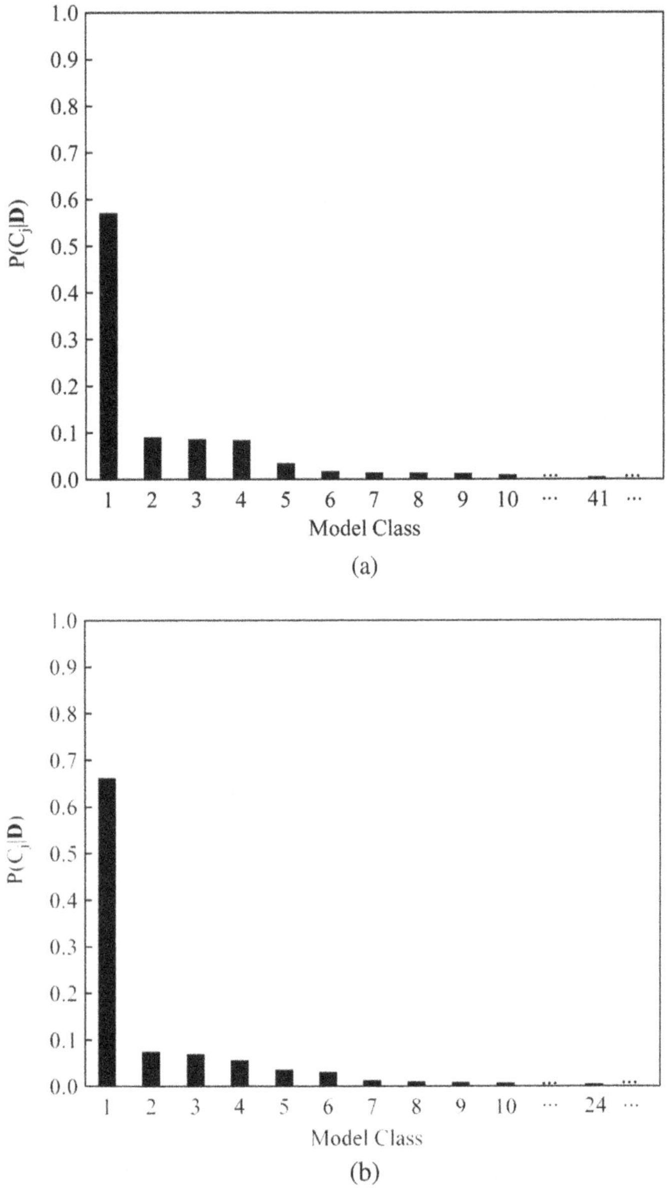

Fig. 3.1 Model class selection results: **a** coarse-grained soils and **b** fine-grained soils

Table 3.4 Optimal parameters of selected predictive model classes and their uncertainties in coarse-grained soils

Number	b_1	b_2	b_3	b_4	b_5	b_6	b_7	b_8
1	7.371 (0.999)	–	−3.115 (0.154)	–	–	−4.404 (0.657)	–	−0.896 (0.001)
2	1.884 (0.065)	–	−2.954 (0.157)	–	–	–	0.922 (0.031)	−1.638 (0.020)
3	10.03 (1.251)	–	–	–	−1.968 (0.065)	−4.629 (0.666)	–	−0.923 (0.001)
4	5.551 (4.46)	–	−3.050 (0.158)	–	–	−2.954 (2.851)	0.359 (0.135)	−1.183 (0.087)
5	12.45 (2.230)	7.094 (9.410)	–	–	−5.781 (2.786)	−4.193 (0.686)	–	−0.861 (0.002)
6	6.636 (1.437)	1.825 (2.740)	−4.586 (1.933)	–	–	−4.183 (0.694)	–	−0.863 (0.002)
7	8.591 (1.139)	−3.414 (0.227)	–	–	–	−4.832 (0.687)	–	−0.939 (0.002)
8	8.039 (4.968)	–	–	–	−1.924 (0.067)	−3.089 (2.900)	0.380 (0.136)	−1.226 (0.088)
9	4.112 (0.286)	–	–	–	−1.852 (0.066)	–	0.968 (0.032)	−1.699 (0.021)
10	7.373 (0.999)	–	−3.088 (0.190)	−0.020 (0.019)	–	−4.422 (0.673)	–	−0.899 (0.002)
41	27.41 (942.84)	25.08 (508.61)	13.57 (692.56)	−0.241 (0.152)	−23.83 (851.71)	−2.237 (2.871)	0.502 (0.138)	−1.268 (0.092)

Note The values in parentheses are the posterior variances of parameters

Table 3.5 Optimal parameters of selected predictive model classes and their uncertainties in fine-grained soils

Number	b_1	b_2	b_3	b_4	b_5	b_6	b_7	b_8
1	−2.219 (0.036)	–	–	–	–	–	3.508 (0.027)	−2.731 (0.014)
2	−0.688 (0.358)	–	−1.709 (0.402)	–	–	–	3.302 (0.032)	−2.693 (0.014)
3	−3.903 (3.629)	–	–	–	–	1.251 (1.984)	3.768 (0.112)	−2.925 (0.062)
4	−0.680 (0.504)	−1.574 (0.490)	–	–	–	–	3.359 (0.031)	−2.704 (0.014)
5	0.475 (1.125)	–	–	–	−1.036 (0.161)	–	3.320 (0.032)	−2.697 (0.014)
6	−2.180 (0.036)	–	–	−0.557 (0.082)	–	–	3.417 (0.029)	−2.717 (0.014)
7	−96.83 (894.49)	–	−87.00 (702.06)	−11.29 (12.60)	66.69 (429.43)	–	3.265 (0.032)	−2.695 (0.014)
8	−16.74 (26.72)	30.60 (93.66)	−16.68 (23.09)	−6.610 (4.615)	–	–	3.302 (0.032)	−2.710 (0.014)
9	−2.571 (3.806)	–	−1.737 (0.402)	–	–	1.417 (1.955)	3.593 (0.115)	−2.911 (0.061)
10	2.181 (3.737)	36.19 (133.02)	–	−5.252 (3.334)	−15.17 (20.06)	–	3.311 (0.032)	−2.714 (0.014)
24	−66.68 (3.98e3))	15.25 (585.25)	−58.63 (3.08e3)	−9.787 (20.426)	38.59 (2.68e3)	1.817 (1.940)	3.652 (0.144)	−2.982 (0.060)

Note The values in parentheses are the posterior variances of parameters

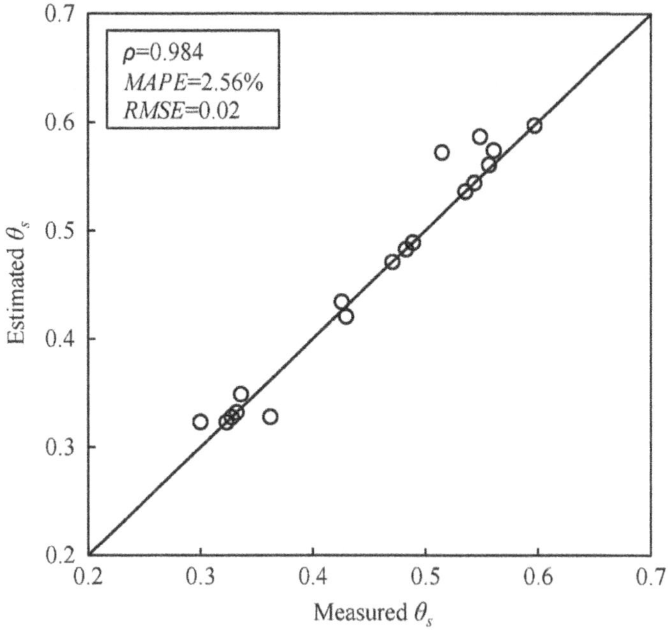

Fig. 3.2 Comparison between measured values of saturated volumetric water content and those estimated by Eq. (3.7)

$$\rho = \frac{\sum_{n=1}^{N}\left(y_n - \mu_y\right)\left(\bar{y}_n - \mu_{\bar{y}}\right)}{\sqrt{\sum_{n=1}^{N}\left(y_n - \mu_y\right)^2}\sqrt{\sum_{n=1}^{N}\left(\bar{y}_n - \mu_{\bar{y}}\right)^2}} \tag{3.27}$$

$$\text{MAPE} = \frac{1}{N}\sum_{n=1}^{N}\left|\frac{y_n - \bar{y}_n}{y_n}\right| \times 100\% \tag{3.28}$$

$$\text{RMSE} = \sqrt{\frac{1}{N}\sum_{n=1}^{N}(y_n - \bar{y}_n)^2} \tag{3.29}$$

where N is the number of measurements and y_n and $\bar{y}_n$ are the nth measurement and prediction, respectively.

The saturated volumetric water contents of all coarse- and fine-grained soil samples are estimated by Eq. (3.7) and shown in Fig. 3.2. The calculated values of ρ and *RMSE* are equal to 0.984 and 0.02, respectively, suggesting that Eq. (3.7) is capable of estimating the volumetric water content. Figure 3.3 shows a comparison

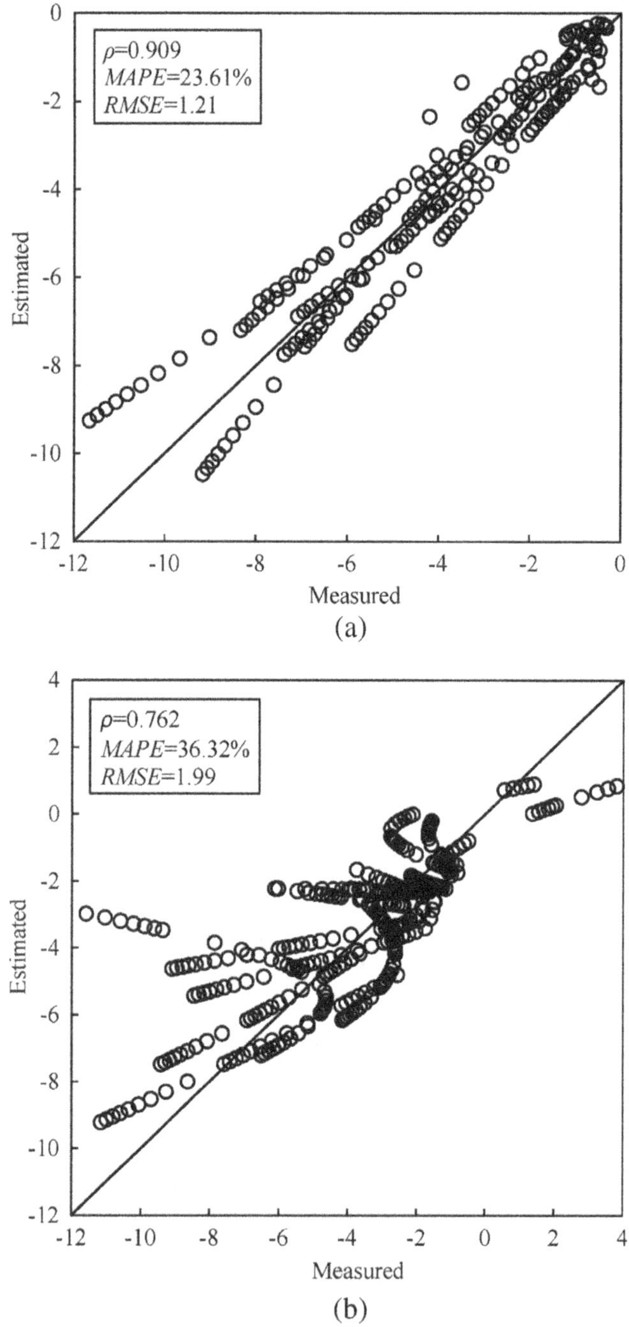

Fig. 3.3 Comparison of measured and estimated adjustment parameters: **a** coarse-grained soils and **b** fine-grained soils

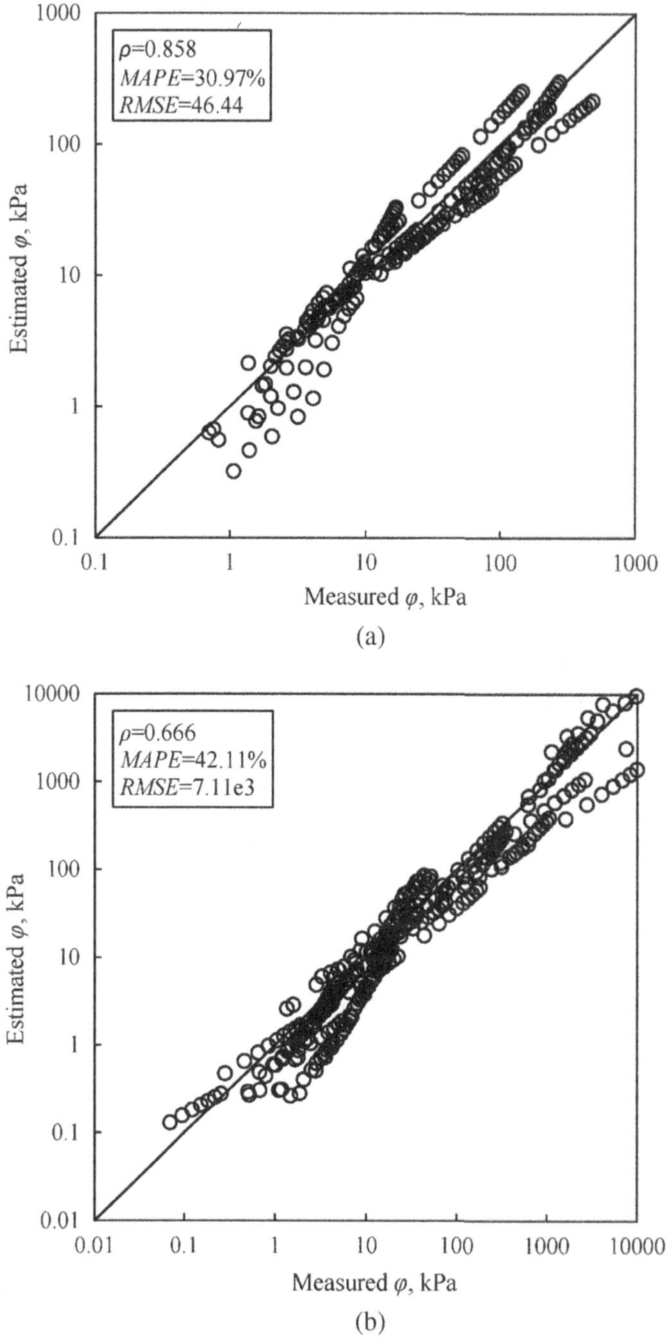

Fig. 3.4 Comparison of measured and estimated matric suctions: **a** coarse-grained soils and **b** fine-grained soils

of measured and estimated ln β results for coarse- and fine-grained soils. The correlation coefficients ρ are 0.909 and 0.762 for coarse-grained soils and fine-grained soils, respectively. Combining Eqs. (3.14) and (3.25) or Eq. (3.26), the matric suction of the same-textured soil with a different value of the initial void ratio can be estimated. Figure 3.4 presents the matric suction estimated using the values of β according to Eqs. (3.25) and (3.26), and the correlation coefficients between the estimated and measured matric suctions are 0.858 and 0.666 for coarse- and fine-grained soils, respectively. Thus, the estimated matric suctions are generally in agreement with the measured data.

3.4.2 Comparison of SWCCs Predicted by Different Methods

In this section, the SWCCs estimated using the proposed method are compared with the experimental data, as well as the estimations from the AP model and the modified Fredlund and Xing (MFX) equation proposed in Chap. 2. In the following comparative analysis, the MFX equation is used to estimate the SWCCs of both coarse-grained and fine-grained soils.

Figures 3.5 and 3.6 show comparisons of the SWCCs based on measured data and estimations provided by different predictive models for coarse-grained soils and fine-grained soils, respectively. The solid curve is obtained using the proposed method, that is, Eqs. (3.7) and (3.14), and the adjustment parameter is estimated using the proposed method in Eqs. (3.25) and (3.26), for coarse-grained soil and fine-grained soil, respectively. For comparison, the results using the proposed method with β as a constant, that is, the mean value of the measured results, are also plotted. In Figs. 3.5 and 3.6, β is defined as 0.214 and 1.520 for coarse- and fine-grained soils, respectively. The results of constant β can be used as a reference to evaluate the effectiveness of estimating SWCC using the proposed method with variable β, that is, with Eqs. (3.25) and (3.26). The dashed curve denotes the estimation using the MFX equation proposed in Chap. 2. These estimated SWCCs are obtained using a measured SWCC of the same-textured soil in the corresponding group. The figures show that the SWCCs estimated by the proposed method with variable β exhibit the best fit to the measured data. The estimations of the proposed method with constant β exhibit significant differences from the measurements, especially for fine-grained soils. Additionally, larger errors may occur when the suction is smaller than the air entry value, as observed in Figs. 3.5c and 3.6b. It is difficult to distinguish between the errors resulting from the estimations of water content and matric suction using the proposed method. Although the MFX equation was proposed based on SWCC data associated with granular soils, the MFX equation generally obtained better estimations than the proposed method with constant β for fine-grained soils in this study. One should be cautious when using the proposed relationships of the SWCC variables at high matric suctions because the effect of the initial void ratio at high matric suctions diminishes and the SWCC is dominated by a surface adsorption mechanism.

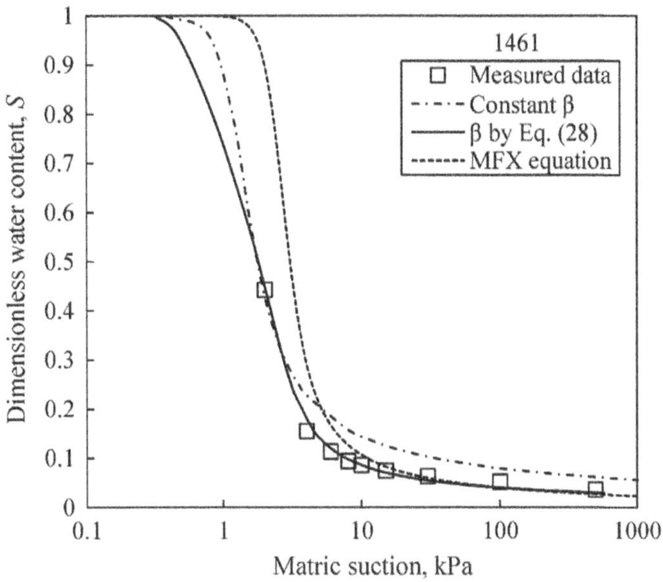

(a)

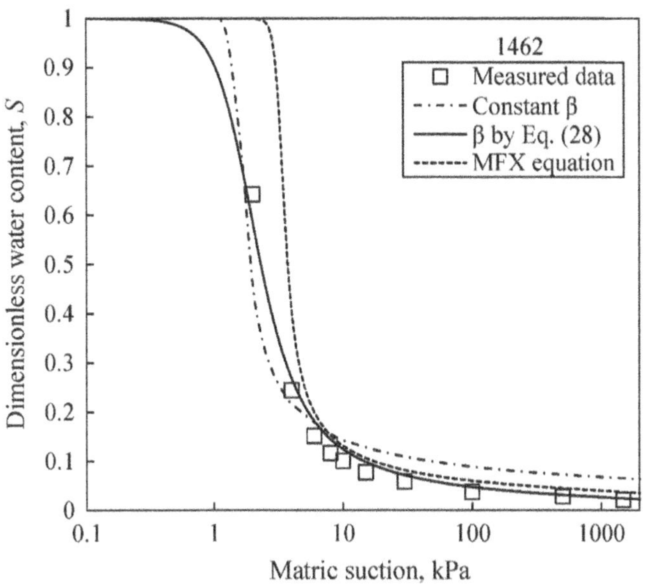

(b)

Fig. 3.5 Comparisons of SWCCs based on measured data and the estimates of different models for coarse-grained soils: **a** soil code 1461, **b** soil code 1462 and **c** soil code 4441

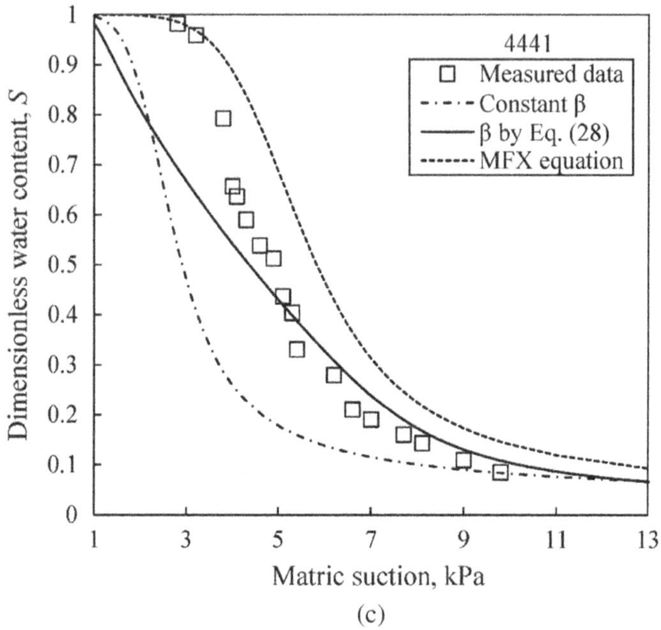

Fig. 3.5 (continued)

Table 3.6 summarizes the statistics of the estimations provided by the AP model, the proposed method with constant β, the proposed method with variable β and the MFX equation. For the AP model, a value of $\alpha = 1.38$ was adopted [14]. Comparing these statistics, the correlation coefficients ρ obtained from these four methods are similar. However, the MAPE and RMSE values of the proposed method with variable β are 5.02–35.4% and 0.009–0.032, respectively, and are the smallest among the four methods. For the proposed method with constant β, the MAPE and RMSE values are 26.57–69.78% and 0.015–0.101, respectively. For the MFX equation, those values are 10.06–55.42% and 0.036–0.071, respectively. For the AP model, those values are 27.74–385.66% and 0.062–0.233, respectively. Thus, the accuracy of the estimated results improves when the adjustment parameter is considered as a variable in both coarse-grained and fine-grained soil analyses.

3.4.3 Verification and Uncertainty Analysis

The reliability of a predicted SWCC can be evaluated by considering the posterior PDFs of uncertain parameters (b_1, b_3, b_6 and b_8 for coarse-grained soils and b_1, b_7 and b_8 for fine-grained soils). Here, 10,000 samples of these parameters, that is, $[b_1,\ b_3,\ b_6,\ b_8]^{\mathrm{T}}_{(j)}$, $j = 1, 2, \ldots, 10,000$ for coarse-grained soils and

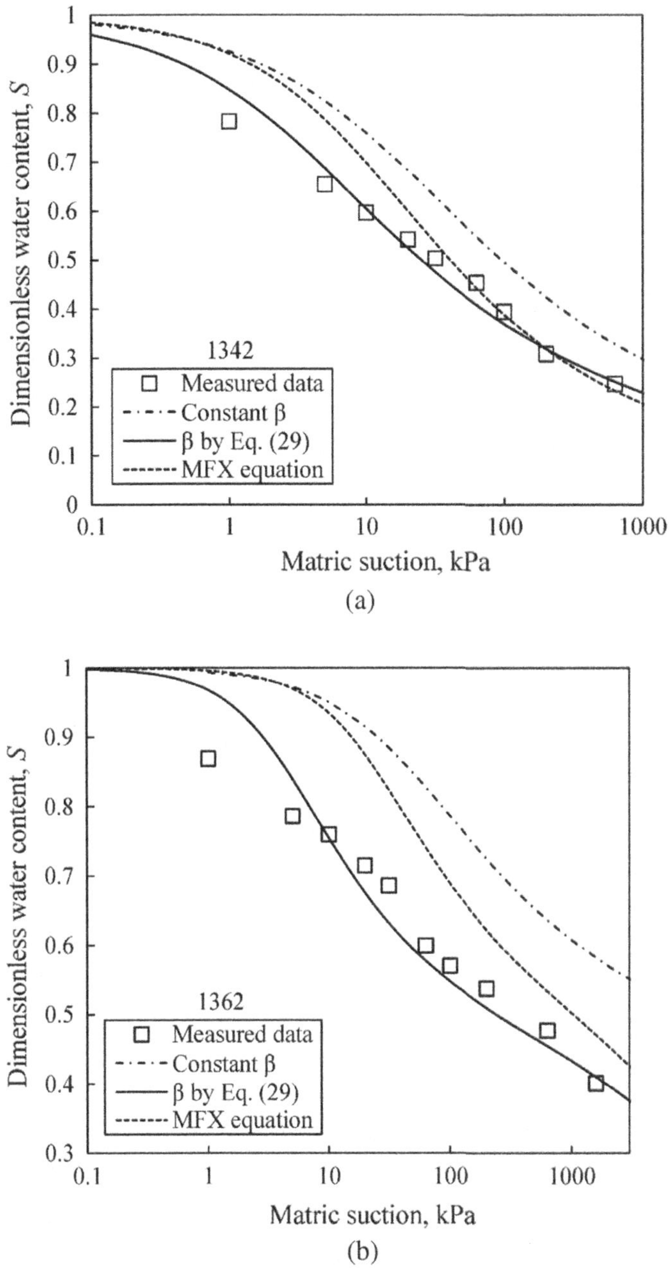

Fig. 3.6 Comparisons of SWCCs based on measured data and the estimates of different models for fine-grained soils: **a** soil code 1342, **b** soil code 1362 and **c** soil code 2550

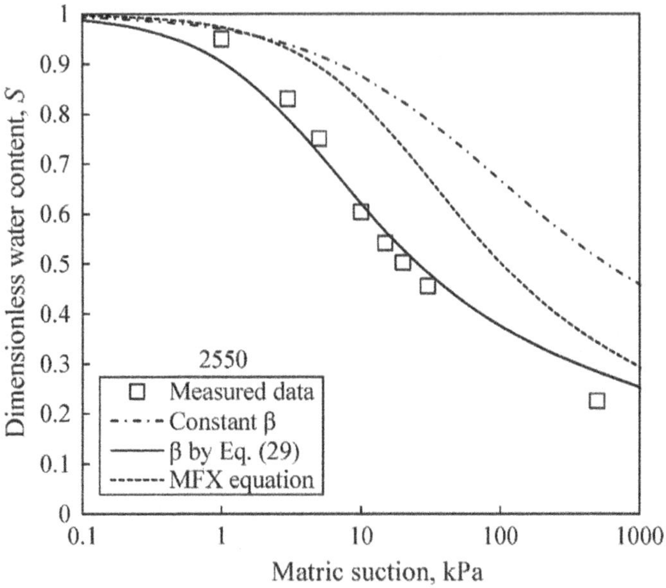

(c)

Fig. 3.6 (continued)

$[b_1, b_7, b_8]^{\mathrm{T}}_{(j)}$, $j = 1, 2, \ldots, 10,000$ for fine-grained soils, are generated by LHS [27]. For each sample, the SWCC of the proposed method with variable β can be obtained. Thus, 10,000 SWCC samples are predicted, and the SWCCs at different percentiles can be obtained by statistical analysis. To obtain continuous and smooth SWCCs, Eq. (3.1) is utilized to fit the SWCC at a percentile. Then, a set of fitting parameters (a, n, m) can be determined. Note that in this uncertainty analysis, the water contents predicted by Eq. (3.7) are assumed accurate, and the uncertainty in predicted matric suctions are evaluated. Therefore, the uncertainty and prediction errors of SWCCs are only related to the predicted matric suctions in this analysis.

In the verification and uncertainly analysis, six additional groups of soil samples, which are different from the samples in Table 3.1, are selected from the literature, as shown in Table 3.7. "Measured soil sample" denotes a soil sample with a known SWCC, and "Estimated soil sample" represents a soil sample whose SWCC is predicted. Table 3.8 shows the values of fitting parameters $(a, n, m)^{\mathrm{T}}$ at different percentiles of SWCC samples.

Figures 3.7 and 3.8 show the upper and lower boundaries of the 95% credibility interval (CI) of the predicted SWCCs (2 coarse-grained soils and 2 fine-grained soils as shown in Table 3.7) and the mean curves for coarse- and fine-grained soils, respectively. The predicted mean curves, that is, the predictions of SWCC using the optimal models of β, are in good agreement with the measured data. Comparing the 95% CIs of SWCCs predicted for the four soil samples, the accuracy of the proposed method decreases with decreasing water content in fine-grained soils; however, it

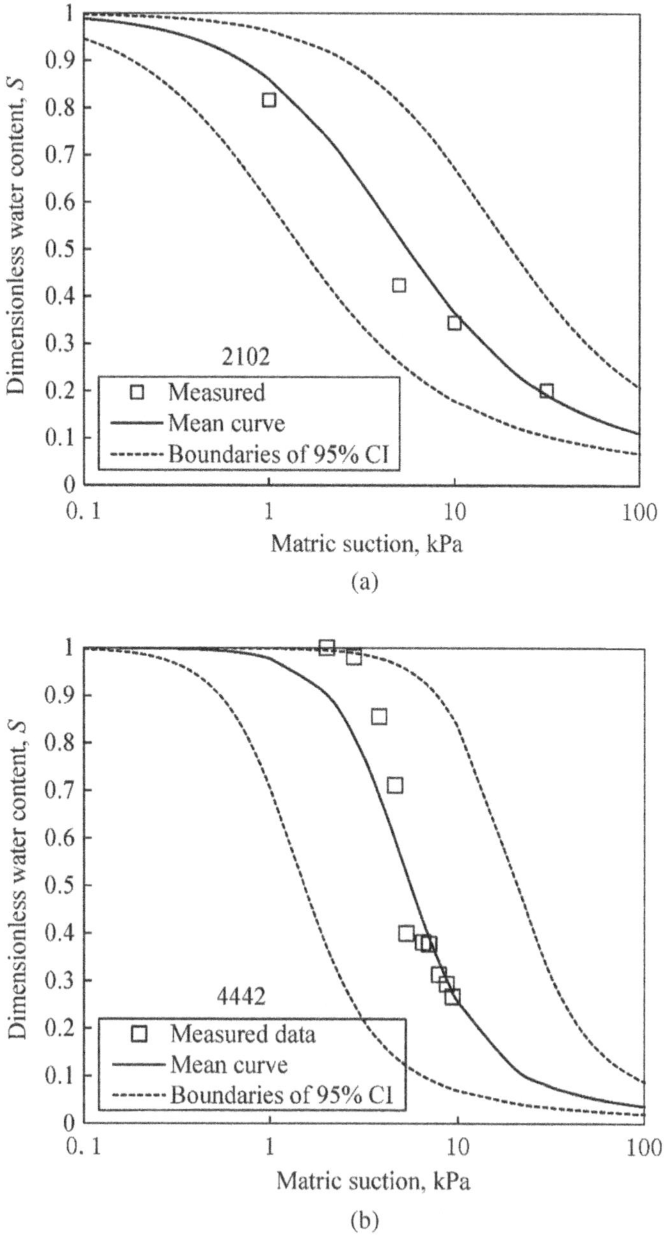

Fig. 3.7 95% CIs of predicted SWCCs for coarse-grained soils: **a** soil code 2102 and **b** soil code 4442

Table 3.6 Comparison of the results estimated by four methods

		Soil code for coarse-grained soil			Soil code for fine-grained soil		
		1461	1462	4441	1342	1362	2550
AP model	ρ	0.902	0.885	0.802	0.910	0.958	0.982
	MAPE (%)	109.49	117.87	385.66	45.91	27.74	48.15
	RMSE	0.062	0.101	0.233	0.100	0.100	0.071
MFX equation	ρ	0.998	0.998	0.961	0.995	0.990	0.959
	MAPE (%)	40.13	48.53	55.42	10.06	19.32	36.18
	RMSE	0.058	0.055	0.051	0.036	0.071	0.057
Constant β	ρ	0.979	0.979	0.943	0.996	0.982	0.935
	MAPE (%)	42.37	42.37	48.49	26.57	31.02	69.78
	RMSE	0.015	0.015	0.083	0.061	0.102	0.094
Variable β	ρ	0.980	0.980	0.988	0.991	0.974	0.999
	MAPE (%)	25.78	25.78	35.40	5.02	5.30	13.21
	RMSE	0.009	0.009	0.032	0.015	0.024	0.016

Table 3.7 Soil samples for verification and uncertainty analysis

Number of figure	Measured soil sample	Estimated soil sample	Soil type	Data source
Figure 3.7a	Soil code 2103	Soil code 2102	Coarse-grained	UNSODA [30]
Figure 3.7b	Soil code 4444	Soil code 4442		
Figure 3.8a	Soil code 1351	Soil code 1341	Fine-grained	
Figure 3.8b	Soil code 2723	Soil code 2722		
Figure 3.9a	$e_0 = 0.55$	$e_0 = 0.68$	Coarse-grained	Salager et al. [32]
	$e_0 = 0.68$	$e_0 = 0.86$		
Figure 3.9b	$e_0 = 0.50$	$e_0 = 0.54$	Fine-grained	Tarantino [33]
	$e_0 = 0.54$	$e_0 = 0.62$		

does not significantly change in coarse-grained soils. This is consistent with the fact that the effect of the initial void ratio gradually diminishes at high suction levels (i.e., low water content) when the water retention is dominated by a surface adsorption mechanism, as stated in Chap. 2.

To evaluate SWCCs with different initial void ratios, more measured SWCCs are collected for verification, as shown in Table 3.7. Three SWCCs with different initial void ratios of 0.55, 0.68 and 0.86 are obtained from Salager et al. [32] for coarse-grained soil, and three SWCCs with different initial void ratios of 0.50, 0.54 and 0.62 are obtained from Tarantino [33] for fine-grained soil. Figure 3.9 shows the predictions of SWCCs with different initial void ratios, and Table 3.9 presents

Table 3.8 Fitting parameters for different percentiles of SWCC samples

Soil code	Fitting parameters	Percentiles				
		2.5	25	50	75	97.5
2102	*a*	0.894	1.903	3.022	4.639	11.037
	n	1.142	1.128	1.133	1.139	1.125
	m	1.616	1.584	1.583	1.571	1.586
4442	*a*	1.180	2.757	4.297	6.732	15.976
	n	2.202	2.240	2.255	2.270	2.318
	m	1.715	1.694	1.685	1.678	1.658
1341	*a*	3.635	4.728	6.315	8.227	11.901
	n	0.725	0.665	0.610	0.563	0.511
	m	1.030	1.015	1.012	1.006	1.003
2722	*a*	1.32e4	2.23e4	4.2e4	9.26e4	2.40e5
	n	0.492	0.466	0.438	0.409	0.381
	m	14.061	13.339	12.950	12.653	12.600

the corresponding fitting parameters of SWCCs at different percentiles. The dashed curves and the dotted-dashed curves are the boundaries of 95% CIs at different initial void ratios, while the solid curve and the large dashed curve represent the mean values. The locations of 95% CIs change with the initial void ratio, and the mean curves agree well with the measured data. The predicted SWCCs with larger initial void ratios are always below the corresponding predictions with smaller initial void ratios. The initial void ratio has no apparent influence on the reliability of the predicted SWCCs. For a measured SWCC, the SWCCs of same-textured soils with different initial void ratios can be predicted at a certain significance level.

3.5 Summary

To model the SWCCs of same-textured soils at different initial void ratios, a new method is introduced by establishing the relationships of the SWCC variables between two different soil samples with the same texture. The method requires a set of SWCC data and the initial void ratios of the soil samples. Both coarse-grained and fine-grained soils are analyzed. In total, 10 coarse-grained soils and 24 fine-grained soils are analyzed by Bayesian model class selection to obtain the optimal estimation equation of the adjustment parameter. Then, the proposed method is verified using six additional soils. The conclusions of the study can be summarized as follows:

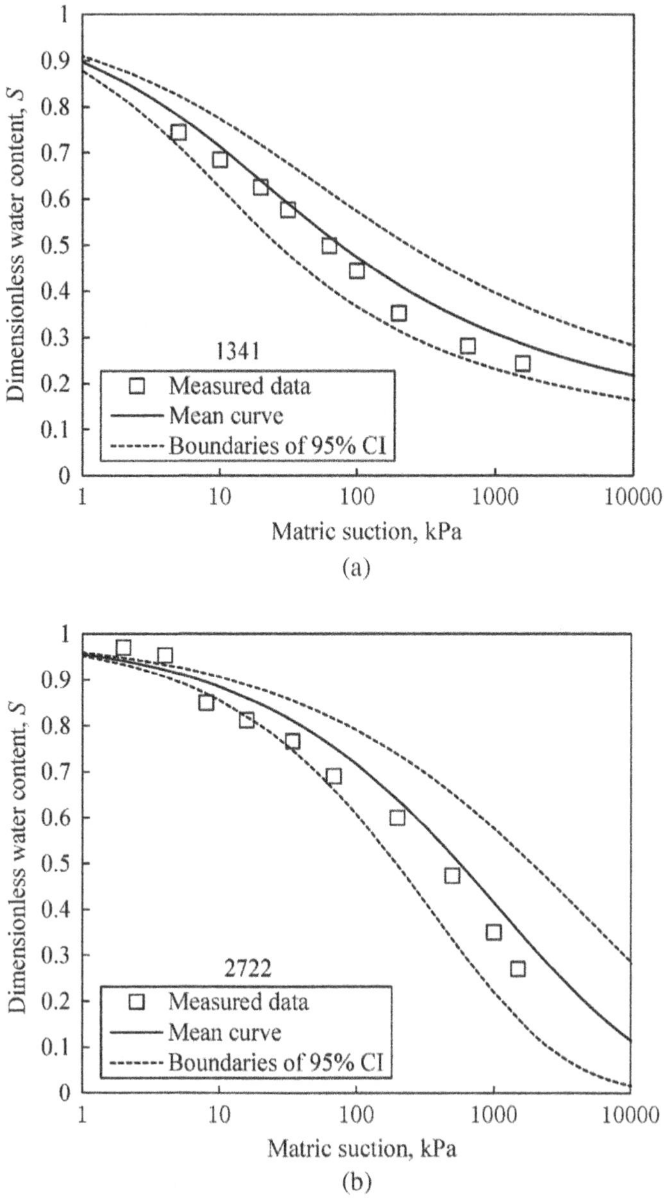

Fig. 3.8 95% CIs of predicted SWCCs for fine-grained soils: **a** soil code 1341 and **b** soil code 2722

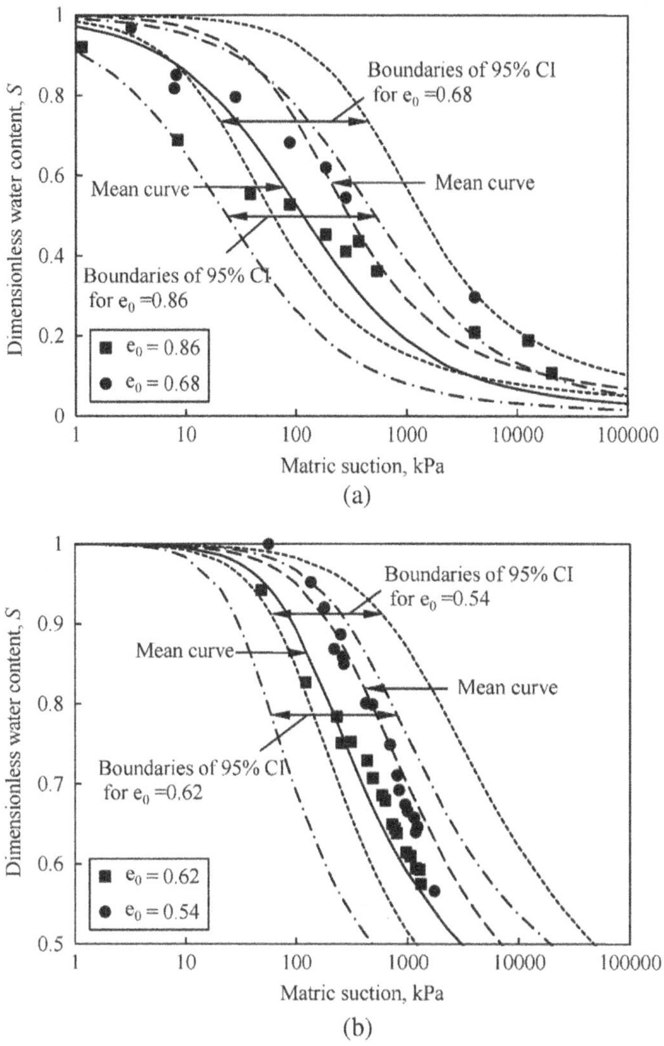

Fig. 3.9 95% CIs of predicted SWCCs at different initial void ratios for different soil types: **a** coarse-grained soil (data from Salager et al. [32]) and **b** fine-grained soil (data from Tarantino [33])

(1) For different soil samples with the same texture, the relationship between their volumetric water contents depends only on their initial void ratios.
(2) The adjustment parameter β as a function of different initial void ratios, matric suctions, and volumetric water contents can better describe the relationship between the matric suctions of two soil samples.

Table 3.9 Fitting parameters for different percentiles of SWCC samples with different initial void ratios

Data source	Initial void ratio	Fitting parameters	Percentiles				
			2.5	25	50	75	97.5
Salager et al. [32]	0.68	a	39.64	82.12	152.73	270.77	513.20
		n	1.076	1.069	1.080	1.086	1.145
		m	1.354	1.348	1.317	1.287	1.232
	0.86	a	18.27	40.44	85.88	160.25	330.21
		n	0.731	0.734	0.738	0.737	0.768
		m	2.238	2.183	2.138	2.095	2.017
Tarantino [33]	0.54	a	87.47	141.01	262.80	490.01	960.30
		n	1.444	1.323	1.197	1.104	1.015
		m	0.496	0.500	0.502	0.500	0.499
	0.62	a	32.53	54.78	94.06	174.58	331.54
		n	1.812	1.650	1.499	1.355	1.234
		m	0.432	0.428	0.428	0.429	0.428

(3) The SWCCs estimated using the proposed method with variable β agree better with the measured data than do SWCCs produced by other models in both coarse-grained and fine-grained soils.
(4) The sensitivity analysis reveals that the contributions of the model parameters in the two optimal models to the sensitivity of the predicted matric suction are different, and there is positive correlation between the predicted matric suction and all model parameters.
(5) The uncertainty analysis shows that the reliability of predicted SWCC decreases with decreasing water content in fine-grained soils, and the initial void ratio has no apparent influence on the reliability of predicted SWCCs in either type of soils.

References

1. Gupta SC, Larson WE (1979) Estimating soil-water retention characteristics from particle size distribution, organic matter percent, and bulk density. Water Resour Res 16(6):1633–1635
2. Saxton KE, Rawls WJ (2006) Soil water characteristic estimations by texture and organic matter for hydrologic solution. Soil Sci Soc Am J 70:1569–1578
3. Walczak RT, Moreno F, Sławiński C, Fernandez E, Arrue JL (2006) Modeling of soil water retention curve using soil solid phase parameters. J Hydrol 329:527–533
4. Yang F, Zhang GL, Yang JL, Li DC, Zhao YG, Liu F, Yang RM, Yang F (2014) Organic matter controls of soil water retention in an alpine grassland and its significance for hydrological processes. J Hydrol 519:3086–3093

5. Saxton KE, Rawls WJ, Romberger JS, Papendick RI (1986) Estimating generalized soil water characteristics from texture. Trans ASAE 50:1031–1035
6. Vereecken JHM, Feyen J, Maes J, Darius P (1989) Estimating the soil moisture retention characteristic from texture, bulk density, and carbon content. Soil Sci 148:389–403
7. Schaap MG, Bouten W (1996) Modelling water retention curves of sandy soils using neural networks. Water Resour Res 32:3033–3040
8. Scheinost AC, Sinowski W, Auerswald K (1997) Regionalization of soil-water retention curves in a highly variable soilscape, Developing a new pedo-transfer function. Geoderma 78:129–143
9. Huang G, Zhang R (2005) Evaluation of soil water retention curve with the pore-solid fractal model. Geoderma 127:52–61
10. Russell AR (2010) Water retention characteristics of soils with double porosity. Eur J Soil Sci 61:412–424
11. Russell AR (2014) How water retention in fractal soils depends on particle and pore sizes, shapes, volumes and surface areas. Géotechnique 64(5):379–390
12. Chin KB, Leong EC, Rahardjo H (2010) A simplified method to estimate the soil-water characteristic curve. Can Geotech J 47:1382–1400
13. Chiu CF, Yan WM, Yuen KV (2012) Estimation of water retention curve of granular soils from particle size distribution—a Bayesian probabilistic approach. Can Geotech J 49(9):1024–1035
14. Arya LM, Paris JF (1981) A physic-empirical model to predict the soil moisture characteristic form particle size distribution and bulk density data. Soil Sci Soc Am J 45:1023–1030
15. Fredlund MD, Fredlund DG, Wilson GW (1997) Prediction of the soil-water characteristic curve from grain-size distribution and volume-mass properties. In: Proceedings of the 3rd Brazilian symposium on unsaturated soils, Rio de Janeiro, 22–25 April 1997, vol 1, pp 13–23
16. Hwang SI, Powers S (2003) Using particle-size distribution models to estimate soil hydraulic properties. Soil Sci Soc Am J 67:1103–1112
17. Fredlund DG, Xing A (1994) Equations for the soil-water characteristic curve. Can Geotech J 31:521–532
18. Fredlund MD, Fredlund DG, Wilson GW (2000) An equation to represent grain-size distribution. Can Geotech J 37:817–827
19. Fredlund MD, Wilson GW, Fredlund DG (2002) Use of the grain-size distribution for estimation of the soil-water characteristic curve. Can Geotech J 39:1103–1117
20. Liu SY, Yasufuku N, Liu Q, Omine K (2013) Bimodal and multimodal descriptions of soil-water characteristic curves for structure soils. Water Sci Technol 67(8):1740–1747
21. Nasta P, Kamai T, Chirico GB, Hopmans JW, Romano N (2009) Scaling soil water retention functions using particle-size distribution. J Hydrol 374:223–234
22. Kosugi K, Hopmans JW (1998) Scaling water retention curves for soils with lognormal pore-size distribution. Soil Sci Soc Am J 62:1496–1505
23. Birle E, Heyer D, Vogt Norbert (2008) Influence of the initial water content and dry density on the soil-water retention curve and the shrinkage behavior of a compacted clay. Acta Geotech 3:191–200
24. Beckett CTS, Augarde CE (2012) Prediction of soil water retention properties using pore-size distribution and porosity. Can Geotech J 50:435–450
25. Li X, Zhang LM (2009) Characterization of dual-structure pore-size distribution of soil. Can Geotech J 46:129–141
26. Zhang LM, Li X (2010) Microporosity structure of coarse granular soils. J Geotech Geoenviron Eng 136(10):1425–1436
27. Mckay M, Beckman R, Conover W (1979) A comparison of three methods for selecting values of input variables in the analysis of output from a computer code. Technometrics 21(2):239–245
28. Basil A, D'Urso G (1997) Experimental corrections of simplified methods for predicting water retention curves in clay-loamy soils from particle-size determination. Soil Technol 10:261–272

29. Vaz CMP, Iossi MF, Naime JM (2005) Validation of the Arya and Paris water retention model for Brazilian soils. Soil Sci Soc Am J 69:577–583
30. Leij FJ, Alves WJ, van Genuchten MT, Williams JR (1996) The UNSODA unsaturated hydraulic database. National Risk Management Research Laboratory, Office of Research and Development, U.S. Environmental Protection Agency, Cincinnati, Ohio. EPA/600/R-96/095
31. American Society for Testing and Materials (ASTM) (2004) Annual Book of ASTM Standards. American Society for Testing and Materials, Sec. 4, Vol. 04.08, West Conshohocken, PA
32. Salager S, Youssoufi MSEI, Saix C (2010) Definition and experimental determination of a soil-water retention surface. Can Geotech J 47:609–622
33. Tarantino A (2009) A water retention model for deformable soils. Géotechnique 59(9):751–762

Chapter 4
Model Updating and Uncertainty Analysis for Creep of Clay

Abstract Many researchers have proposed various constitutive models for the purpose of capturing the complex physical mechanisms governing the creep behavior of soft soils. However, the more complex the model, the larger the number of associated uncertain parameters it has, and the less robust it is against modeling/measurement error. In this chapter, the Bayesian model class selection approach is applied to select the most plausible/suitable model describing the creep behavior of soft soil using laboratory measurements. In total, nine 1-D time-dependent constitutive models for the analysis of creep of clay are chosen for the assessment. Consolidated data from the intact samples of Vanttila clay and reconstituted samples of Hong Kong Marine Clay were adopted in the case study. All unknown model parameters are identified simultaneously by adopting the transitional Markov Chain Monte Carlo (TMCMC) method, and their uncertainty is quantified through the posterior probability density functions (PDFs). Engineers can adopt the present probabilistic method to determine the most suitable model and its associated model parameters for a given soft soil for the prediction of long-term creep behavior. The strategy also provides uncertainty evaluation of the model prediction based on the given data.

4.1 Introduction

Over the past few decades, researchers have conducted numerous laboratory studies to investigate the time-dependent stress-strain behavior for soft soils [1–9]. To describe the viscous features of soft soils, the strain rates were used in the models [5, 10, 11]. Based on laboratory observations, mathematical models have been proposed to consider the creep behavior of soft soil under one-dimensional (1-D) straining [1, 5, 10, 12–15], triaxial stress states, and general stress states [16–19]. Advanced elastic viscoplastic (EVP) models have also been developed to incorporate the anisotropy and destructuration effects [20, 21]. Due to different physical, chemical and biological processes over time, soils from different places can perform very differently under loading. The main physical mechanism controlling the creep behaviors may

W. Zhou et al., *Practice of Bayesian Probability Theory in Geotechnical Engineering*,
https://doi.org/10.1007/978-981-15-9105-1_4

be very different for different soft soils. It is difficult for engineers or researchers to select a suitable model in the analysis of creep behavior for a certain soil.

When analyzing the creep behavior of soft soil using an EVP model, the associated model parameters should be determined in advance. Traditionally, curve-fitting techniques are commonly adopted for separate parts of the experimental data to determine the model parameters [12, 22, 23]. Yin and Graham [12] used this technique to delineate the model parameters, which were separated based on the experimental data at different consolidation stages. The studies by Yin [22] and Yin et al. [23] suggested that the reference equivalent time (t_0) can be approximately calculated by referring to the time at which the excess pore water pressure has dissipated completely, while the other two parameters, i.e., creep coefficient and creep strain limit, are obtained by fitting the measured data after the primary consolidation stage. However, this procedure is not rigorous. The measured total strain includes the elastic and viscoplastic strain, although the elastic and plastic parameters are determined using the data on total strain. On the other hand, certain parameters are determined subjectively, such as reference equivalent time (t_0). To overcome the limitation of the traditional curve-fitting technique, Le et al. [24] adopted the trust-region reflective least-squares method to simultaneously obtain all model parameters based on the test data under different loading, while the time parameter t_0 was assumed to have the unit of minute. The recent studies [25, 26] also used the aforementioned optimization method to conduct the analysis with the same assumption on the time parameter. Additionally, certain researchers [27–29] used the enhanced genetic algorithm to identify the model parameters, and Jin et al. [27] even conducted selection work on the sand models.

From our best knowledge, Buisman [30] was the first to model the effect of time on the compression of clay by introducing the term secondary compression. Thereafter, researchers conducted a significant number of studies on one-dimensional secondary compression [1, 5, 10, 12, 13, 16, 18, 19, 21, 29, 31–37]. In general, a complex model with more parameters can better fit the measured creep deformation of soft soil. However, in engineering practice, a simple model with adequate accuracy is preferred. It should be mentioned that all these works adopted the deterministic method to determine the model parameters. The best fitting between the predictions and the observations is the only concern in these studies. This ignores the uncertainties associated with statistical uncertainty, measurement error, and model uncertainty and its effect on the model predictions. Recently, the probabilistic methods for back analysis have been widely conducted in geotechnical engineering, e.g. maximum likelihood method and Bayesian probabilistic method [38]. The Bayesian probabilistic method is considered to be more effective for updating the soil parameters [39–42], and it can quantify the uncertainty of soil parameters by the obtained posterior probabilistic distributions. It has been successfully applied to various geotechnical problems [43–50].

In this chapter, we investigate the efficiency of several well-known time-dependent models in the predictions of long-term 1-D straining behavior of soft soils. Model updating is realized using a Bayesian method, namely the transitional Markov Chain Monte Carlo (TMCMC) method. Two groups of 1-D oedometer test data from intact and reconstituted soft soil samples are employed for the model updating. A total of nine models and their associated model class selections are conducted respectively. The analysis results indicate that the most suitable model should be selected to capture the different characteristics of deformation for different soil samples. This method can evaluate the corresponding uncertainty during the model updating and hence the prediction with 95% (or any other target level) credibility interval (CI) can be computed.

4.2 Review of Time-Dependent Models for Soft Soils

Many researchers have proposed models to describe the viscoplastic strain rate for the time-dependent behavior of soft soils. The total strain rate is composed of the elastic strain rate $\dot{\varepsilon}_z^e$ and viscoplastic strain rate $\dot{\varepsilon}_z^{vp}$:

$$\dot{\varepsilon}_z = \dot{\varepsilon}_z^e + \dot{\varepsilon}_z^{vp} \tag{4.1}$$

The elastic strain rate can be written as follows:

$$\dot{\varepsilon}_z^e = \frac{\kappa}{1+e_0}\frac{\dot{\sigma}'_z}{\sigma'_z} \tag{4.2}$$

where e_0 is the initial void ratio; κ is used to denote the elastic behavior of soil; and σ'_z is the effective vertical stress. In the following, several well-known time-dependent 1-D compression models for soft soils are briefly reviewed. We will assess their performance in the simulation of different soft soil samples, using probabilistic approach, in the later sections.

(1) Yin and Graham [10, 12] developed the idea of "equivalent time" and deduced a general EVP constitutive model for the 1-D straining condition. This model uses a logarithmic function with a constant creep coefficient to describe the creep compression:

$$\dot{\varepsilon}_z^{vp} = \frac{\psi}{(1+e_0)t_0} \times \exp\left[-(\varepsilon_z - \varepsilon_{z0}^r)\frac{1+e_0}{\psi}\right]\left(\frac{\sigma'_z}{\sigma'_{zr}}\right)^{\lambda/\psi} \tag{4.3}$$

where λ is used to denote the elastic-plastic behavior of soil; ψ is a creep parameter; t_0 is the reference equivalent time, which is different from the real test time; ε_z is the current total vertical strain; and ε_{z0}^r is the vertical strain at $\sigma'_z = \sigma'_{zr}$.

(2) Yin [22] further proposed a non-linear function to model the creep strain, which contains the creep coefficient ψ'_0 and creep strain limit ε_{lm}^{vp}. The use of creep

strain limit can avoid obtaining an unrealistic infinite creep strain when the equivalent time approaches infinite, and the creep strain rate is given as follows:

$$\dot{\varepsilon}_z^{vp} = \frac{\psi_0'}{t_0}\left(1 + \frac{\varepsilon_z^r - \varepsilon_z}{\varepsilon_{lm}^{vp}}\right)^2 \times \exp\left(\frac{\varepsilon_z^r - \varepsilon_z}{1 + \frac{\varepsilon_z^r - \varepsilon_z}{\varepsilon_{lm}^{vp}}}\frac{1}{\psi_0'}\right) \tag{4.4}$$

where ε_z^r is the elastic-plastic strain and is computed by $\varepsilon_z^r = \varepsilon_{z0}^r + \lambda/(1 + e_0) \times ln(\sigma_z'/\sigma_{zr}')$. Yin [22] proposed that the creep strain limit and creep coefficient could be considered to reduce linearly with the logarithmic growth of the vertical stress:

$$\varepsilon_{lm}^{vp} = a - b\ln(\sigma_z'/\sigma_u') \tag{4.5}$$

$$\psi_0' = c - d\ln\left(\sigma_z'/\sigma_u'\right) \tag{4.6}$$

From the analysis of the consolidation data of remoulded sample, Le et al. [24] indicated that the creep strain limit can be considered as a constant for the entire consolidation process. Certain researchers [19, 51] have considered the creep coefficient as a constant. In this regard, no research has been conducted to evaluate different models of varying complexities in the analysis of the creep behavior of soft soil. Selection of a suitable model and identification of the associated parameters for a specific soft soil, given a limited test data, are of significant interest to engineers and researchers.

(3) Vermeer et al. [52] proposed a 1-D model based on the secondary compression coefficient $C_{\alpha e}$:

$$\dot{\varepsilon}_z^{vp} = \frac{C_{\alpha e}}{(1 + e_0)}\left\{\frac{\sigma_z'}{\sigma_p' \exp\left(\frac{1+e_0}{\lambda-\kappa}\varepsilon_z^{vp}\right)}\right\}^{\lambda-\kappa/C_{\alpha e}} \tag{4.7}$$

where τ is the reference time and usually takes the duration of the load increment in the conventional oedometer test; σ_p' is the preconsolidation pressure.

(4) Kim and Leroueil [45] developed a viscoplastic strain rate equation based on the stress-strain-rate behavior proposed by Leroueil et al. [5]:

$$\dot{\varepsilon}_z^{vp} = 10^{\left[\left(log\sigma_z' - \Gamma - \varepsilon_{0i} - C_\varepsilon \varepsilon_z^{vp}\right)/C_p\right]} \tag{4.8}$$

where Γ is the value of $\log\sigma'_{p0}$ in a constant rate-of-strain test with a strain rate $\dot{\varepsilon}_z^{vp} = 1\ \text{s}^{-1}$; ε_{oi} is the intercept of the line in the $\log\left(\sigma'_z / \sigma'_{p0}\right) - \varepsilon_z^{vp}$ plane; C_ε is the compression index, i.e., the slope of the line in the $\log\left(\sigma'_z / \sigma'_{p0}\right) - \varepsilon_z^{vp}$ plane; and C_p is the preconsolidation index, i.e., the slope of the line in the $\log\left(\sigma'_{p0}\right) - \dot{\varepsilon}_z^{vp}$ plane.

(5) Yin et al. [21] proposed a 1-D EVP model to describe the creep behavior of natural structured clays. The destructuration effect on the deformation of said clays was considered by introducing two structure parameters, χ_0 and ρ, and the corresponding creep strain rate was given by:

$$\dot{\varepsilon}_z^{vp} = \frac{C_{\alpha e}}{(1+e_0)\tau}\left\{\frac{\sigma_z'(1+\chi_0)}{\left[1+\chi_0\exp(-\rho\varepsilon_z^{vp})\right]\sigma'^{r}_{p0}\exp\left(\frac{1+e_0}{\lambda_i-\kappa}\varepsilon_z^{vp}\right)}\right\}^{\lambda_i-\kappa/C_{\alpha e}} \tag{4.9}$$

where σ'^{r}_{p0} is the initial reference pre-consolidation pressure and λ_i is the intrinsic slope of the compression line.

All these models can be used to describe the long-term creep behavior of soft soil albeit in different ways. Different complexities are associated with different models. When it comes to practical applications, it is of significant interest to engineers to identify the optimal model that would allow good predictions for the long-term settlement of the soil with the necessary complexity. In the later section, an elastoplastic model by Yang et al. [54] is also chosen for comparison in the case of intact samples.

(6) Yang et al. [54] proposed a nonlinear normal compression line to consider the structure effect, in which the plastic strain rate can be expressed as:

$$\dot{\varepsilon}_z^{p} = \frac{[\lambda_i^*(1+\chi_0^* e^{(-\rho^*\varepsilon_z^p)})]-\kappa}{1+e_0}\frac{\dot{\sigma}'_z}{\sigma'_z} \tag{4.10}$$

where χ_0^* and ρ^* are two parameters that have similar meanings to χ_0 and ρ in Eq. (4.9) for the description of the variation of λ with the destructuration; λ_i^* is the compression coefficient obtained at large strain where the state of the soil structure stabilizes under ongoing normal consolidation stress paths.

To compute the vertical strain development with time during the 1-D consolidation tests, coupled analyses should be performed. The mass continuity equation for the consolidation process is thus adopted as follows:

$$\frac{k}{\gamma_w}\frac{\partial^2 u}{\partial z^2} = -\frac{\partial \varepsilon_z}{\partial t} \tag{4.11}$$

where u is the pore water pressure; γ_w is the unit weight of water; t is the elapsed time; and k is the hydraulic conductivity of soil, which is assumed to conform to the Carman-Kozeny equation [55]. To consider its variation with developed vertical strains, k can be calculated during the consolidation process as follows:

$$k = \frac{k_0}{1 - \varepsilon_z}\left(1 - \varepsilon_z \frac{1 + e_0}{e_0}\right)^3 \tag{4.12}$$

In this study, a total of 9 well-known 1-D compression models proposed by the above researchers [10, 21, 22, 33, 52–54] for soft soils will be assessed. Table 4.1 summarizes the 9 models and the equations of their viscoplastic or plastic strain rate. The parameters or symbols involved in the table are explained in the notations. The candidates include the six models with the vertical viscoplastic/plastic strain

Table 4.1 Summary of vertical strain rates by the 9 models

EVP models	Viscoplastic strain rate $\dot{\varepsilon}_z^{vp}$
YG_1989	$\frac{\psi}{(1+e_0)t_0} \times \exp\left[-(\varepsilon_z - \varepsilon_{z0}^r)\frac{1+e_0}{\psi}\right]\left(\frac{\sigma_z'}{\sigma_{zr}'}\right)^{\lambda/\psi}$
Yin_1999.a	$\frac{\psi_0'}{t_0}\left(1 + \frac{\varepsilon_z^r - \varepsilon_z}{\varepsilon_{lm}^{vp}}\right)^2 \times \exp\left(\frac{\varepsilon_z^r - \varepsilon_z}{1 + \frac{\varepsilon_z^r - \varepsilon_z}{\varepsilon_{lm}^{vp}}}\frac{1}{\psi_0'}\right)$
Yin_1999.b	$\frac{\psi_0'}{t_0}\left(1 + \frac{\varepsilon_z^r - \varepsilon_z}{a - b\ln\frac{\sigma_z'}{\sigma_u'}}\right)^2 \times \exp\left(\frac{\varepsilon_z^r - \varepsilon_z}{1 + \frac{\varepsilon_z^r - \varepsilon_z}{a - b\ln\frac{\sigma_z'}{\sigma_u'}}}\frac{1}{\psi_0'}\right)$
Yin_1999.c	$\frac{c - d\ln\frac{\sigma_z'}{\sigma_u'}}{t_0}\left(1 + \frac{\varepsilon_z^r - \varepsilon_z}{\varepsilon_{lm}^{vp}}\right)^2 \times \exp\left(\frac{\varepsilon_z^r - \varepsilon_z}{1 + \frac{\varepsilon_z^r - \varepsilon_z}{\varepsilon_{lm}^{vp}}}\frac{1}{c - d\ln\frac{\sigma_z'}{\sigma_u'}}\right)$
Yin_1999.d	$\frac{c - d\ln\frac{\sigma_z'}{\sigma_u'}}{t_0}\left(1 + \frac{\varepsilon_z^r - \varepsilon_z}{a - b\ln\frac{\sigma_z'}{\sigma_u'}}\right)^2 \times \exp\left(\frac{\varepsilon_z^r - \varepsilon_z}{1 + \frac{\varepsilon_z^r - \varepsilon_z}{a - b\ln\frac{\sigma_z'}{\sigma_u'}}}\frac{1}{c - d\ln\frac{\sigma_z'}{\sigma_u'}}\right)$
Vermeer et al._1997	$\frac{C_{\alpha e}}{(1+e_0)\tau}\left\{\frac{\sigma_z'}{\sigma_p' \exp\left(\frac{1+e_0}{\lambda-\kappa}\varepsilon_z^{vp}\right)}\right\}^{\lambda-\kappa/C_{ae}}$
KL_2001	$10^{[(log\sigma_z' - \Gamma - \varepsilon_{oi} - C_\varepsilon \varepsilon_z^{vp})/C_p]}$
Yin et al._2011	$\frac{C_{\alpha e}}{(1+e_0)\tau}\left\{\frac{\sigma_z'(1+\chi_0)}{[1+\chi_0\exp(-\rho\varepsilon_z^{vp})]\sigma_p'^r \exp\left(\frac{1+e_0}{\lambda_i-\kappa}\varepsilon_z^{vp}\right)}\right\}^{\lambda_i-\kappa/C_{ae}}$
EP model	Plastic vertical strain $\dot{\varepsilon}_z^p$
Yang et al._2015	$\frac{[\lambda_i^*(1+\chi_0^* e^{(-\rho^*\varepsilon_z^p)})] - \kappa}{1+e_0}\frac{\dot{\sigma}'_z}{\sigma_z'}$

*$\dot{\varepsilon}_z$ is total vertical strain rate ($\dot{\varepsilon}_z = \partial\varepsilon_z/\partial t$); $\dot{\sigma}_z'$ is vertical effective stress rate ($\dot{\sigma}_z' = \partial\sigma_z'/\partial t$)

rate defined by Eqs. (4.3)–(4.4) and Eqs. (4.7)–(4.10), respectively. Models 2–5 in Table 4.1 are the variations on Yin's model [22]. The YG_1989 model [10, 33] uses a logarithmic function with a constant creep coefficient to model creep compression. The Yin_1999 model [22] overcomes the limitations of the YG_1989 model in terms of obtaining unrealistic infinite creep strain using a nonlinear function to model the creep strain, which contains the creep strain limit and creep coefficient. However, some researchers have found that these two creep parameters play different roles in different cases. The study by Le et al. [24] shows that the variation of the creep strain limit with time can be ignored for the entire consolidation process of remolded soils. Some researchers [19, 51] considered the creep coefficient to be a constant. In this regard, a series of Yin_1999 models [22] of different complexity that can explore different situations are investigated in this study, and their performance in the prediction of deformation is evaluated. The model name postfixes a, b, c and d for said models denote that the creep function, respectively, considers constant ε_{lm}^{vp} and constant ψ_0', Eq. (4.5) for ε_{lm}^{vp} and constant ψ_0', constant ε_{lm}^{vp} and Eq. (4.6) for ψ_0', and Eq. (4.5) for ε_{lm}^{vp} and Eq. (4.6) for ψ_0'.

The Vermeer et al._1997 model [52] has many similar features to the YG_1989 model, except the physical interpretation of the time parameter. The KL_2001 model [53] was derived based on the stress-strain-strain rate model by Leroueil et al. [5] and Kabbaj et al. [37] and can model the viscoplastic behavior of clay under both laboratory and field conditions. The Yin et al._2011 model [21] introduces two structure parameters to take into consideration the destructuration effect on the deformation of natural clays. Additionally, an elastic-plastic model, i.e., the Yang et al._2015 model [54], which can incorporate the structure effect of intact soft soil, is also investigated in this study.

The solution can be easily obtained under proper boundary and initial conditions, using the finite difference method. The reader may refer to Yin and Graham [56] and Le et al. [24] for a detailed solution process of the coupled equations for 1-D consolidation of soft soil. A MATLAB code for different soft soil models with finite difference method is shown in Appendix B.

4.3 Model Updating with Bayesian Method

The Bayesian probabilistic method can be used to quantify the uncertainty in the parameters of a model and select the most suitable model class from a family of competitive model classes based on measurements of the system. The uncertainty of the model parameters can be quantified through derived posterior probability density functions (PDFs). In the following, the Bayesian probabilistic framework and transitional Markov Chain Monte Carlo algorithm are employed for the model updating for the analysis of creep of clay.

Using $\bar{\varepsilon}_z(\boldsymbol{\theta}_{\mathrm{m}}; C)$ to denote the predicted vertical strains given model class C, its relationship with the measured data ε_z can be assumed to satisfy the following equation:

Table 4.2 The 9 model class candidates and their parameters

Number	Model class	Uncertain parameter vector $\boldsymbol{\theta}$
1	YG_1989	$\left[\lambda/V, \kappa/V, \psi/V, \sigma'_{zr}, t_0, k_0, \sigma_\varepsilon^2\right]^T$
2	Yin_1999.a	$\left[\lambda/V, \kappa/V, \psi'_0, \varepsilon_{lm}^{vp}, \sigma'_{zr}, t_0, k_0, \sigma_\varepsilon^2\right]^T$
3	Yin_1999.b	$\left[\lambda/V, \kappa/V, \psi'_0, a, b, \sigma'_{zr}, t_0, k_0, \sigma_\varepsilon^2\right]^T$
4	Yin_1999.c	$\left[\lambda/V, \kappa/V, c, d, \varepsilon_{lm}^{vp}, \sigma'_{zr}, t_0, k_0, \sigma_\varepsilon^2\right]^T$
5	Yin_1999.d	$\left[\lambda/V, \kappa/V, a, b, c, d, \sigma'_{zr}, t_0, k_0, \sigma_\varepsilon^2\right]^T$
6	Vermeer et al._1997	$\left[\lambda/V, \kappa/V, C_{\alpha e}, \sigma'_p, k_0, \sigma_\varepsilon^2\right]^T$
7	KL_2001	$\left[C_\varepsilon, \kappa/V, C_p, \Gamma, \varepsilon_{oi}, k_0, \sigma_\varepsilon^2\right]^T$
8	Yin et al._2011	$\left[\lambda/V, \kappa/V, C_{\alpha e}, \chi_0, \rho, \sigma'^{r}_{p0}, k_0, \sigma_\varepsilon^2\right]^T$
9	Yang et al._2015	$\left[\lambda/V, \kappa/V, \chi_0^*, \rho^*, \sigma'_p, k_0, \sigma_\varepsilon^2\right]^T$

$$\varepsilon_z = \bar{\varepsilon}_z(\boldsymbol{\theta}_{\text{m}}; C) + \varepsilon \tag{4.13}$$

where ε is the zero-mean Gaussian random variable with variance σ_ε^2 referred hereafter as prediction error variance; and σ_ε^2 is another unknown parameter besides the soil model parameters. Thus, the uncertain parameter vector $\boldsymbol{\theta}$ includes the model parameters and the prediction-error variance σ_ε^2, i.e., $\boldsymbol{\theta} = \left[\boldsymbol{\theta}_{\text{m}}^T, \sigma_\varepsilon^2\right]^T$. Table 4.2 lists the nine model class candidates and their associated uncertain parameter vector $\boldsymbol{\theta}$.

Given the measured data D, the posterior PDF of $\boldsymbol{\theta}$ in the model class C can be computed through the Bayes' theorem [57]:

$$f(\boldsymbol{\theta}|D, C) = \frac{f(\boldsymbol{\theta}|C) \cdot f(D|\boldsymbol{\theta}, C)}{f(D|C)} \tag{4.14}$$

where $f(\boldsymbol{\theta}|C)$ is the prior PDF of $\boldsymbol{\theta}$, representing the user's adjustment before the data D are observed; $f(D|C) = \int f(D|\boldsymbol{\theta}, C) f(\boldsymbol{\theta}|C) d\boldsymbol{\theta}$ is the evidence of model class C, provided by the data D and important for the model class selection; and $f(D|\boldsymbol{\theta}, C)$ is the likelihood function expressing the level of data fitting, which can be computed by:

$$f(D|\boldsymbol{\theta}, C) = (2\pi\sigma_\varepsilon^2)^{-\frac{N}{2}} \exp\left\{ -\frac{\sum_{j=1}^{N} \left[(\varepsilon_z)_j - (\bar{\varepsilon}_z(\boldsymbol{\theta}_{\text{m}}|C))_j\right]^2}{2\sigma_\varepsilon^2} \right\} \tag{4.15}$$

where N is the number of data points.

The posterior PDF $f(\boldsymbol{\theta}|D, C)$ represents the updated belief about the parameter vector $\boldsymbol{\theta}$ after obtaining the evidence D. The optimal parameter vector $\boldsymbol{\theta}^*$ can be acquired by maximizing the posterior $f(\boldsymbol{\theta}|D, C)$, or equivalently $f(\boldsymbol{\theta}|C)\cdot f(D|\boldsymbol{\theta}, C)$. Considering that the model classes involve the high-dimensional nonlinear functions, the transitional TMCMC [58] method was used to quantify the uncertainty of model parameters and conduct the model class selection. The detailed steps of TMCMC method is introduced in Sect. 1.2.3.3 and Fig. 4.1 shows a flowchart of the TMCMC method. A matlab code for the implementation of TMCMC is shown in Appendix C.

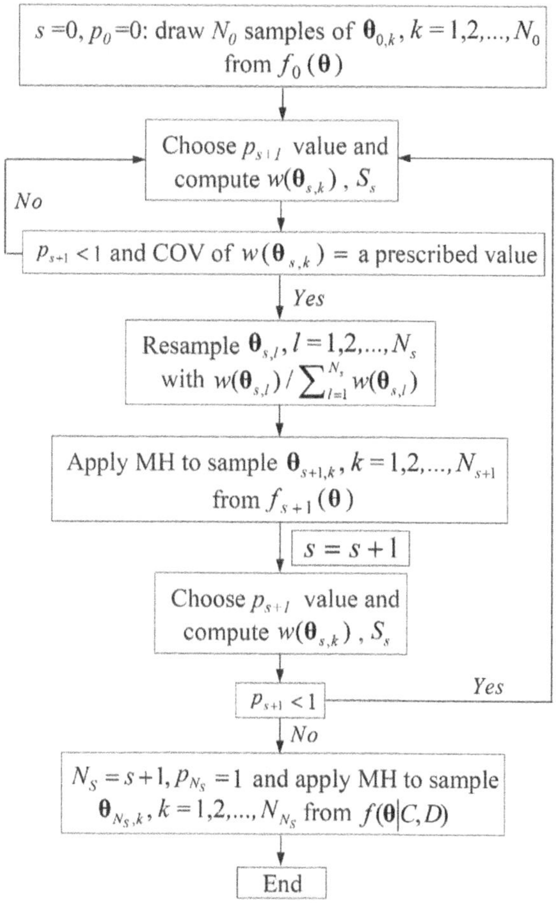

Fig. 4.1 Flowchart of the TMCMC method based on Ching and Chen' theory [58]

4.4 Case Study

As described in Sects. 4.2 and 4.3, a total of 9 different model candidates (shown in Table 4.1) were chosen to denote the 1-D compression behavior of soft soil. To investigate the simulation performance of these various models, two groups of test data from conventional oedometer tests were collected from the literature [21, 59]. In the following, two case studies are presented on intact soft soil and reconstituted soft soil, respectively.

4.4.1 Case Study 1: Intact Soft Soil Sample of Vanttila Clay

The multi-stage consolidation data of two intact samples of Vanttila clay were collected from the study of Yin et al. [21] for the model updating and verification. The test information is displayed in Table 4.3. Two intact samples (IS1 and IS2) were consolidated under 6 and 7 stages, respectively. For IS1, loading was maintained for 1 day for each loading stage while for IS2, each loading step lasted for 10 days. The initial height of soil samples is $H_0 = 15$ mm. In this case study, the 1-day oedometer test data of IS1 were used for training, i.e., identifying the model parameters. At the same time, the most plausible models were obtained via Bayesian model class selection. Then, the 10-day oedometer test data of IS2 were used for testing, i.e., verifying the performance of the models.

Table 4.3 Test information and initial values for intact samples

Sample	Loading duration t/day	Loading stage	Initial vertical strain $(\varepsilon_z)_{i,0}$/%	Vertical surcharge $(\sigma_z)_{i,0}$/kPa	Initial excess pore water pressure $u_{i,0}$/kPa
IS1	1	IS1_1	0.74	9.4	8.3
		IS1_2	1.40	18.75	9.35
		IS1_3	2.59	37.5	18.75
		IS1_4	13.08	75	37.5
		IS1_5	29.59	150	75
		IS1_6	37.51	300	150
IS2	10	IS2_1	0.46	6.25	5.15
		IS2_2	1.00	12.5	6.25
		IS2_3	2.04	25	12.5
		IS2_4	5.05	50	25
		IS2_5	25.45	100	50
		IS2_6	34.6	200	100
		IS2_7	41.55	400	200

Table 4.4 Model class selection results based on the consolidation data of intact sample

Model class	Average ln$p(D\|\boldsymbol{\theta}$, C)	ln$S$	$P(C\|D)$
YG_1989	−241.39	−253.91	1.22e-81
Yin_1999.a	−241.57	−256.39	1.02e-82
Yin_1999.b	−241.63	−256.22	1.21e-82
Yin_1999.c	−240.93	−255.57	2.32e-82
Yin_1999.d	−240.86	−255.90	1.67e-82
Vermeer et al._1997	−240.27	−253.08	2.80e-81
KL_2001	−241.33	−253.96	1.16e-81
Yin et al._2011	−36.97	−67.60	1.00
Yang et al._2015	−112.97	−142.68	2.47e-33

Uniform prior plausibility was ssigned to the 9 model class candidates, i.e., $P(C^{(m)}) = 1/9\,(m = 1, 2, \ldots, 9)$, and the evaluation of model classes can be conducted by comparing their evidence (S). The model class with the largest S value is the optimal model class among the 9 model classes. Table 4.4 presents the model class selection results based on the consolidation data of IS1. The observer may clearly discern that the Yin et al._2011 model class outperformed the other model classes with a posterior probability value $P(C|D) \approx 1$. This points to the importance of a model capturing the destructuration effect in the creep analysis of the intact samples.

During the process of TMCMC simulation, the sample numbers of all stages were fixed to 10,000; then, 10,000N_S samples were generated in the simulation. To guarantee their quality, the initial 2,000 samples, as burn-in samples, were discarded at the final stage. The corresponding posterior statistics of the model parameters were then obtained. Table 4.5 shows the updated parameter values of the first two model

Table 4.5 Posterior statistics of model parameters updated based on the consolidation data of IS1

Model class	Model parameter							
Yin et al._2011	λ/V	κ/V	$C_{\alpha e}$	χ_0	ρ	σ'^{r}_{p0}	k_0	σ_ε^2
	0.0712 (7.57e-4)	0.0088 (1.35e-4)	0.0039 (1.45e-4)	82.85 (5.71)	13.50 (0.15)	24.71 (0.27)	8.04e-8 (7.09e-9)	0.115 (0.011)
Yang et al._2015	λ/V	κ/V	–	χ_0^*	ρ^*	σ'_p	k_0	σ_ε^2
	0.0861 (3.63e-4)	0.0095 (0.0062)	–	6.91 (0.512)	11.78 (1.09)	36.25 (0.094)	2.21e-8 (1.38e-9)	0.478 (0.050)

Note The values outside the parentheses are the updated mean values, while those in the parentheses are the corresponding standard deviations

classes, i.e., Yin_et al._2011 and Yang_et al._2015. The values outside the parentheses are the updated mean values, and those within the parentheses are the corresponding standard deviations. The prediction error $\sigma_{\varepsilon}^2 = 0.115$ by Yin et al._2011 is smaller than that by Yang et al._2015 ($\sigma_{\varepsilon}^2 = 0.478$), which means that the Yin et al._2011 model can fit better the deformation of the intact sample IS1. The estimated pre-consolidation pressures in Model Yin_et al._2011 and Yang_et al._ 2015 are 24.7 kPa and 36.25 kPa, respectively. The values are close to the pre-consolidation pressure (29 kPa) reported by Yin et al. [21]. In addition, it is worth noting that with the proposed method, all the unknown parameters can be obtained simultaneously, and the optimal values are estimated based on all the test data.

The vertical strain estimations at the six loading stages by the nine updated models, based on IS1 consolidation data, were computed, and the corresponding statistic indices are shown in Table 4.6. The values outside the parentheses are the root mean square errors (*RMSEs*), while those in the parentheses are the mean absolute percentage errors (*MAPEs*). The model with smaller *RMSE* and *MAPE* gives better fittings with the measurements of the IS1 sample. The average values of *RMSE* and *MAPE* for the last four loading stages are compared in Fig. 4.2. It can clearly seen that the Yin et al._2011_IS1 model has the smallest values of *RMSE* and *MAPE*. The

Table 4.6 Statistic indices for vertical strains computed by the nine updated models based on the consolidation data of IS1

Model	Loading stage of IS1					
	1	2	3	4	5	6
YG_1989_IS1	0.63 (50.68)	0.43 (19.57)	2.49 (35.90)	2.80 (9.58)	2.67 (7.30)	2.38 (4.80)
Yin_1999. a_IS1	0.66 (51.70)	0.51 (23.59)	2.60 (37.21)	2.67 (9.28)	2.66 (7.17)	2.45 (4.93)
Yin_1999. b_IS1	0.71 (56.60)	0.53 (24.91)	2.62 (37.30)	2.71 (9.83)	2.27 (6.06)	2.59 (5.47)
Yin_1999. c_IS1	0.64 (51.95)	0.48 (21.28)	2.46 (36.80)	2.80 (9.66)	2.62 (7.14)	2.42 (5.10)
Yin_1999. d_IS1	0.65 (52.66)	0.47 (20.70)	2.40 (36.21)	2.87 (9.81)	2.66 (7.29)	2.37 (5.01)
Vermeer et al._1997_IS1	0.68 (54.35)	0.59 (26.15)	3.20 (47.27)	2.27 (9.83)	1.93 (4.88)	3.00 (5.65)
KL_2001_IS1	0.67 (51.14)	0.55 (25.36)	3.21 (46.17)	2.11 (8.70)	2.02 (4.80)	3.03 (5.86)
Yin et al._2011_IS1	0.96 (71.47)	0.87 (42.15)	0.63 (8.23)	0.68 (3.31)	0.86 (2.33)	0.65 (1.56)
Yang et al._2015_IS1	0.89 (53.97)	0.97 (47.11)	0.88 (11.48)	1.15 (3.63)	0.87 (1.79)	0.69 (1.35)

Note The values outside the parentheses are the *RMSE*, and those in the parentheses are the *MAPE* (%)

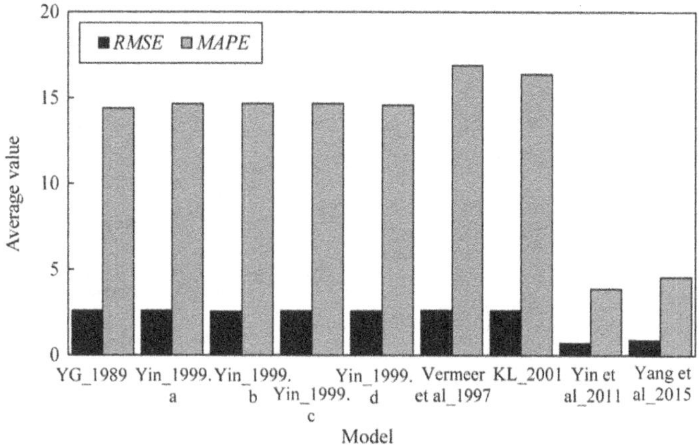

Fig. 4.2 Comparison of average statistical indices for the models updated using the consolidation data of IS1

other 1-D EVP models proposed by Yin and his coworkers [10, 12, 22] have similar performance in the fitting with the measurement data of IS1.

The variations of vertical strain with time, estimated by the top two models (i.e., Yin et al._2011_IS1 and Yang et al._2015_IS1; see Table 4.5) are plotted in Fig. 4.3a, b together with the measured data of the IS1 sample, respectively. Figure 4.3a indicates that the Yin et al._2011_IS1 model, considering both the destructuration effect and creep effect, has good fitting ability throughout the entire consolidation process. The Yang et al._2015_IS1 model does not take the creep effect into account. Correspondingly, Fig. 4.3b shows some disparities between the model predictions by Yang et al._2015_IS1 and the test data, particularly in the later period of each loading stage during which the creep effect becomes more significant.

According to the identified parameters shown in Table 4.5, the uncertainty regarding predictions of vertical strain can be evaluated. Here, 95% CIs (Credibility Interval) of predictions for the IS2 sample were computed. Table 4.7 summarizes the statistic indices between the estimated mean values and the oedometer test data for IS2 samples. It is noted that the Yin et al._2011_IS1 model exhibits a distinct advantage in terms of fitting ability, especially for the last four loading stages over the Yang et al._2015_IS1 model. It is because this model has a lack of description of the rate-dependent effect. Figure 4.4 plots the variations of vertical strain with time by the optimal model, i.e., the Yin et al._2011_IS1 model. The two dash curves in the figure are the upper and lower boundaries of 95% CI, while the solid curve represents the mean values of the predictions. It can be found that the measured data of IS2 at the final four loading stages are located near or between the areas of 95% CI, while the predicted mean values are close to the measured data.

Based on the previous results, it was found that the deformations of the analyzed intact soft soil sample should be simulated using an EVP model which can capture both the destructuration and time effect. Based on a set of measurements, the proposed

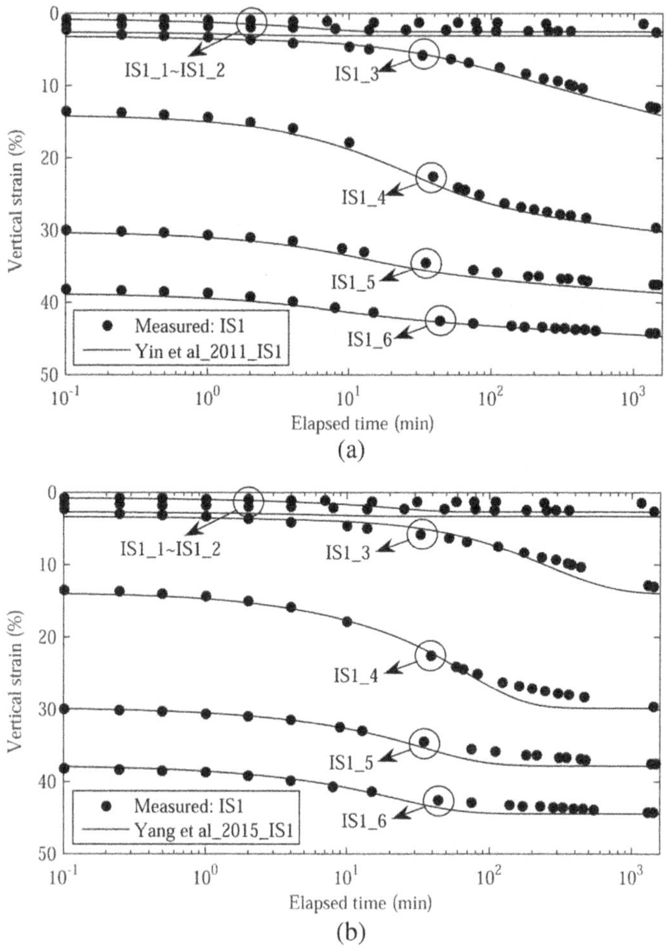

Fig. 4.3 Comparisons of vertical strains between the measurements and estimations by: **a** Yin et al._2011_IS1 model and **b** Yang et al._2015_IS1 model

Table 4.7 Prediction results for IS2 from two updated models based on the consolidation data of IS1

Model	Statistical indice	Loading stage of IS2						
		1	2	3	4	5	6	7
Yin_2011_IS1	*RMSE*	0.99	0.79	1.27	3.72	2.56	1.35	1.58
	MAPE (%)	121.02	49.8	25.31	21.51	7.48	3.31	3.37
Yang et al._2015_IS1	*RMSE*	1.03	0.9	1.23	11.19	15.4	16.08	16.17
	MAPE (%)	117.18	56.47	24.06	54.53	47.38	40.85	35.00

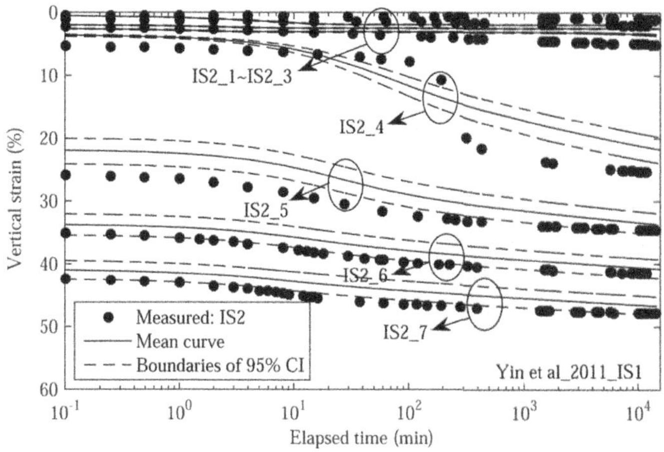

Fig. 4.4 IS2 measured data and 95% CI of vertical strains estimated by the Yin et al._2011_IS1 model

Bayesian procedure can be employed to effectively identify the optimal model with its associated parameters. Furthermore, the uncertainty analysis for the prediction of soil deformation can be carried out.

4.4.2 *Case Study 2: Reconstituted Sample of Hong Kong Marine Clay*

In this case study, the consolidation data of two reconstituted samples of Hong Kong Marine Clay (HKMC) were analyzed. The samples with a height of $H_0 = 19$ mm under four monotonic loading stages [59] were collected for this case study. Table 4.8

Table 4.8 Test information and initial values for reconstituted samples

Sample	Loading stage	Initial vertical strain $(\varepsilon_z)_{i,0}$/%	Vertical surcharge $(\sigma_z)_{i,0}$/KPa	Initial excess pore water pressure $u_{i,0/kPa}$
RS1	RS1_1	3.12	50	48.9
	RS1_2	26.63	100	50
	RS1_3	31.96	250	150
	RS1_4	38.61	500	250
RS2	RS2_1	23.76	100	50
	RS2_2	29.87	250	150
	RS2_3	37.61	500	250
	RS2_4	43.53	1000	500

Table 4.9 Model class selection results based on the consolidation data of the reconstituted sample

Model class	Average $\ln p(D\|\boldsymbol{\theta}, C)$	ln*S*	$P(C\|D)$
YG_1989	32.64	9.59	4.00e-3
Yin_1999. a	30.89	6.69	2.21e-4
Yin_1999. b	32.58	7.21	3.73e-4
Yin_1999. c	41.56	14.96	0.859
Yin_1999. d	35.67	13.09	0.133
Vermeer et al._1997	20.34	2.12	2.28e-6
KL_2001	27.84	5.68	8.05e-5
Yin et al._2011	27.05	1.73	1.55e-6
Yang et al._2015	-64.55	-84.34	6.44e-44

displays more detailed test information. Each loading stage of the two reconstituted samples was maintained for 1 day. The 1-day oedometer test data of RS1 were used for training, and the most plausible model was obtained via Bayesian model class selection. Then, the 1-day oedometer test data of RS2 were used for testing.

Table 4.9 presents the Bayesian model class selection results. It is found that the Yin_1999.c model class is the most plausible one ($P(C|D) = 0.859$), with the optimal tradeoff between the data fitting and model robustness, and the Yin_1999.d model class ranks the second. These are followed by the other EVP model classes proposed by Yin and coworkers [10, 12, 22]. Two structural model classes, i.e., Yin et al._2011 and Yang et al._2015, have the smallest values of posterior plausibilities. These results indicate that the model classes with consideration of the nonlinear creep effect are the most capable of capturing the deformation characteristics of reconstituted HKMC samples, while the destructuration effect is not necessary to be considered in the prediction model.

Table 4.10 shows the posterior statistics of model parameters in the top two model classes, i.e., Yin_1999.c and Yin_1999.d. By comparing their mean values, it is observed that the two updated models have close values for all parameters. This means that both will have similar performance in the prediction of the deformation behavior of the reconstituted samples. Their prediction error values ($\sigma_\varepsilon^2 = 0.031$ for Yin_1999.c and 0.032 for Yin_1999.d) further illustrate that both are almost equally accurate in the terms of prediction.

The estimations of the vertical strains at four loading stages by the 9 updated predictive models were computed, and the corresponding statistic indices are shown in Table 4.11. Figure 4.5 shows the average values of the statistic indices of the four stages. It is clearly seen that the first five EVP models have close average values for *RMSE* and *MAPE*, as well as having similar accuracy in fitting to the training data. The variations of vertical strain with time, estimated by the top two models, i.e., Yin_1999.c_RS1 and Yin_1999.d_RS1, are clearly shown in Fig. 4.6. Both models show reasonable agreement with the training data.

Table 4.10 Posterior statistics of model parameters updated based on the consolidation data of RS1

Model class	Model parameter									
Yin_1999.c	k/*V*	k/*V*	ε_{lm}^{vp}	–	$\psi_0^{'} = c - d\ \ln(\sigma_z^{'}/\sigma_u^{'})$		$\sigma_{zr}^{'}$	t_o	k_O	σ_{zr}^2
					c	d				
	0.0820 (0.0019)	0.0627 (1.12e-4)	0.286 (0.0708)	–	0.0195 (0.0027)	0.0027 (4.38e-4)	2.827 (0.370)	42.86 (12.93)	5.58e-8 (4.03e-10)	0.031 (0.002)
Yin_1999.d	λ/V	k/*V*	$\varepsilon_{lm}^{vp} = a - b\ \ln\left(\sigma_z^{'}/\sigma_u^{'}\right)$		$\psi_0^{'} = c - d\ \ln(\sigma_z^{'}/\sigma_u^{'})$					
			a	b	c	d	$\sigma_{zr}^{'}$	t_o	k_O	σ_{zr}^2
	0.0808 (0.0014)	0.0646 (9.77e-5)	0.236 (0.0103)	0.0252 (0.0030)	0.0201 (0.0033)	0.0025 (6.73e-4)	2.703 (0.239)	50.45 (10.00)	5.62e-8 (2.69e-9)	0.032 (0.0029)

Note The values outside the parentheses are the updated mean values, while those in the parentheses are the corresponding standard deviations

Table 4.11 Comparison of the estimated results by the updated models based on the consolidation data of RS1

Model	Statistical indice: *RMSE* (*MAPE*, %)			
	RS1_1	RS1_2	RS1_3	RS1_4
YG_1989_RS1	0.26 (2.08)	0.17 (0.37)	0.12 (0.29)	0.17 (0.28)
Yin_1999. a_RS1	0.26 (2.10)	0.17 (0.38)	0.13 (0.32)	0.17 (0.29)
Yin_1999. b_RS1	0.26 (2.08)	0.18 (0.41)	0.13 (0.28)	0.17 (0.25)
Yin_1999. c_RS1	0.24 (2.02)	0.16 (0.40)	0.14 (0.32)	0.16 (0.26)
Yin_1999. d_RS1	0.24 (2.02)	0.17 (0.40)	0.13 (0.28)	0.19 (0.25)
Vermeer et al._1997_RS1	0.30 (2.37)	0.22 (0.70)	0.17 (0.45)	0.07 (0.10)
KL_2001_RS1	0.29 (2.23)	0.18 (0.58)	0.12 (0.25)	0.10 (0.20)
Yin et al._2011_RS1	0.30 (2.31)	0.22 (0.68)	0.17 (0.43)	0.09 (0.12)
Yang et al._2015_RS1	0.29 (2.26)	0.48 (1.28)	0.47 (0.96)	0.41 (0.77)

Note The values outside the parentheses are the *RMSE*, while those in the parentheses are the *MAPE* (%)

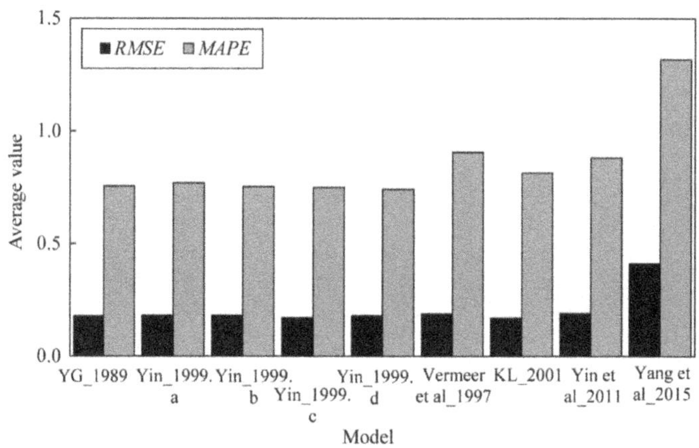

Fig. 4.5 Comparison of average statistical indices for the models updated using the consolidation data of RS1

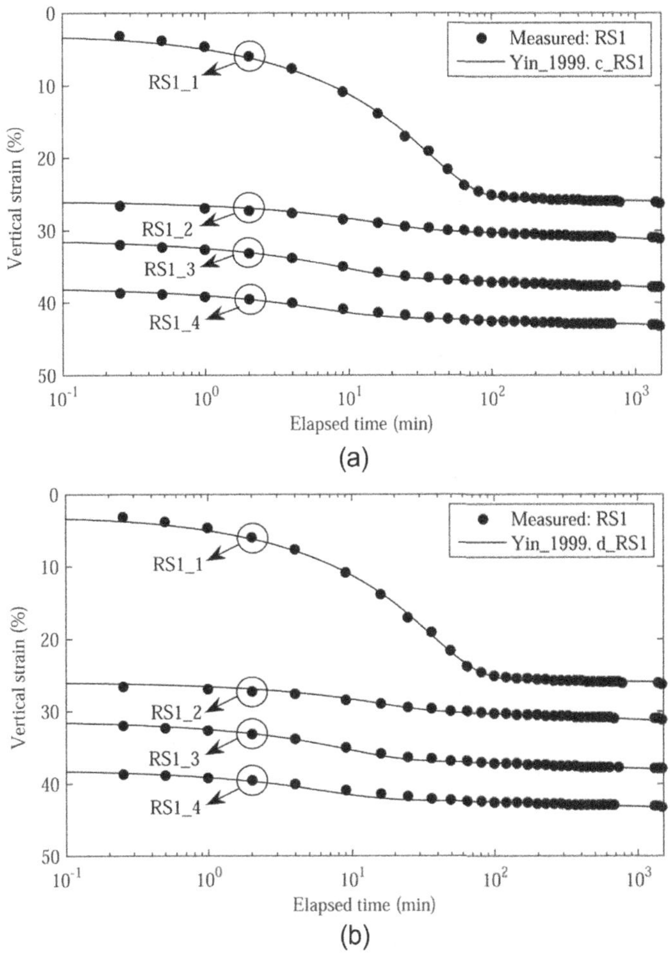

Fig. 4.6 Comparisons of vertical strains between the measurements and estimations by: **a** the Yin_1999.c_RS1 model and **b** the Yin_1999.d_RS1 model

Using the identified parameters shown in Table 4.10, the 95% CIs of predictions for the RS2 sample were computed. Table 4.12 presents the statistic indices of the testing results from the Yin_1999.c_RS1 and Yin_1999.d_RS1 models. An onlooker may discern that the former model yields more accurate predictions than the latter. Figure 4.7 plots the variations of vertical strain with time, as estimated by the Yin_1999.c_RS1 model. It is seen that most of the measured data are located in the areas of 95% CIs for the entire consolidation process. Especially for the final loading stage, the predicted mean values are in good agreement with the measured data. It can be considered that the optimal model, Yin_1999.c_RS1, has sufficient accuracy in predicting the creep deformation of the same type of soft soil. However, if higher accuracy is required, the consolidation data in longer loading duration are needed for

Table 4.12 Verification results of two updated models based on the consolidation data of RS1

Model	Statistical indice	Loading stage			
		RS2_1	RS2_2	RS2_3	RS2_4
Yin_1999.c_RS1	*RMSE*	1.66	1.23	0.54	0.47
	MAPE (%)	5.54	3.45	1.17	0.82
Yin_1999.d_RS1	*RMSE*	1.67	1.27	0.62	0.45
	MAPE (%)	5.59	3.60	1.38	0.86

Note The values outside the parentheses are the *RMSE*, while those in the parentheses are the *MAPE* (%)

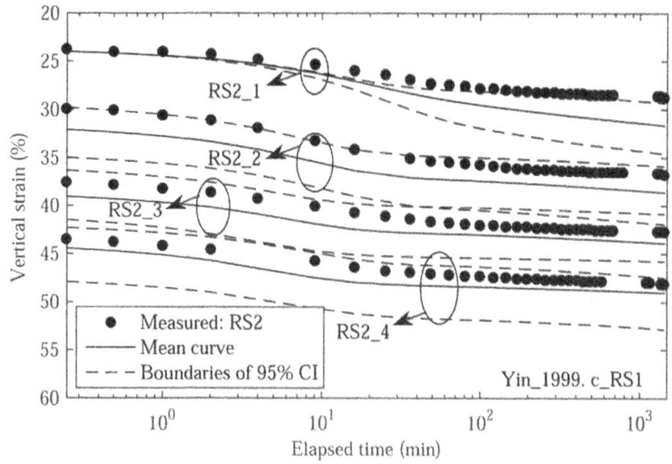

Fig. 4.7 RS2 measured data and 95% CI of vertical strains estimated by Yin_1999.c_RS1 model

model updating. Additionally, it is also observed that the prediction uncertainty can be quantified. Given a set of measured data, one can predict the creep deformation using the updated model with certain credibility.

4.5 Conclusions

In this chapter, a probabilistic method of identifying the model parameters and investigating the efficiency of different models in the prediction of the creep behavior of clay, under 1-D straining, is presented. Consolidated data from the intact samples of Vanttila clay and reconstituted samples of Hong Kong Marine Clay were collected in the case study. In the current approach, all model parameters can be identified simultaneously using the TMCMC method, without artificially determination of any parameter in advance.

From the investigation, we found that the presented approach is efficient for identifying the most plausible model and its associated optimal parameters for different types of soils. This approach is useful in engineering practice; engineers can adopt the present probabilistic method to determine the most suitable model class and its associated model parameters for a given soft soil for the prediction of long-term creep behavior. The strategy also provides uncertainty evaluation of the model prediction based on the given data.

References

1. Bjerrum L (1967) Seventh Rankine Lecture: Engineering geology of Norwegian normally-consolidated marine clays as related to settlement of buildings. Géotechnique 17(2):81–118
2. Mesri G, Godlewski PM (1977) Time and stress-compressibility interrelationship. J Geotech Eng ASCE 103(5):417–430
3. Vaid YP, Campanella RG (1977) Time-dependent behavior of undisturbed clay. J Geotech Geoenviron Eng ASCE 103(7):693–709
4. Graham J, Crooks JHA, Bell AL (1983) Time effects on the stress-strain behavior of natural soft clays. Géotechnique 33(3):327–340
5. Leroueil S, Kabbaj M, Tavenas F, Bouchard R (1985) Stress-strain-strain rate relationship for the compressibility of sensitive natural clays. Géotechnique 35(2):159–180
6. Nash DFT, Sills GC, Davison LR (1992) One-dimensional consolidation testing of soft clay from Bothkennar. Géotechnique 42(2):241–256
7. Sheahan TC, Ladd CC, Germaine JT (1996) Rate-dependent undrained shear behavior of saturated clay. J Geotech Geoenviron Eng ASCE 122(2):99–108
8. Yin JH, Cheng CM (2006) Comparison of strain-rate dependent stress-strain behavior from K0-consolidated compression and extension tests on natural Hong Kong Marine deposits. Mar Georesour Geotechnol 24(2):119–147
9. Yin ZY, Yin JH, Huang HW (2015) Rate-dependent and long-term yield stress and strength of soft Wenzhou marine clay: experiments and modeling. Mar Georesour Geotechnol 33(1):79–91
10. Yin JH, Graham J (1989) Viscous-elastic-plastic modelling of one-dimensional time-dependent behavior of clays. Can Geotech J 26:199–209
11. Yin ZY, Chang CS, Karstunen M, Hicher PY (2010) An anisotropic elastic viscoplastic model for soft clays. Int J Solids Struct 47(5):665–677
12. Yin JH, Graham J (1994) Equivalent times and one-dimensional elastic viscoplastic modelling of time-dependent stress-strain behavior of clays. Can Geotech J 31:42–52
13. Yin ZY, Zhu QY, Yin JH, Ni Q (2014) Stress relaxation coefficient and formulation for the soft soils. Géotechnique Lett 4(1):45–51
14. Zhu QY, Yin ZY, Hicher PY, Shen SL (2016) Nonlinearity of one-dimensional creep characteristics of soft clays. Acta Geotech 11(4):887–900
15. Yin ZY, Zhu QY, Zhang DM (2017) Comparison of two creep degradation modeling approaches for soft structured soils. Acta Geotech. https://doi.org/10.1007/s11440-017-0556-y
16. Borja RI, Kavazanjian E (1985) A constitutive model for the stress-strain-time behavior of 'wet' clays. Géotechnique 35(3):283–298
17. Kutter BL, Sathialingam N (1992) Elastic-viscoplastic modelling of the rate-dependent behavior of clays. Géotechnique 42(3):427–441
18. Yin JH, Graham J (1999) Elastic viscoplastic modelling of the time-dependent stress-strain behaviour of soils. Can Geotech J 36:736–745
19. Vermeer PA, Neher H (1999) A soft soil model that accounts for creep. In: Proceedings of the international symposium beyond 2000 in computational Geotechnics, pp. 249–261

20. Kimoto S, Oka F (2005) An elasto-viscoplastic model for clay considering destructuralization and consolidation analysis of unstable behavior. Soils Found 45(2):29–42
21. Yin ZY, Karstunen M, Chang CS, Koskinen M, Lojander M (2011) Modeling time-dependent behavior of soft sensitive clay. J Geotech Geoenviron Eng ASCE 137(11):1103–1113
22. Yin JH (1999) Non-linear creep of soils in oedometer tests. Géotechnique 49(5):699–707
23. Yin JH, Zhu JG, Graham J (2002) A new elastic viscoplastic model for time-dependent behavior of normally and overconsolidated clays: theory and verification. Can Geotech J 39:157–173
24. Le TM, Fatahi B, Khabbaz H (2015) Numerical optimisation to obtain elastic viscoplastic model parameters for soft clay. Int J Plast 65:1–21
25. Fatahi B, Le TM (2016) Trust-region reflective optimisation to obtain soil visco-plastic properties. Eng Comput: Int J Comput Aided Eng Softw 32(2):410–442
26. Le TM, Fatahi B, Khabbaz H, Sun WJ (2017) Numerical optimization applying trust-region reflective least squares algorithm with constraints to optimize the non-linear creep parameters of soft soil. Appl Math Model 41:236–256
27. Jin YF, Yin ZY, Shen SL, Hicher PY (2016) Selection of sand models and identification of parameters using an enhanced genetic algorithm. Int J Numer Anal Meth Geomech 40(8):1219–1240
28. Jin YF, Yin ZY, Shen SL, Zhang DM (2017) A new hybrid real-coded genetic algorithm and its application to parameters identification of soils. Inverse Problems Sci Eng 25(9):1343–1366
29. Yin ZY, Jin YF, Shen SL, Huang HW (2017) An efficient optimization method for identifying parameters of soft structured clay by an enhanced genetic algorithm and elastic-viscoplastic model. Acta Geotech 12(4):849–867
30. Buisman AS (1936) Results of long duration settlement tests. In: Proceedings 1st international conference on soil mechanics and foundation engineering, Cambridge, Mass, 1, pp. 103–107
31. Garlanger JE (1972) The consolidation of soils exhibiting creep under constant effective stress. Géotechnique 22(1):71–78
32. Magnan JP (1984) Modélisation numériques du comportement des argiles molles naturelles. DSc thesis, Université Pierre et Marie Curie, Paris
33. Yin JH, Graham J (1989) "General elastic viscous plastic constitutive relationships for 1-D straining in clays. In: Proceedings 3rd. International symposium numerical models geomechanics, Elsevier Applied Science, pp. 108–117
34. Hu YY, Zhou WH, Cai YQ (2014) Large strain elastic visco-plastic consolidation analysis of very soft clay layers with vertical drains under preloading. Can Geotech J 51(2):144–157. https://doi.org/10.1139/cgj-2013-0200
35. Yin ZY, Zhu QY, Zhang DM (2017) Comparison of two creep degradation modeling approaches for soft structured soils. Acta Geotech. https://doi.org/10.1007/s11440-017-0556-y
36. Mesri G, Vardhanabhuti B (2005) Secondary compression. J Geotech Geoenviron Eng 131:398–401
37. Kabbaj M, Oka F, Leroueil S, Tavenas F (1986) Consolidation of natural clays and laboratory testing. In: Yong RN, Townsend FC (eds) Consolidation of soils: testing and evaluation, ASTM STP 892. Am Soc Testing Mater, Philadelphia, pp 378–404
38. Wang L, Luo Z, Xiao JH, Juang CH (2014) Probabilistic inverse analysis of excavation-induced wall and ground responses for assessing damage potential of adjacent buildings. Geotech Geol Eng 32:273–285
39. Tang WH (1971) A Bayesian evaluation of information for foundation engineering design. Statistics and probability in civil engineering 1971; P. Lumb, ed., Hong Kong University Press, Hong Kong
40. Beck JL, Au SK (2002) Bayesian updating of structural models and reliability using Markov chain Monte Carlo simulation. J Eng Mech 128(4):380–391
41. Ching J, Phoon KK, Chen YC (2010) Reducing shear strength uncertainties in clays by multivariate correlations. Can Geotech J 47(1):16–33
42. Zhang J, Tang WH, Zhang LM (2010) Efficient probabilistic back-analysis of slope stability model parameters. J Geotech Geoenviron Eng 136(1):99–109

43. Yan WM, Yuen KV, Yoon GL (2009) Bayesian probabilistic approach for the correlations of compression index for marine clays. J Geotech Geoenviron Eng 135(12):1932–1940
44. Chiu CF, Yan WM, Yuen KV (2012) Estimation of water retention curve of granular soils from particle-size distribution—a Bayesian probabilistic approach. Can Geotech J 49(9):1024–1035
45. Park JH, Kim D, Chung CK (2012) Implementation of Bayesian theory on LRFD of axially loaded driven piles. Comput Geotech 42:73–80
46. Zhou WH, Yuen KV, Tan F. Estimation of maximum pullout shear stress of grouted soil nails using Bayesian probabilistic approach. Int J Geomech ASCE 13(5):659–664
47. Zhou WH, Yuen KV, Tan F (2014) Estimation of soil-water characteristic curve and relative permeability for granular soils with different initial dry densities. Eng Geol 179:1–9
48. Huang J, Kelly R, Li D, Zhou C, Sloan S (2016) Updating reliability of single piles and pile groups by load tests. Comput Geotech 73:221–230
49. Tan F, Zhou WH, Yuen KV (2016) Modeling soil water retention property of the same textured soil with different initial void ratios. J Hydrol 542:731–743
50. Gamse S, Zhou WH, Tan F, Yuen KV, Oberguggenberger M (2018) Hydrostatic-season-time model updating using Bayesian model class selection. Reliab Eng Syst Saf 169:40–50
51. Bodas Freitas TM, Potts DM, Zdravkovic L (2011) A time dependent constitutive model for soils with isotach viscosity. Comput Geotech 38:809–820
52. Vermeer PA, Stolle DFA, Bonnier PG. From the classical theory of secondary compression to modern creep analysis. In: Proceedings of the 9th International Conference on Computer Methods and Advances in Geomechanics, Wuhan, China, Vol. 4, Rotterdam: Balkema, pp 2469–2478
53. Kim YT, Leroueil S (2001) Modeling the viscoplastic behavior of clays during consolidation: Application to Berthierville clay in both laboratory and field conditions. Can Geotech J 38:484–497
54. Yang C, Liu XL, Liu XF, Yang CS, Carter JP (2015) Constitutive modelling of Otaniemi soft clay in both natural and reconstituted states. Comput Geotech 70:83–95
55. Das BM (2010) Principles of geotechnical engineering. 7th ed. Cengage. ISBN: 9780495411321
56. Yin JH, Graham J (1996) Elastic visco-plastic modelling of one-dimensional consolidation. Géotechnique 46(3):515–527
57. Yuen KV (2010) Bayesian methods for structural dynamics and civil engineering. Wiley (Asia) Pte Ltd, Singapore
58. Ching JY, Chen YC (2007) Transitional Markov Chain Monte Carlo method for Bayesian model updating, model class selection, and model averaging. J Eng Mech 133(7):816–832
59. Feng WQ, Lalit B, Yin ZY, Yin JH (2017) Long-term non-linear creep and swelling behavior of Hong Kong marine deposits in oedometer condition. Comput Geotech 84:1–15

Chapter 5
Effect of Loading Duration on Uncertainty in Creep Analysis for Clay

Abstract Following the work presented in the previous chapter, we further investigate the effect of loading duration on the uncertainty of the creep behavior prediction for clay analysis. Consolidation data from intact samples of Vanttila clay was used in the analysis. The results show that sufficient creep test data is crucial for reliable models and the associated parameters. The appropriate amount of test data used for updating can be determined by the Bayesian probabilistic approach, and this can avoid conducting an unnecessary long-term consolidation test. The proposed method can quantify the uncertainty of model prediction to facilitate better judgment by engineers for the loading duration and model uncertainty in their predictions.

5.1 Introduction

The previous studies show a consensus that the updated creep parameters are influenced by the test data used. The creep parameters should be updated using more creep test data to capture the deformation behavior of soft soils. However, different types of soft soils have different characteristics of deformation behavior, and their primary consolidations may end at different times. Therefore, it is unreasonable to conduct consolidation tests with the same loading duration to update the model parameters for different types of soft soils. Additionally, as mentioned in the previous chapter that a deterministic method only concerns the best fitting between the predictions and the observations in said determination. It ignores the uncertainties associated with statistical uncertainty, measurement error, and model uncertainty and its effect on the model predictions.

In the previous chapter, we adopt the Bayesian probabilistic method to identify the model parameters of nine well-known 1-D compression models, and select the optimal models to describe the creep behavior of a certain soft soil. In this chapter, the effect of loading duration on the identified model parameters is investigated based on two groups of 1-D oedometer test data. The analyses indicate that the form of the optimal models is unique for two intact soil samples under different loading durations, and they can better describe the creep behavior of intact soft soils. The appropriate amount of test data used for updating can be determined by the method

W. Zhou et al., *Practice of Bayesian Probability Theory in Geotechnical Engineering*,
https://doi.org/10.1007/978-981-15-9105-1_5

adopted, and this can avoid conducting an unnecessary long-term consolidation test. The uncertainty of updated model parameters decreases when more creep data is used. The corresponding uncertainty of predicted deformation is also evaluated by computing the 95% credible interval (CI).

5.2 Data for Training and Testing

To identify the model parameters and investigate the effect of loading duration on updated results, compression data for Vanttila clay yielded by conventional oedometer tests was collected from Yin et al. [1]. In this study, consolidation data for four intact samples with the same height ($H_0 = 15$ mm) at 3 different loading durations (1 day, 10 days, and 100 days) was used for analysis. Table 5.1 shows the initial conditions for different samples. Each loading stage is maintained for 1 day and 100 days, respectively, for S1 and S4 samples, and 10 days for S2 and S3 samples. The 1-day oedometer test data for S1 and the 10-day oedometer test data for S2, respectively, were used for training, i.e., identifying the model parameters. At the same time, the most plausible models were obtained by Bayesian model class selection. Then the 10-day and 100-day oedometer test data for S3 and S4 were used for testing, i.e., verifying the performance of the models. By comparing the results obtained using the data for S1 and S2, the effect of the amount of data at the creep stage on the model parameters can be investigated.

5.3 Results and Discussion

5.3.1 Results by TMCMC Method

The prior PDFs for $\boldsymbol{\theta}$ were assumed to follow the lognormal distribution, and their distribution parameters should assure that the generated samples in the initial stage are spread widely. To ensure the low computational cost and the smooth transition between the adjacent PDFs, the scaling parameter β was set to 0.2, and the prescribed threshold for the coefficient of variation (COV) of plausibility weight was set to 100% [2]. During the TMCMC simulation process, the sample numbers of all stages were fixed to 10,000, and then 10,000 N_S total samples were generated in the simulation. The nine model class candidates are assumed to have uniform prior plausibility in this study, i.e., $P\left(C^{(m)}\right) = 1/9\,(m = 1, 2, \ldots, 9)$, and the evaluation of the model classes can be conducted by comparing the computed evidence (S). The model class with the larger S value is the optimal model class among the 9 chosen model classes.

The log-scaled S estimations of the nine model classes, which are obtained based on two groups of consolidation test data from the S1 and S2 samples, respectively, are presented in Table 5.2a, b. It is clear to see that the optimal model classes are the

Table 5.1 Test information of intact soil samples

Sample	Loading duration t/day	Loading stage	Initial vertical strain $(\varepsilon_z)_{i,0}$/%	Vertical surcharge $(\sigma_z)_{i,0}$/kPa	Initial excess pore water pressure $u_{i,0}$/kPa
S1	1	S1_1	0.74	9.4	8.3
		S1_2	1.40	18.75	9.35
		S1_3	2.59	37.5	18.75
		S1_4	13.08	75	37.5
		S1_5	29.59	150	75
		S1_6	37.51	300	150
S2	10	S2_1	0.61	9.4	8.3
		S2_2	1.39	18.75	9.35
		S2_3	4.00	37.5	18.75
		S2_4	21.73	75	37.5
		S2_5	31.62	150	75
		S2_6	39.13	300	150
S3	10	S3_1	0.46	6.25	5.15
		S3_2	1.00	12.5	6.25
		S3_3	2.04	25	12.5
		S3_4	5.05	50	25
		S3_5	25.45	100	50
		S3_6	34.6	200	100
		S3_7	41.55	400	200
S4	100	S4_1	1.61	9.4	8.3
		S4_2	2.73	18.75	9.35
		S4_3	7.58	37.5	18.75
		S4_4	22.81	75	37.5
		S4_5	33.84	150	75
		S4_6	42.13	300	150

same for the two intact samples under different loading durations. The Yin et al_2011 model class always outperformed the other model classes with a posterior probability value $P(C|D) \approx 1.0$, and the Yang et al_2015 model class is second. This indicates that it is important for a model to capture the destructuration effect in the creep analysis of intact samples. With the exception of these two model classes, the other model classes when considering only the time effect show no significant difference in the magnitude of the posterior probability based on the 1-day oedometer test data for S1. Among these model classes, the Vermeer et al_1997 model class with the smallest number of uncertain parameters (6) has a larger possibility of yielding better predictions than the other model classes. The YG_1989 model class with a

Table 5.2 Model class selection results based on the consolidation data for (a) S1 and (b) S2

Model class	Average $\ln p(D\|\boldsymbol{\theta}, C)$	$\ln S$	$P(C\|D)$
(a)			
YG_1989	−241.39	−253.91	1.22e-81
Yin_1999.a	−241.57	−256.39	1.02e-82
Yin_1999.b	−241.63	−256.22	1.21e-82
Yin_1999.c	−240.93	−255.57	2.32e-82
Yin_1999.d	−240.86	−255.90	1.67e-82
Vermeer_et al_1997	−240.27	−253.08	2.80e-81
KL_2001	−241.33	−253.96	1.16e-81
Yin_et al_2011	−36.97	−67.60	1.00
Yang_et al_2015	−112.97	−142.68	2.47e-33
(b)			
YG_1989	−572.77	−587.61	4.10e-225
Yin_1999.a	−528.18	−547.85	7.60e-208
Yin_1999.b	−513.66	−544.21	2.89e-206
Yin_1999.c	−521.97	−547.69	8.91e-208
Yin_1999.d	−519.02	−553.88	1.83e-210
Vermeer_et al_1997	−565.43	−580.60	4.54e-222
KL_2001	−567.95	−582.96	4.29e-223
Yin_et al_2011	−35.58	−70.94	1.00
Yang_et al_2015	−247.41	−286.30	2.95e-94

linear creep function and 7 uncertain parameters is slightly better than the other four model classes, i.e., a series of Yin_1999 model classes with nonlinear creep functions. These results indicate that when the 1-day oedometer test data for S1 are used to train the model classes, the model classes considering only the time effect may behave similarly in the prediction of vertical strain. When employing the 10-day oedometer test data for S2 to determine the model parameters, several model classes considering only the time effect show a distinct difference in the values of posterior probability. Among them, the Yin_1999 model classes outperform the others in their predictions. Of special note, the Yin_1999.b model class, which considers the variation of creep strain limit with effective vertical stress, performs better than the other three Yin_1999 model classes. This means that the Yin_1999.b model class can reflect the destructuration effect of soil deformation by considering the variation of creep strain limit with effective stress.

Figures 5.1 and 5.2 show the evolution samples of two of the same parameters (κ/V and σ_ε^2) for the nine model classes at selected stages in two cases. Among the four subgraphs in each column of figures, the first and last correspond to the initial stage and final stage, respectively. For all the model classes, it is clear to see that the samples gradually converge to a small region. The samples of the parameters at the final stage are asymptotically distributed as the posterior PDFs $f(\boldsymbol{\theta}|C, D)$; the corresponding posterior statistics, i.e., the mean values and the standard deviations, can be respectively obtained based on the two groups of consolidation data. Here, only the results of the top two model classes are shown in Table 5.3a, b. The values in the parentheses are the updated mean values and the corresponding standard deviations, respectively, and those outside the parentheses are the COV values for the parameters. Comparing the COV values in Table 5.3a, b, it can be observed that the COV values of all updated parameters based on the 10-day oedometer test data for S2 are generally less than those based on the 1-day oedometer test data for S1. Moreover, the obtained value of prediction-error variances σ_ε^2 is found to decrease in the case of S2. The

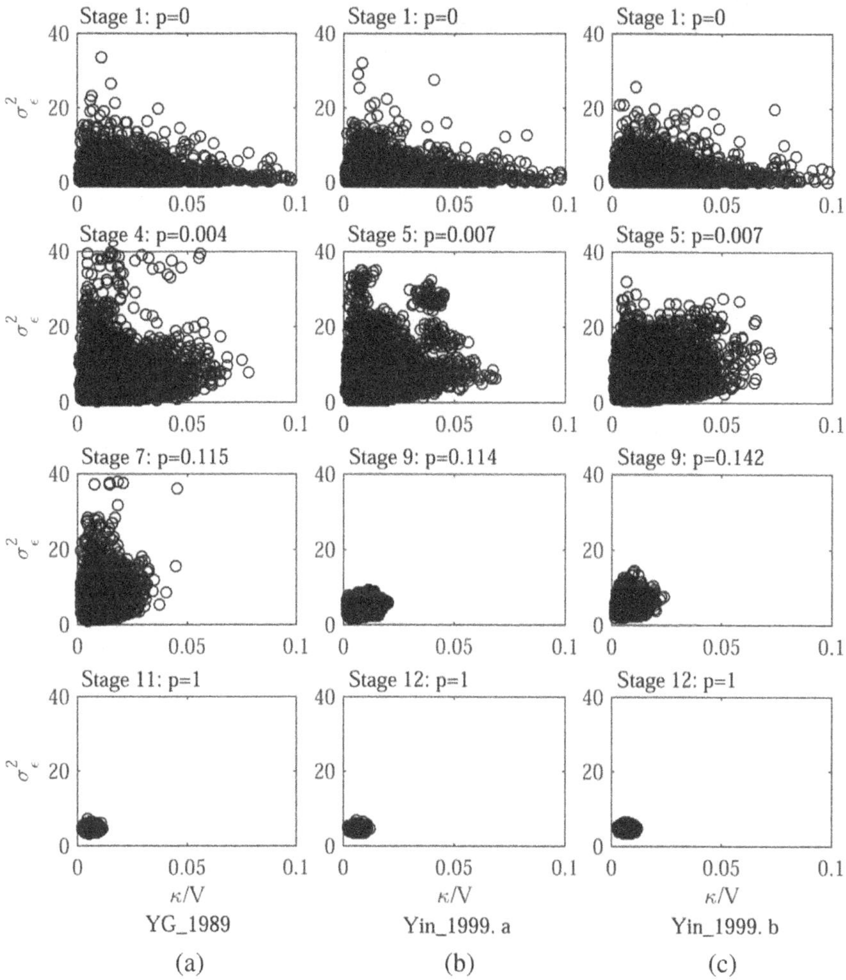

Fig. 5.1 Evolution of samples from TMCMC at selected stages for nine model classes: S1

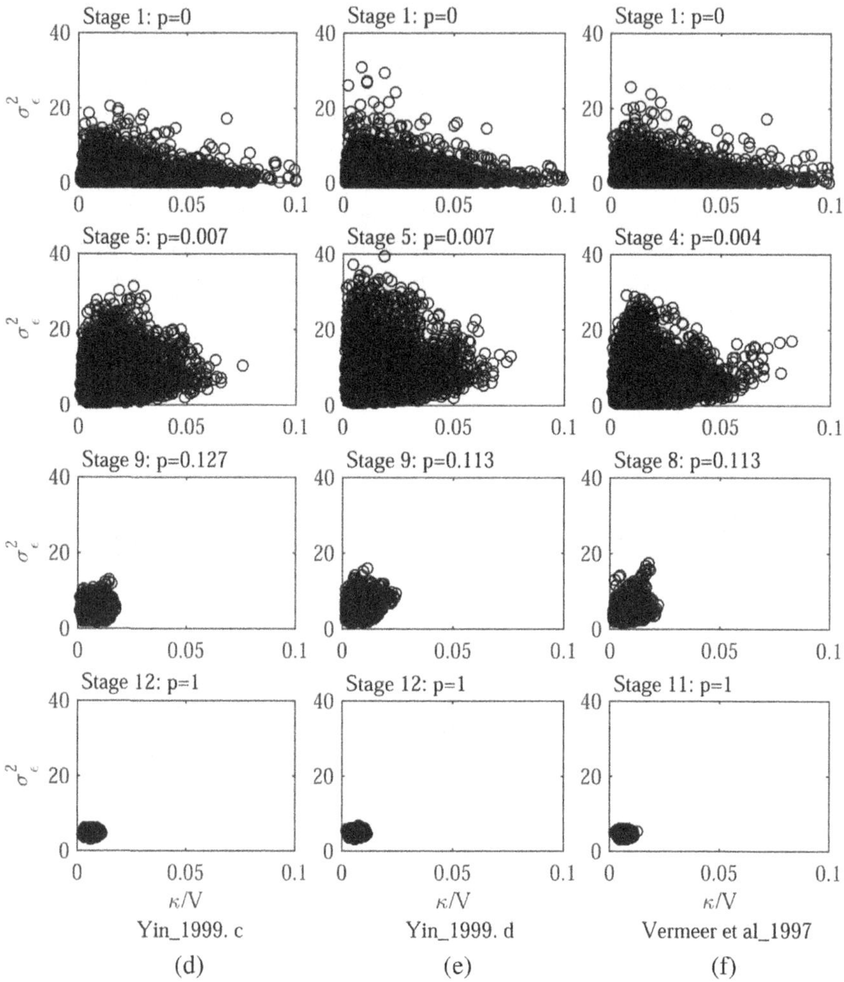

Fig. 5.1 (continued)

changes in COV and σ_ε^2 values indicate that the uncertainty of model parameters and model predictions generally decreases with increasing the amount of creep data used for updating.

Based on the two groups of results under different loading durations, it can be concluded that the determination of the optimal model is not only related to the mechanical properties of the soft soil, but also to the experimental data. To capture the creep deformation of the intact soft soil, the optimal model and its associated parameters should be updated using the experimental data with sufficient loading duration.

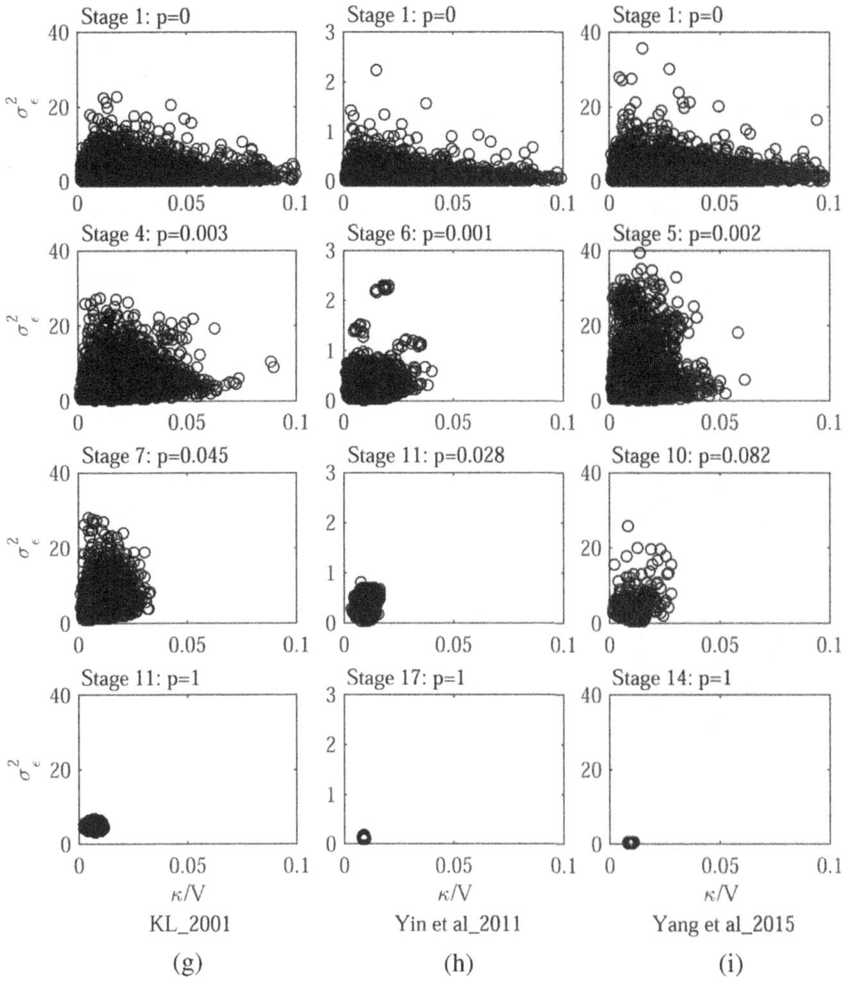

Fig. 5.1 (continued)

5.3.2 Estimations by Updated Models

The estimation of vertical strain at each loading stage by the nine predictive models can be computed using the mean values of parameters, e.g., the values shown in Table 5.3. Table 5.4a, b summarizes the statistical indices between the estimations and the oedometer test data from the S1 and S2 samples, respectively. The average values of *RMSE* and *MAPE* for the last 4 loading stages are compared in Fig. 5.3a, b. It is clear to see that the updated Yin et al_2011 models have the smallest values for *RMSE* and *MAPE* and yield more accurate predictions than the other models.

When employing the 1-day oedometer test data to determine the model parameters, Table 5.4a shows no obvious difference among the *MAPE* and *RMSE* values of the five models proposed by Yin and his coworkers [3, 5, 7]. The Yin et al_2011_S1 and Yang et al_2015_S1 models perform very well in the prediction of vertical strain at the last 4 loading stages. However, when the soil sample is loaded with the small vertical surcharge in a short time, these two models cannot capture the deformation behavior. This implies that when the applied loading is smaller than the preconsolidation pressure and loading duration is short, the models that consider only the time effect may show better performance in their predictions. When the applied

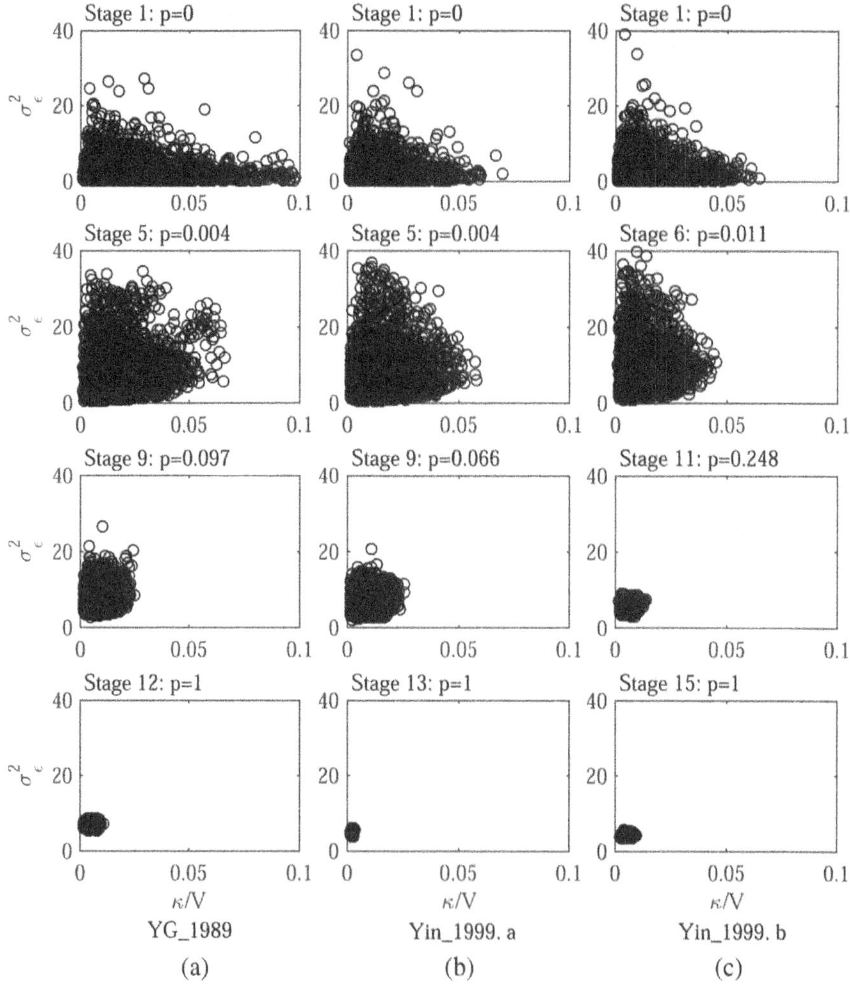

Fig. 5.2 Evolution of samples from TMCMC at selected stages for nine model classe

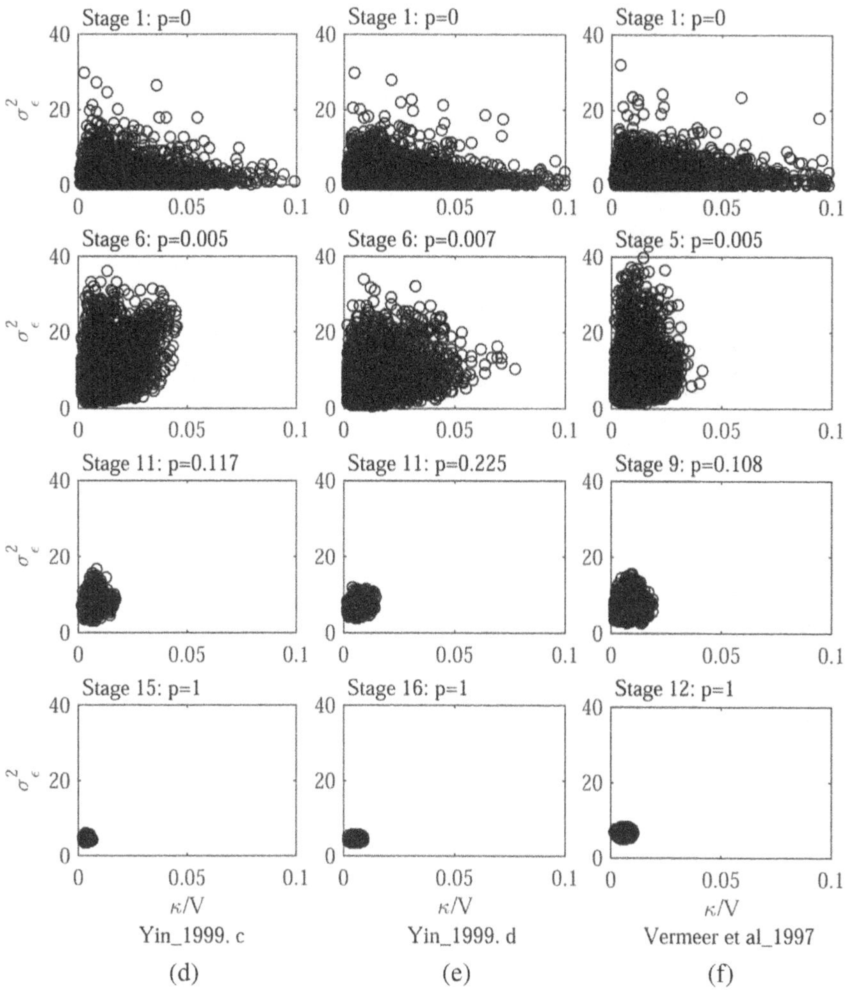

Fig. 5.2 (continued)

loading is larger than the preconsolidation pressure, the destructuration effect should be considered even under a shorter loading duration.

When the 10-day oedometer test data was employed to determine the model parameters, the Yin et al_2011_S2 and Yang et al_S2 models also exhibited a distinct advantage regarding the fitting ability (Fig. 5.3b). Compared to the histograms in Fig. 5.3a, their *MAPE* and *RMSE* values are found to decrease for all models. This means that when more creep data is used in updating a specific model, accuracy will be improved. Comparing the *MAPE* and *RMSE* values in Table 5.4b, it can be concluded that, even if the intact sample is loaded with a smaller vertical surcharge,

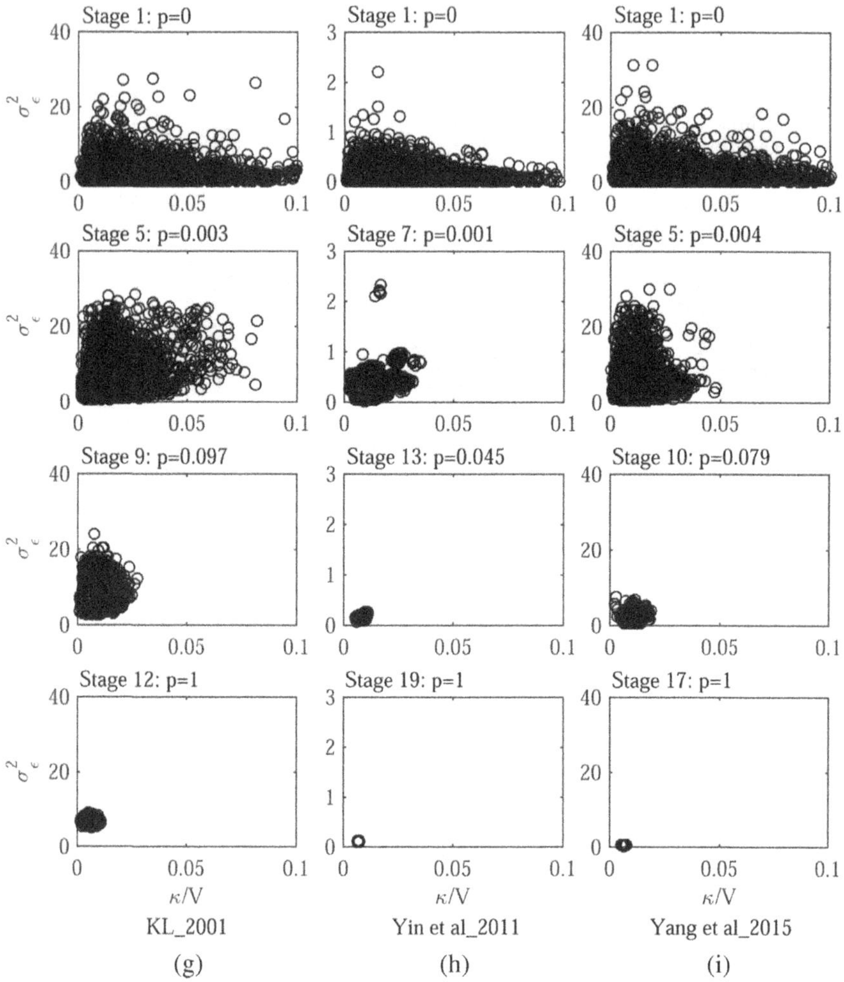

Fig. 5.2 (continued)

its deformation over a long duration should be described using models that consider the destructuration effect.

The variations of vertical strain with time estimated by the top two models [i.e., Yin et al_2011_S1 and Yang et al_2015_S1, see Table 5.3a] are plotted in Fig. 5.4a, b together with the measured data for S1, respectively. It is observed in Fig. 5.4a that the Yin et al_2011_S1 model, considering both the destructuration effect and creep effect, fits well during the whole consolidation process. The Yang et al_2015_S1 model does not take the creep effect into account. Correspondingly, Fig. 5.4b shows some disparities between the model predictions by Yang et al_2015_S1 and the test data, particularly in the later period of each loading stage, during which the creep effect becomes more significant. Similar results are also found in Fig. 5.5; the

Table 5.3 Posterior statistics of model parameters updated based on the consolidation data for (a) S1 and (b) S2

Model class	Model parameter							
(a)								
Yin et al_2011	λ/V	κ/V	$C_{\alpha e}$	χ_0	ρ	σ'^{r}_{p0} (kPa)	k_0 (m/min)	σ^2_ε
	(0.071, 7.57e-4) 0.011	(0.0088, 1.35e-4) 0.015	(0.0039, 1.45e-4) 0.037	(82.85, 5.71) 0.069	(13.50, 0.15) 0.011	(24.71, 0.27) 0.011	(8.04e-8, 7.09e-9) 0.088	(0.115, 0.011) 0.096
Yang et al_2015	λ/V	κ/V	–	χ^*_0	ρ^*	σ'_p (kPa)	k_0 (m/min)	σ^2_ε
	(0.086, 0.0062) 0.072	(0.0095, 3.63e-4) 0.038	–	(6.91, 0.512) 0.074	(11.78, 1.09) 0.093	(36.25, 0.094) 0.0026	(2.21e-8, 1.38e-9) 0.062	(0.478, 0.050) 0.105
(b)								
Yin et al_2011	λ/V	κ/V	$C_{\alpha e}$	χ_0	ρ	σ'^{r}_{p0} (kPa)	k_0 (m/min)	σ^2_ε
	(0.078, 7.39e-5) 9.47e-3	(0.0068, 5.34e-5) 7.85e-3	(0.0040, 1.69e-5) 4.23e-3	(30.42, 0.117) 3.85e-3	(11.44, 0.020) 1.75e-3	(16.17, 0.026) 1.61e-3	(5.79e-8, 1.83e-9) 0.032	(0.110, 0.003) 0.027
Yang et al_2015	λ/V	κ/V	–	χ^*_0	ρ^*	σ'_p (kPa)	k_0 (m/min)	σ^2_ε
	(0.052, 0.0032) 0.069	(0.0064, 3.56e-4) 0.027	–	(7.05, 0.417) 0.059	(4.70, 0.164) 0.035	(18.54, 0.015) 8.09e-4	(1.29e-8, 5.10e-10) 0.039	(0.469, 0.032) 0.068

Note The values in the parentheses are the mean values and standard deviations, while those values outside the parentheses are the COV values

Table 5.4 Statistic indices for vertical strains computed by the nine updated models based on the consolidation data for (a) S1 and (b) S2

Model	Loading stage of S1					
	1	2	3	4	5	6
(a)						
YG_1989_S1	0.63 (50.68)	0.43 (19.57)	2.49 (35.90)	2.80 (9.58)	2.67 (7.30)	2.38 (4.80)
Yin_1999. a_S1	0.66 (51.70)	0.51 (23.59)	2.60 (37.21)	2.67 (9.28)	2.66 (7.17)	2.45 (4.93)
Yin_1999. b_S1	0.71 (56.60)	0.53 (24.91)	2.62 (37.30)	2.71 (9.83)	2.27 (6.06)	2.59 (5.47)
Yin_1999. c_S1	0.64 (51.95)	0.48 (21.28)	2.46 (36.80)	2.80 (9.66)	2.62 (7.14)	2.42 (5.10)
Yin_1999. d_S1	0.65 (52.66)	0.47 (20.70)	2.40 (36.21)	2.87 (9.81)	2.66 (7.29)	2.37 (5.01)
Vermeer et al_1997_S1	0.68 (54.35)	0.59 (26.15)	3.20 (47.27)	2.27 (9.83)	1.93 (4.88)	3.00 (5.65)
KL_2001_S1	0.67 (51.14)	0.55 (25.36)	3.21 (46.17)	2.11 (8.70)	2.02 (4.80)	3.03 (5.86)
Yin et al_2011_S1	0.96 (71.47)	0.87 (42.15)	0.63 (8.23)	0.68 (3.31)	0.86 (2.33)	0.65 (1.56)
Yang et al_2015_S1	0.89 (53.97)	0.97 (47.11)	0.88 (11.48)	1.15 (3.63)	0.87 (1.79)	0.69 (1.35)
(b)						
YG_1989_S2	0.54 (47.20)	2.89 (79.57)	3.41 (24.12)	3.32 (11.98)	1.58 (3.34)	2.84 (5.37)
Yin_1999. a_S2	0.20 (16.72)	2.53 (67.56)	3.03 (21.58)	2.44 (8.74)	1.67 (4.11)	1.65 (3.21)
Yin_1999. b_S2	0.49 (42.40)	2.54 (63.39)	2.74 (19.59)	2.43 (8.76)	1.05 (2.35)	1.82 (3.61)
Yin_1999. c_S2	0.28 (23.22)	1.97 (48.95)	3.25 (21.58)	2.19 (7.70)	2.11 (5.48)	1.73 (3.48)
Yin_1999. d_S2	0.52 (45.23)	2.64 (66.21)	2.77 (20.12)	2.47 (8.89)	1.08 (2.33)	1.88 (3.73)
Vermeer et al_1997_S2	0.68 (59.72)	3.23 (89.79)	2.99 (23.35)	2.64 (9.55)	1.21 (2.65)	3.42 (6.83)
KL_2001_S2	0.56 (48.80)	3.22 (90.14)	2.99 (22.79)	2.68 (9.66)	1.24 (2.43)	3.50 (6.96)
Yin et al_2011_S2	0.75 (63.96)	0.46 (13.44)	0.56 (4.21)	0.82 (2.74)	0.67 (1.72)	0.47 (0.89)
Yang et al_2015_S2	0.61 (48.33)	0.65 (19.10)	1.17 (5.29)	0.85 (2.46)	0.96 (2.12)	0.89 (1.68)

Note The values outside the parentheses are the *RMSE*, while those in the parentheses are the *MAPE* (%)

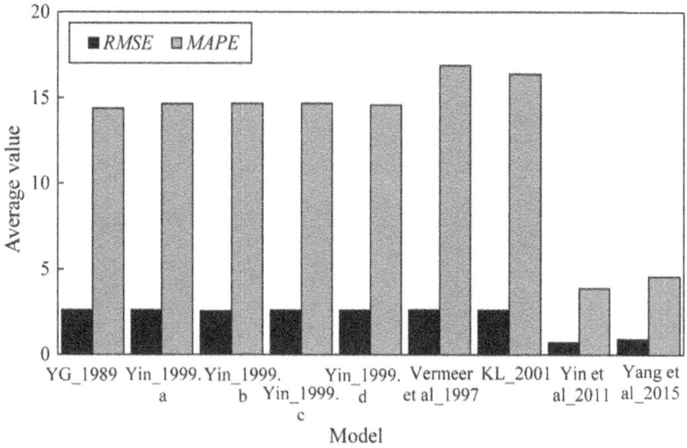

(a)

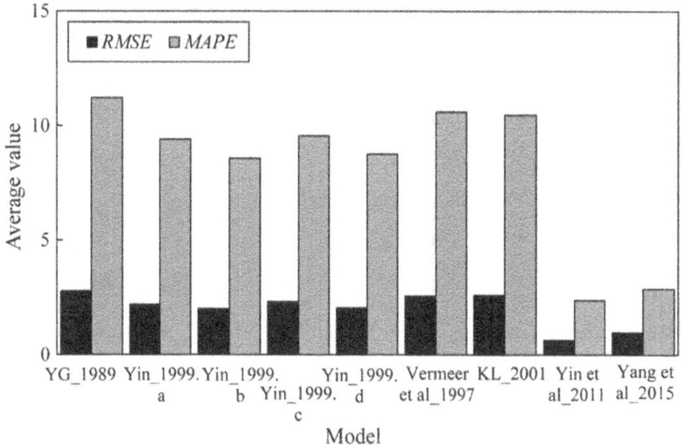

(b)

Fig. 5.3 Comparison of average statistical indices for the model updated using 1-day test data for S1 and 10-day test data for S2 samples

Yin et al_2011_S2 model is observed to yield a better estimation for vertical strain during the whole consolidation process at each loading stage. It should be noted that the above conclusions are based on an intact sample of Vanttila clay, and one should be cautious about the application of these findings on other types of soft soils.

When employing the 1-day oedometer test data to determine the model parameters, Table 5.4a shows no obvious difference among the *MAPE* and *RMSE* values of the five models proposed by Yin and his coworkers [3–5]. The Yin et al_2011_S1 and Yang et al_2015_S1 models perform very well in the prediction of vertical strain at the last 4 loading stages. However, when the soil sample is loaded with the small vertical surcharge in a short time, these two models cannot capture the deformation

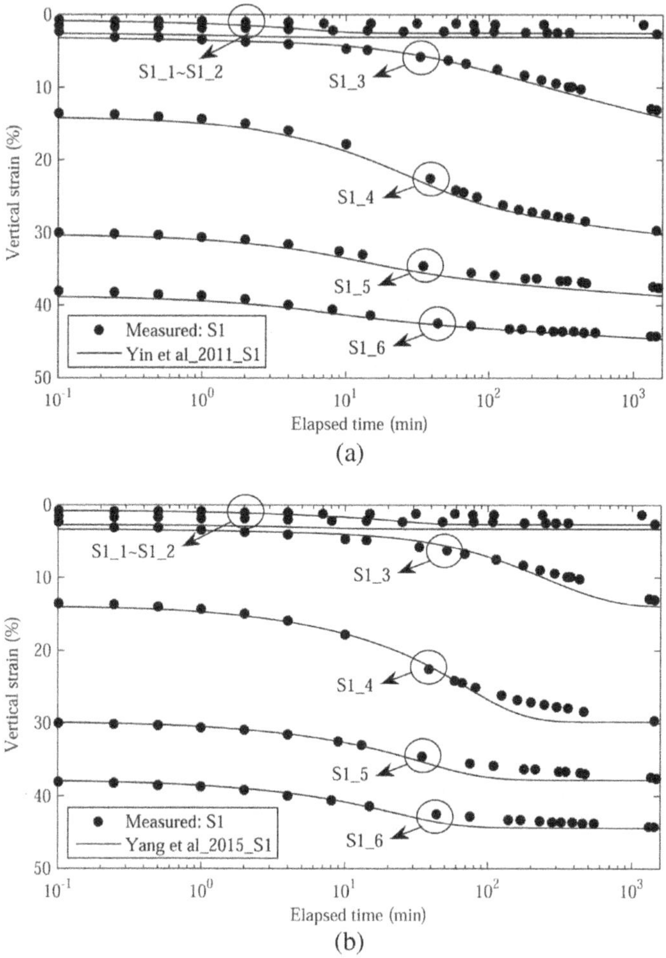

Fig. 5.4 Comparisons of vertical strain between the measured and estimations by: **a** Yin et al_2011_S1 model and **b** Yang et al_2015_S1 model

behavior. This implies that when the applied loading is smaller than the preconsolidation pressure and loading duration is short, the models that consider only the time effect may show better performance in their predictions. When the applied loading is larger than the preconsolidation pressure, the destructuration effect should be considered even under a shorter loading duration.

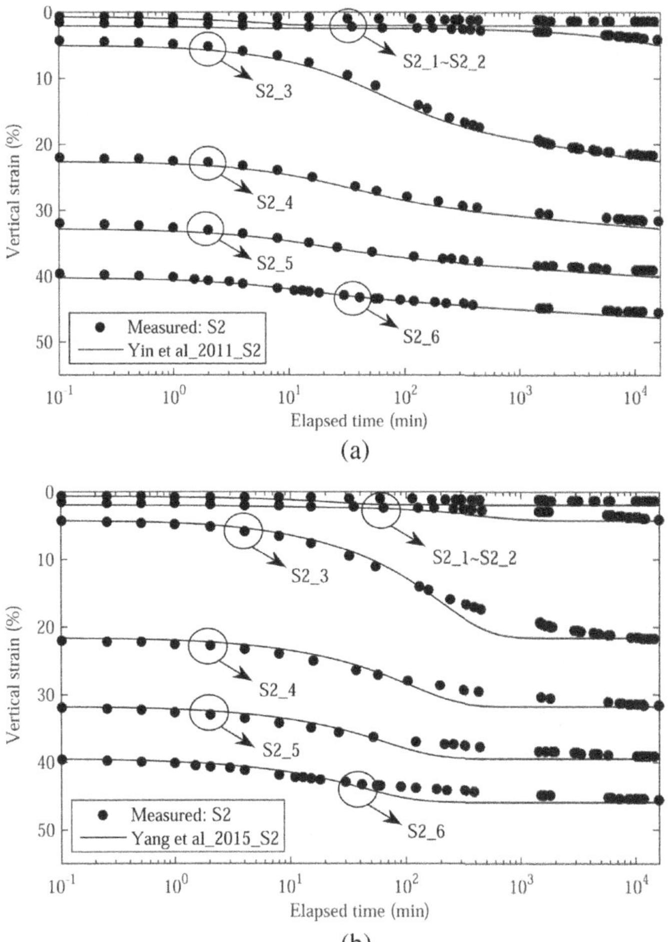

Fig. 5.5 Comparisons of vertical strain between the measured and estimations by: **a** Yin et al_2011_S2 model and **b** Yang et al_2015_S2 model

5.3.3 Verification and Uncertainty Analysis

If the updated model can appropriately capture the creep behavior of soft soil, practical engineers can employ the model to predict soil deformation in the long term. In this section, two groups of newly measured data for intact soil samples, i.e., 10-day oedometer test data for sample S3 and 100-day oedometer test data for sample S4, were used to test the efficiency of the updated models. This testing can show whether more data is needed for the identification of model parameters. Test information for samples S3 and S4 is shown in Table 5.1.

According to the identified parameters shown in Table 5.3a, the vertical strain for samples S3 and S4 is predicted by the Yin et al_2011_S1 and Yang et al_2015_S1 models. Table 5.5a, b summarizes the corresponding statistical results between the measured data and the mean values of predictions. It is seen that the Yin et al_2011_S1 model exhibits a distinct advantage regarding the fitting ability for the last 4 loading stages of S3 over the Yang et al_2015_S1 model. When predicting the deformations for S4, there is no significant difference between the predictions of the two models. When using the Yin et al_2011_S2 and Yang et al_2015_S2 models to predict the vertical strain for S4, the associated *RMSE* and *MAPE* values shown in Table 5.6 are mostly smaller than those by the Yin et al_2011_S1 and Yang et al_2015_S1 models. The *RMSE* and *MAPE* for the final 4 loading stages of S4 by Yin et al_2011_S2 are 11.03%–58.73% and 12.42%–66.82%, respectively, of those yielded by the Yin et al_2011_S1 model. The corresponding *RMSE* and *MAPE* by Yang et al_2015_S2 are 18.18%–50.14% and 14.30%–46.25%, respectively, of those yielded by the Yang

Table 5.5 Prediction results for S3 and S4 from two updated models based on the consolidation data for S1

Model	Statistical indices	Loading stage of S3						
		1	2	3	4	5	6	7
(a) S3								
Yin et al_2011_S1	*RMSE*	0.99	0.79	1.27	3.72	2.56	1.35	1.58
	MAPE (%)	121.02	49.8	25.31	21.51	7.48	3.31	3.37
Yang et al_2015_S1	*RMSE*	1.03	0.90	1.23	11.19	15.40	16.08	16.17
	MAPE (%)	117.18	56.47	24.06	54.53	47.38	40.85	35.00

Model	Statistical indices	Loading stage of S4					
		1	2	3	4	5	6
(b) S4							
Yin et al_2011_S1	*RMSE*	1.11	2.06	6.53	3.76	2.89	3.24
	MAPE (%)	45.26	28.12	34.15	11.19	6.8	6.63
Yang et al_2015_S1	*RMSE*	1.06	2.20	8.91	5.20	3.19	3.61
	MAPE (%)	46.44	29.56	46.19	16.78	7.97	7.59

Table 5.6 Prediction results for S4 from two updated models based on the consolidation data for S2

Model	Statistical indices	Loading stage of S4					
		1	2	3	4	5	6
Yin et al_2011_S2	*RMSE*	0.70	1.59	0.72	0.72	1.33	2.12
	MAPE (%)	30.52	23.15	4.24	2.05	3.13	4.43
Yang et al_2015_S2	*RMSE*	1.31	1.82	1.62	0.95	1.46	1.81
	MAPE (%)	52.3	27.25	8.74	2.40	3.56	3.51

et al_2015_S1 model. Additionally, the *RMSE* and *MAPE* for the first 5 loading stages of S4 by Yin et al_2011_S2 are 44.44%–91.10% and 48.51%–87.92%, respectively, of those yielded by the Yang et al_2015_S2 model. These results imply that the Yin et al_2011_S1 and Yin et al_2011_S2 models generally perform better than Yang et al_2015_S1 and Yang et al_2015_S2 regarding predictions of vertical strain. When more creep data is used in updating, the updated models display better performance in the predictions related to the longer loading period.

Based on the obtained posterior PDFs of parameters shown in Table 5.3a, b, we draw 10,000 samples of these parameters by Latin Hypercube sampling [6]. For each sample, the vertical strains at all loading stages can be predicted by the proposed model. Thus, 10,000 groups of predictions are calculated, and the vertical strains at different percentiles can be obtained by statistical analysis. Herein, we just show the 95% CIs of predictions by Yin et al_2011_S1 and Yin et al_2011_S2 models, and show in Figs. 5.6a, b. The two broken curves in the figures are the upper and lower boundaries of 95% CI, and the solid curve represents the mean values of the predictions. It is found that the measured data for S3 at the last 4 loading stages is located near or between the areas of 95% CI by Yin et al_2011_S1. In comparison, it can be observed that the area of 95% CI by Yin et al_2011_S2 for S4 is smaller than that in Fig. 5.6a, and most of the predictions are closer to the measured data. This means that the accuracy of the updated model in the predictions increases, and its uncertainty is reduced, as more creep data are used for updating.

It is noted that from the present analysis, the uncertainty of the model prediction under different loading durations can be quantified so that engineers can make better judgment on whether the loading duration is sufficient to obtain an updated model with associated parameters for a reliable prediction. In addition, one can predict the creep deformation using the updated models with any designated range of credibility. Considering the different creep characteristics of the soils and engineering requirements, one can easily recommend and justify the testing program, e.g., loading durations or loading stages, based on the method presented in this study.

5.4 Conclusions

We adopted the Bayesian probabilistic method to identify model parameters and investigate the effect of loading duration on updated results. Consolidation data from intact samples of Vanttila clay was used in the analysis. The results show that sufficient creep test data is crucial for reliable models and the associated parameters. The uncertainty of the updated model and its parameters reduces as a greater quantity of creep test data is used. The optimal models can describe the creep deformation of intact soft soils. Moreover, the results also indicate that the proposed method can quantify the uncertainty of model prediction to facilitate better judgment by engineers for the loading duration and model uncertainty in their predictions.

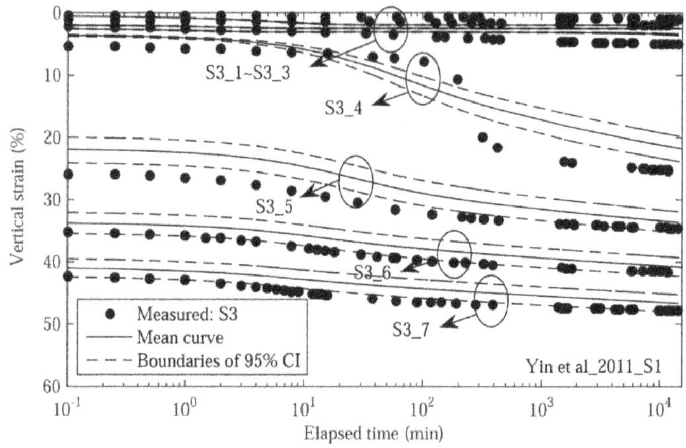

(a)

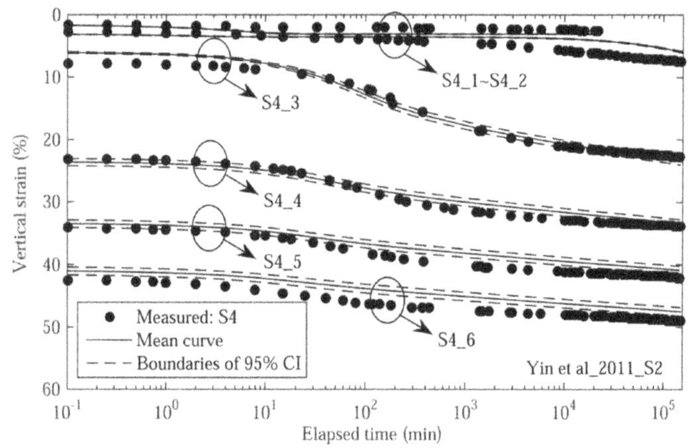

(b)

Fig. 5.6 S3 and S4 measured data and 95% CI of estimated vertical strain for **a** S3 by Yin et al_2011_S1 model and **b** S4 by Yin et al_2011_S2 model

Although this study is based on one type of soft soil, the proposed method can be applied to other types of soft soils. Additionally, other, more complex constitutive models, which incorporate creep, swelling, soil anisotropy, and structural effect, can also be evaluated in the framework of this analysis.

References

1. Yin ZY, Karstunen M, Chang CS, Koskinen M, Lojander M (2011) Modeling time-dependent behavior of soft sensitive clay. J Geotech Geoenviron Eng 137(11):1103–1113
2. Ching JY, Chen YC (2007) Transitional Markov Chain Monte Carlo method for Bayesian model updating, model class selection, and model averaging. J Eng Mech 133(7):816–832
3. Yin JH, Graham J (1989) Viscous-elastic-plastic modelling of one-dimensional time-dependent behavior of clays. Can Geotech J 26:199–209
4. Yin JH, Graham J (1999) Elastic viscoplastic modelling of the time-dependent stress-strain behaviour of soils. Can Geotech J 36:736–745
5. Yin JH, Graham J (1994) Equivalent times and one-dimensional elastic viscoplastic modelling of time-dependent stress-strain behavior of clays. Can Geotech J 31:42–52
6. Mckay M, Beckman R, Conover W (1979) A comparison of three methods for selecting values of input variables in the analysis of output from a computer code. Technometrics 21(2):239–245
7. Yin JH (1999) Non-linear creep of soils in oedometer tests. Géotechnique 49(5):699–707

Chapter 6
Model Class Selection for Sand with Generalization Ability Evaluation

Abstract Current studies have been focused on constitutive models selection using optimization methods, or simple formulas and models selection using Bayesian methods. In contrast, this chapter presents the Bayesian selection for advanced soil models accounting for soil uncertainty. Six representative sand model classes from elastic perfect plastic to critical-state-based are considered. Triaxial tests on Hostun sand are selected as training and testing data. The Bayesian model class selection is performed using an enhanced transitional Markov chain Monte Carlo method, whereby the generalization ability for each model is simultaneously evaluated. The most plausible/suitable model in terms of predictive ability, generalization ability and model complexity is selected.

6.1 Introduction

Reliable predictions depend heavily on plausible constitutive models with reasonable parameters in geotechnical engineering [1–7]. To achieve this purpose, many constitutive models have been proposed for soils [8–11]. However, different constitutive models render dissimilar results in the process of numerical simulations, prompting variations in engineering decisions that affect the levels of safety, economy and risk in construction [12–14]. Therefore, selecting the appropriate model is a critical issue in the application of constitutive models to practical use. Lack of attention to the problem of model class selection has become a primary source of risk for accident [15].

Model class selection is the task of choosing a model having the correct inductive bias from a set of candidate models. In practice, selection of a constitutive model often depends on the user's preferences and experiences, both of which are subjective. Accordingly, an efficient approach plays an important role in conducting model class selection. Among current studies, model class selection primarily relies on optimization methods and the Bayesian approach. However, model class selection based on optimization methods is less attractive due to the lack of considering soil uncertainty [8, 16]. The Bayesian approach appears to be primarily responsible for this

W. Zhou et al., *Practice of Bayesian Probability Theory in Geotechnical Engineering*,
https://doi.org/10.1007/978-981-15-9105-1_6

problem, in which the ability to select the most plausible/suitable model while simultaneously obtaining the posterior uncertainty of parameters would be of considerable use to engineers [17–28]. Unfortunately, the Bayesian methods are only applied to selecting some simple formula or simple soil models in geotechnical engineering [17, 19, 22, 23, 29]. Accordingly, an investigation of Bayesian model class selection for advanced soil models is desirable.

Furthermore, a reliable soil model class should offer a reasonable tradeoff between predictive capability and robustness. The predictive capability is usually evaluated using the magnitude of difference between predictions and experiments for optimization-based model class selection [8, 11, 16, 30–34] or by the maximum likelihood value for Bayesian model class selection [18, 19, 22, 23, 25, 35, 36]. For a given soil, simple models with insufficient features are likely to miss some of the characteristics of soil behavior. In contrast, advanced models with a large number of adjustable parameters will be more powerful to capture different soil behaviors, but they are therefore more prone to over-fitting the data, thereby reducing generalization ability. Accordingly, it is desirable to select a soil model that offers an outstanding predictive ability and generalization ability so that both efficient and robust application performance can be expected in practice [26].

This chapter conducts soil model class selection using the Bayesian method while evaluating the model's generalization ability. For this purpose, a set of model classes is selected that includes six representative sand models: (1) the elastic perfectly plastic Mohr–Coulomb model (MC), (2) the nonlinear Mohr–Coulomb model (NLMC), (3) the critical-state-based sand model (SIMSAND), (4) the simple anisotropic sand model (SANISAND), (5) the critical-state-based hypoplastic sand model (HYPOSAND) and (6) the Cam-clay-based sand model (MCCSAND). Three drained triaxial tests on Hostun sand are selected to make five combinations having different number of tests, making five groups of training data. For each combination, the Bayesian model class selection is performed using the enhanced transitional Markov chain Monte Carlo (TMCMC) method, in which the generalization ability for each selected sand model is simultaneously evaluated by testing data on the same sand. Finally, the most appropriate model in terms of predictive ability, generalization ability and model complexity is selected and validated.

6.2 Representative Advanced Sand Models

An impressive variety of sand models that can be classified as (1) elastic perfectly plastic models (such as the Drucker–Prager model, the Mohr–Coulomb model), (2) nonlinear simple models (such as the nonlinear Mohr–Coulomb (NLMC) [8], hardening soil model [37, 38]), (3) critical-state-based advanced models (such as Nor-Sand model [39]; CSAM model [40]; Severn–Trent sand model [41]; UH models [42–45]; SANISAND model [46]; the critical-state-based simple sand model (SIMSAND) [8, 11, 16, 47]; and (4) hypoplasticity models [48–51] have been developed. Not surprisingly, such models display different performances when modeling sand behaviors. To compare the performance of different models and select the most appropriate model among all typical sand models, a set of model classes that included six representative sand models was chosen for their popularity to perform model class selection: (1) the Mohr–Coulomb model (MC) is the elastic perfectly plastic model representing the basic model for simulating soil behavior so that the stress–strain nonlinearity cannot be described, (2) the NLMC model, in contrast to MC, a nonlinear plastic stress–strain behavior is incorporated, (3) the SIMSAND model, in contrast to NLMC, the critical state and interlocking effect are incorporated so that the stress dilatancy and contraction can be described, (4) the SANISAND model, which incorporates the concept of bounding surface compared to SIMSAND, (5) the critical-state-based hypoplastic sand model (HYPOSAND) by Wang et al. [52], which belongs to the framework of hypoplasticity and (6) the MCCSAND, which belongs to the framework of modified Cam-clay [43, 44]. The basic constitutive equations of some selected sand models are summarized in Tables 6.1 and 6.2.

Although the equations are basic, there are some subtle issues to be addressed. For MC and HYPOSAND, Young's modulus is constant. For NLMC, SIMSAND and SANISAND, Young's modulus is expressed as follows, according to Richart et al. [53]:

Table 6.1 Typical constitutive relations of SIMSAND, SANISAND, and MCCSAND

Model class	SIMSAND	SANISAND	MCCSAND
Elasticity	$d\varepsilon_{ij}^{e}=\frac{1+\upsilon}{E}\sigma'_{ij}-\frac{\upsilon}{E}\sigma'_{kk}\delta_{ij}$		
Yield function	$f=\frac{q}{p'}-H=0$	$f=(q-p\alpha)^2-m^2p^2=0$	$f=\ln\left[\left(1+\frac{q^2}{M^2p^2-\chi q^2}\right)p+p_s\right]-\ln(p_{x0}-p_s)-\frac{1}{c_p}H=0$
Potential function	$\frac{\partial g}{\partial p'}=A_d\left(M_{pt}-\frac{q}{p'}\right)$ $\frac{\partial g}{\partial q}=1$ $M_{pt}=\frac{6\sin\phi_{pt}}{3-\sin\phi_{pt}}$	$D=A_d(M_{pt}-\alpha)$ $M_{pt}=\frac{6\sin\phi_{pt}}{3-\sin\phi_{pt}}$	$g=\ln\frac{p}{p_x}+\ln\left(1+\frac{q^2}{M_c^2p^2}\right)$ $M_c=M\exp(-m\xi)$ $\xi=e_\eta-e$
Hardening law	$H=\frac{M_p\varepsilon_d^p}{k_p+\varepsilon_d^p}$ with $M_p=\frac{6\sin\phi_p}{3-\sin\phi_p}$	$h=k_p\frac{\lvert\mathbf{b}:\mathbf{n}\rvert}{b_{ref}-\lvert\mathbf{b}:\mathbf{n}\rvert}$	$H=\int\frac{M_f^4-\eta^4}{M_c^4-\eta^4}d\varepsilon_v^p$ $M_f=6\left[\sqrt{\frac{k}{R}\left(1+\frac{k}{R}\right)}-\frac{k}{R}\right]$ $k=\frac{M^2}{12(3-M)}, R=\exp\left(-\frac{e_\eta-e}{\lambda-\kappa}\right)$
Critical state	$e_c=e_{ref}\exp\left[-\lambda\left(\frac{p'}{p_{at}}\right)^{\xi}\right]$	$e_c=e_{ref}\exp\left[-\lambda\left(\frac{p'}{p_{at}}\right)^{\xi}\right]$	$e_\eta=Z-\lambda\ln\left(\frac{p+p_s}{1+p_s}\right)-(\lambda-\kappa)\ln\left[\frac{\left(1+\eta^2/(M^2-\chi\eta^2)\right)p+p_s}{p+p_s}\right]$
Inter-locking	$\tan\phi_p=\left(\frac{e_c}{e}\right)^{n_p}\tan\phi_\mu$; $\tan\phi_{pt}=\left(\frac{e_c}{e}\right)^{-n_d}\tan\phi_\mu$	$\tan\phi_p=\left(\frac{e_c}{e}\right)^{n_p}\tan\phi_\mu$; $\tan\phi_{pt}=\left(\frac{e_c}{e}\right)^{-n_d}\tan\phi_\mu$	$m=-\frac{1}{\xi_c}\ln\left(\frac{M_c^{\xi_c}}{M}\right)$ $\xi_c=(e_\eta-e)_c$ corresponding to $M_c^{\xi_c}$
Number of parameters	10	10	10

Table 6.2 Typical constitutive relations of HYPOSAND

Components	Constitutive equations
Constitutive equation	$\dot{\boldsymbol{\sigma}} = C_1(\mathrm{tr}\boldsymbol{\sigma})\dot{\boldsymbol{\varepsilon}} + C_2(\mathrm{tr}\dot{\boldsymbol{\varepsilon}})\boldsymbol{\sigma} + C_3\frac{\mathrm{tr}(\boldsymbol{\sigma}\dot{\boldsymbol{\varepsilon}})}{\mathrm{tr}\boldsymbol{\sigma}} + C_4(\boldsymbol{\sigma}+\boldsymbol{\sigma}^*)\lVert\dot{\boldsymbol{\varepsilon}}\rVert I_e$
Critical state line	$I_e = (e/e_c)^{\alpha}$ $e_c = e_{ref}\exp\left[-\lambda(p'/p_{at})^{\xi}\right]$
Translated tensor for cohesion	$\boldsymbol{\sigma}_c = \boldsymbol{\sigma} - p_t\delta_{ij}$, and $p_t = c/\tan\varphi$
Number of parameters	8

$$E = E_0 \cdot p_{at}\frac{(2.97 - e)^2}{(1+e)}\left(\frac{p'}{p_{at}}\right)^{\zeta} \tag{6.1}$$

For the MCCSAND, Young's modulus is expressed as:

$$E = \frac{3(1-2\upsilon)(1+e_0)}{\kappa}\left(p' + p_s\right) \text{ with } p_s = \exp\left(\frac{N-Z}{\lambda}\right) - 1 \tag{6.2}$$

where E_0 is the reference value of Young's modulus; e is the void ratio and e_0 is the initial void ratio; p' is the mean effective stress; p_{at} is the atmospheric pressure used as reference pressure ($p_{\mathrm{at}} = 101.3$ kPa); ζ is a constant; υ is Poisson's ratio; κ is the swelling index; λ is the compression index; N and Z are two constants of the MCCSAND.

For SIMSAND, SANISAND and HYPOSAND, a nonlinear relationship of the critical state line (CSL) [54] was adopted:

$$e_c = e_{\mathrm{ref}}\exp\left[-\lambda\left(\frac{p'}{p_{at}}\right)^{\xi}\right] \tag{6.3}$$

where e_{c} is the critical void ratio; e_{ref} is the initial critical void ratio at p' of 0 kPa; λ and ξ are two parameters controlling the shape of CSL in the e-log p' plane.

The parameters of each of these model classes can be divided into: (1) elastic parameters, (2) plastic shear-hardening-related parameters, (3) stress-dilatancy-related parameters and (4) critical-state-related parameters for critical-state-based models.

6.3 Model Class Selection Approach and Generalization Ability Evaluation

6.3.1 Bayesian Model Class Selection

The Bayesian approach for parametric identification and model class selection is briefly outlined in Sect. 1.2 of Chap. 1. Further details can be found in Yuen [25]. Six model class candidates and their uncertain parameters are shown in Table 6.3.

Generally, deformation and stress are two extremely important indicators for soil behavior. The measurements produced by laboratory test usually contain two curves, such as the curves ε_a–q and ε_a–e for the drained triaxial test or the curves ε_a–q and ε_a–u for the undrained triaxial test (where ε_a is axial strain, q is deviatoric stress, e is void ratio and u is excess pore water pressure). Accordingly, a goodness-of-fit function involving these two important indicators is reasonable. According to [8, 16], a normalized goodness-of-fit function is adopted due to the error independent of the magnitude of different variables (e.g., q and e or u), which is expressed as:

$$J_g(\mathbf{b}; D) = \frac{1}{N_0 N} \sum_{j=1}^{N_0} \left[\sum_{i=1}^{N} \left(\frac{U_{\text{obs}}^i - U_{\text{num}}^i}{U_{\text{obs}}^i} \right)^2 \right]_j \tag{6.10}$$

where N is the number of measured values; N_0 is the number of curves for one test; U_{obs}^i is the value of measurement point i; and U_{num}^i is the value of calculation at point i.

With multiple observations and types of observations, the likelihood values for each observation must be combined into an overall value for each candidate parameter set [55]. When the measured data D involve M tests during Bayesian parameter identification, the likelihood function is expressed as:

Table 6.3 Six model class candidates and their uncertain parameters

Model class	Uncertain parameter vector $\boldsymbol{\theta}$	Number of uncertain parameters
MC	$[E, \phi, \psi, \sigma_\varepsilon^2]^T$	4
NLMC	$[\phi, \psi, k_p, \sigma_\varepsilon^2]^T$	4
SIMSAND	$[e_{ref}, \lambda, \xi, \phi, k_p, A_d, n_p, n_d, \sigma_\varepsilon^2]^T$	9
SANISAND	$[e_{ref}, \lambda, \xi, \phi, h_0, A_d, n_p, n_d, \sigma_\varepsilon^2]^T$	9
HYPOSAND	$[e_{ref}, \lambda, \xi, E, \phi, \psi, A_d, \sigma_\varepsilon^2]^T$	8
MCCSAND	$[\phi, \kappa, \lambda, N, Z, \chi, m, \sigma_\varepsilon^2]^T$	8

$$\ln p(D|\boldsymbol{\theta}) = \sum_{i=1}^{M} \ln p(D_i|\boldsymbol{\theta}) \tag{6.11}$$

where M is the number of involved tests and $p(D_i|\boldsymbol{\theta}, C)$ is the likelihood corresponding to the test i.

The posterior PDF $p(\boldsymbol{\theta}|C, D)$ represents the updated belief about the parameter vector $\boldsymbol{\theta}$ after obtaining the evidence D. An accurate estimator of the parameters $\boldsymbol{\theta}$ for the adopted soil model is the maximum a posteriori (MAP) estimation. The MAP parameter vector $\boldsymbol{\theta}^*$ can be obtained by maximizing the posterior $p(\boldsymbol{\theta}|C, D)$, or equivalently, $p(\boldsymbol{\theta}|C)p(D|\boldsymbol{\theta}, C)$. Considering that the model classes involve high-dimensional nonlinear functions, the evidence integral must be evaluated numerically. As the enhanced TMCMC method (see Chap. 7 for details) has been proven more efficient for high-dimensional problems and can also evaluate the evidence for each model class, it was used to quantify the uncertainty of model parameters and conduct the model class selection. After the enhanced TMCMC algorithm is executed, the importance weights produced during the algorithm can be used to estimate the model evidence $p(D|C_j)$, which can be estimated by the enhanced TMCMC algorithm as a by-product:

$$p(D|C_j) \approx S = \prod_{j=0}^{m-1} \left(\frac{1}{N_s} \sum_{k=1}^{N_s} w_{j,k} \right) \tag{6.16}$$

where S is the asymptotically unbiased estimation of the model evidence; $w_{j,k}$ are the importance weights; and m is the total number of transitional stages.

6.3.2 Evaluation of Generalization Ability

Since sand models are always evaluated on the basis of finite samples/tests, the evaluation of a sand model is sensitive to sampling error. As a result, measurements of prediction error for the current data may not provide much information about predictive ability for new data. To track this problem, generalization ability, a measure of how accurately a model is able to predict outcome values for previously unseen data, was adopted for this case.

Due to the difficulties in computing the unknown joint probability distribution for generalization error, generalization ability is usually measured by empirical error, a function of the difference between the actual and predicted results for out-of-sample data. To make the empirical error independent of the type of test and the number of measurement points, a normalized empirical error function was adopted, expressed as:

$$\mathrm{Error}(\boldsymbol{\theta}) = \sqrt{\frac{1}{N}\sum_{i=1}^{N}\left(\frac{U_{\mathrm{obs}}^{i} - U_{\mathrm{num}}^{i}}{U_{\mathrm{obs}}^{i}}\right)^{2}} \tag{6.17}$$

With multiple observations, the combined empirical error is expressed as:

$$\mathrm{Error}(\boldsymbol{\theta}) = \sum_{j=1}^{\mathrm{K}} l_i[\mathrm{Error}(\boldsymbol{\theta})]_i \tag{6.18}$$

where K is the number of observations contributing to the computation of generalization ability, and l_{i} is weight. In this study, $l_{\mathrm{i}} = 1/K$ for a uniform weight.

6.3.3 General Procedure of Model Class Selection

Figure 6.1 shows the procedure for model class selection. In this procedure, a set of candidate model classes was first prescribed. Then, the measured data serving as the training data for model class selection were selected. Next, the likelihood function for multiple observations was determined according to Eq. (6.11). Subsequently, the Bayesian model class selection using the enhanced TMCMC method was carried out, simultaneously giving the posterior distributions of all identified parameters for each model class candidate.

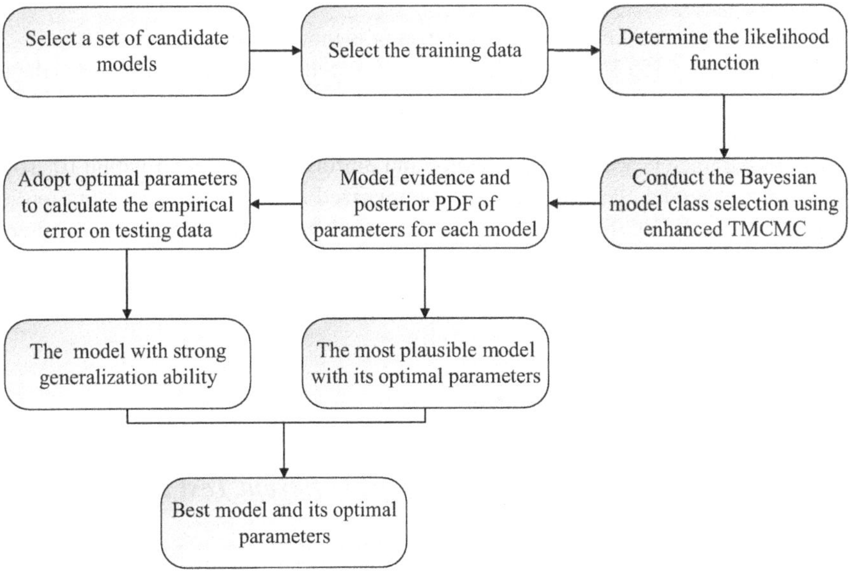

Fig. 6.1 Procedure of Bayesian model class selection with generalization ability evaluation

Continuing the procedure, after obtaining the posterior distributions of all identified parameters for each model class candidate, simulations of the new tests that were different from the training data used in the model class selection were performed using each model class with N optimum sets of parameters drawn from the obtained posterior distributions. The empirical error was simultaneously calculated. The model class showing minimum empirical error was identified as the suitable model having strong generalization ability.

The ideal model class should survive during Bayesian model class selection as well as the evaluation of generalization ability. In fact, all data should be used in Bayesian model class selection in practice for determining the most plausible model. In this case, however, for the purpose of validation few tests were selected to be used in the procedure of generalization ability for validating the performance of the obtained most plausible model. The MATLAB code for model class selection or parametric identification is presented in Appendix D.

6.3.4 Illustration Case of Bayesian Model Class Selection

To show the proposed procedure of model class selection, an illustration case on selecting a polynomial equation is presented. The 31 measured data points were generated by the equation $y = 2x + 5x^2$ with $x \in [0, 3]$. The general model class can be defined as follows:

$$y = a_0 + a_1x + a_2x^2 + a_3x^3 + \varepsilon \tag{6.19}$$

where a_0–a_4 are uncertain model parameters; x is input and y is output.

Then, a total of 15 (2^4−1) models were generated. Based on the measured data, an appropriate model will be finally selected from 15 models. The prior PDF of all uncertain parameters for each model class was assumed in uniform distribution. All uncertain parameters were independently and uniformly distributed within [0, 10]. The results of model class selection are summarized in Table 6.4. According to the values of $\ln p(D|C_j)$, the model 8 is most appropriate and the identified parameters are $a_1 = 2.000009$, $a_2 = 4.9999961$ with the uncertainty of $\varepsilon = 3.63\text{e-}06$. The obtained model approximately equals the predefined equation.

6.4 Selection of Sand Models

6.4.1 Model Class Selection Based on Different Test Sets

A series of laboratory triaxial tests performed on Hostun sand by Liu et al. [56] and Li et al. [57] were chosen for the model class selection. The initial void ratio and

Table 6.4 Model class selection results for illustration case

Model number	Model	ln $p(D\|C_j)$	Model number	Model	ln $p(D\|C_j)$
1	$y = a_0 + \varepsilon$	−4879.1	9	$y = a_1x + a_3x^3 + \varepsilon$	−45.83
2	$y = a_1x^2 + \varepsilon$	−1009.5	10	$y = a_2x^2 + a_3x^3 + \varepsilon$	−48.78
3	$y = a_2x^2 + \varepsilon$	−44.17	11	$y = a_0 + a_1x + a_2x^2 + \varepsilon$	36.62
4	$y = a_3x^3 + \varepsilon$	−394.3	12	$y = a_0 + a_1x + a_3x^3 + \varepsilon$	−49.27
5	$y = a_0x + a_1x + \varepsilon$	−850.2	13	$y = a_0 + a_2x^2 + a_3x^3 + \varepsilon$	−33.81
6	$y = a_0 + a_2x^2 + \varepsilon$	−27.21	14	$y = a_1x + a_2x^2 + a_3x^3 + \varepsilon$	14.59
7	$y = a_0 + a_3x^3 + \varepsilon$	−181.3	15	$y = a_0 + a_1x + a_2x^2 + a_3x^3 + \varepsilon$	−19.60
8	$y = a_1x + a_2x^2 + \varepsilon$	**165.1**			

confining pressure for each test are listed in Table 6.5. Figure 6.2 shows all results of selected triaxial tests. To form a comprehensive group of experiments that can effectively reflect common sand behaviors, a total of three drained triaxial tests including one dilative test (test 1) and two contractive tests (tests 4 and 5) were chosen as training data for model class selection and parameter identification. The exact information of used training data is data points on deviatoric stress–axial strain and void ratio or volumetric strain–axial strain curves (i.e., q–ε_a and e–ε_a). The other tests were considered as testing data used to calculate the empirical error for evaluating the generalization ability of each sand model. Note that the experimental data having strain localization, such as the tests on very dense sand, are not considered since the numerical results are obtained by an element integrated with one single Gauss point in this study. The model class selection when the used data have strain localization can be conducted using finite element method for boundary value problems.

To investigate the effect of a number of tests on model class selection and the evaluation of generalization ability, the model class selection was successively conducted

Table 6.5 Summary of triaxial tests on Hostun sand

Triaxial tests	Drained tests					Undrained tests				
Test number	**1**	2	3	**4**	**5**	6	7	8	9	10
e_0	**0.66**	0.75	0.85	**0.83**	**0.82**	0.69	0.72	0.70	0.72	0.78
p'/kPa	**100**	100	100	**200**	**400**	100	100	400	400	400

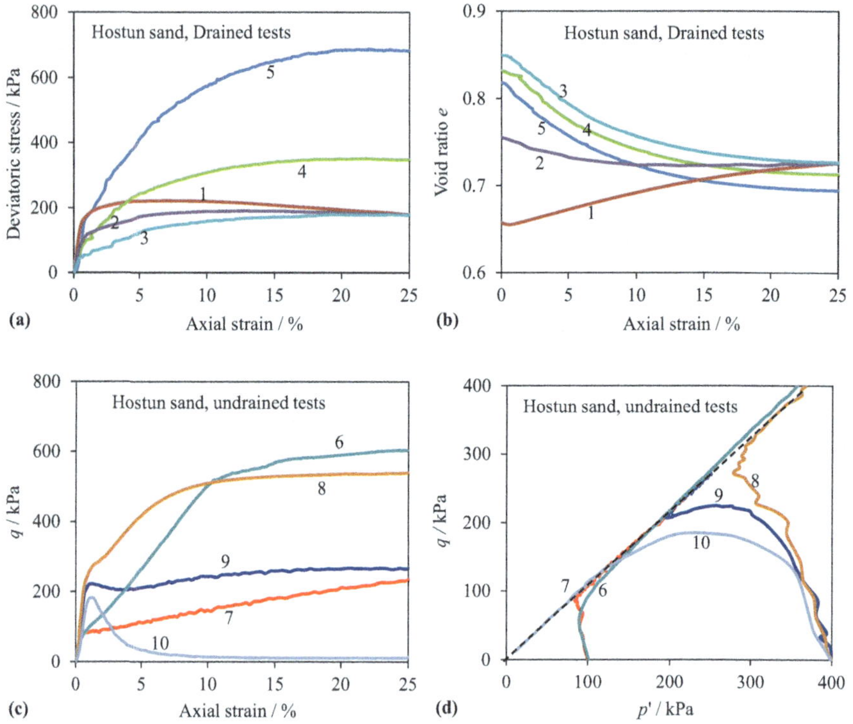

Fig. 6.2 Results of triaxial tests of Hostun sand

based on different number of tests. Based on three selected drained triaxial tests (test 1, 4 and 5), a total of seven combinations were obtained as summarized in Table 6.6. However, combinations 2 and 3 were overlapped because of containing a similar contractive test. Thus, only combination 2 was selected. A similar situation was found for combinations 4 and 5. Thus, only combination 4 was selected. As such, combinations 1, 2, 4, 6 and 7—five cases in total—were finally adopted in the model class selection.

The prior PDF of all uncertain parameters for each model class was assumed to show uniform distribution. All uncertain parameters were independently uniformly distributed [*lower_bound*, *upper_bound*], for example, the friction angle was uniformly distributed [20, 50]. The bounds of each prior PDF for all model classes are summarized in Table 6.7. Note that when the bounds for the uncertain parameters are very narrow, a very small step size is needed in the model class selection.

Table 6.6 Test combinations for model class selection

Test combination	**1**	**2**	3	**4**	5	**6**	**7**
Test number	**1**	**4**	5	**1 and 4**	1 and 5	**4 and 5**	**1, 4 and 5**

For all cases, the enhanced TMCMC with 2000 was implemented, meaning 2000 samples per stage. Therefore, $2000N_s$ total samples were generated in the enhanced TMCMC simulation. According to previous results [23, 58–61], the total number of stages for different model class selections was not always the same.

6.4.2 Results and Discussion

Table 6.8 summarizes the results of evidence obtained by enhanced TMCMC and the results of plausibility using Bayes' theorem based on various test combinations. To intuitively present the results of model class selection while choosing the most plausible model class, all plausibilities are exhibited in Fig. 6.3. The MAP parameters that correspond to the maximum values of their posterior PDFs and the uncertainties for all selected model classes are summarized in Tables 6.9, 6.10, 6.11, 6.12 and 6.13. It can be seen that some parameters with physical meanings (e.g., friction angle, critical-state parameters) identified from three tests for SIMSAND, SANISAND, HYPOSAND and MCCSAND agree well with experimental measurements by Liu et al. [56] and Li et al. [57]. However, the same parameter identified for MC and NLMC are bigger than the experimental values because of defects of these two models themselves.

For model class selection based on test 1 (dense sand) (see Fig. 6.3a), SANISAND was the most plausible model class while the NLMC was the worst one. The results indicate that NLMC had a limitation in capturing the dilatancy behavior of dense sand, and similar results also existed for MC. For selection based on test 4 (loose sand) (see Fig. 6.3b), the most plausible model class was still SANISAND, while the worst was the MC. However, except for MC, little difference in plausibilities for other model classes was found, demonstrating that such model classes show outstanding ability in the simulation of contractive behaviors for sands.

When the number of involved tests in the model class selection increased to 2, a different result for model class selection was found. The most plausible model class was SIMSAND for selection based on tests 1 and 4 (see Fig. 6.3c), while it was SANISAND for selection based on tests 4 and 5 (see Fig. 6.3d). However, the difference of plausibility between SIMSAND and SANISAND was small for both cases, and thus it can sometimes be ignored. The worst model was MC for both cases. Note that the negative values of plausibility in Fig. 6.3c indicate that MC and NLMC cannot simultaneously effectively simulate both the dilatancy and contractive behaviors of sand, which has been highlighted by Jin et al. [8].

Based on preliminary results, it is possible to infer that the most plausible model class is either SIMSAND or SANISAND, which can perform well in capturing the sand behaviors of dilatancy and contractiveness. Furthermore, HYPOSAND and MCCSAND displayed similar performance but were inferior to SIMSAND and SANISAND in simulation.

Unsurprisingly, this surmise is confirmed by the results of model class selection based on three tests, as depicted in Fig. 6.3e. Although the performance of

Table 6.7 Bounds of all uncertain parameters in the prior PDF used in TMCMC

MC	E/kPa	ϕ/°	ψ/°						σ_ε^2
	$[10^3, 5 \times 10^4]$	[20, 50]	[0, 20]						[0, 1]
NLMC	ϕ/°	ψ/°	k_p						
	[20, 50]	[0, 20]	$[10^{-3}, 10^{-1}]$						
SIMSAND	e_{ref}	λ	ξ	ϕ/°	k_p	A_d	n_p	n_d	
	[0.5, 1.5]	$[10^{-3}, 10^{-1}]$	[0.1, 1.0]	[20, 50]	$[10^{-3}, 10^{-1}]$	[0.1, 3]	[0, 10]	[0, 10]	
SANISAND	e_{ref}	λ	ξ	ϕ/°	h_0	A_d	n_p	n_d	
	[0.5, 1.5]	$[10^{-3}, 10^{-1}]$	[0.1, 1.0]	[20, 50]	[1, 100]	[0.1, 3]	[0, 10]	[0, 10]	
HYPOSAND	e_{ref}	λ	ξ	E/kPa	ϕ/°	A_d			
	[0.5, 1.5]	$[10^{-3}, 10^{-1}]$	[0.1, 1.0]	$[10^3, 5 \times 10^4]$	[20, 50]	[0.1, 3]			
MCCSAND	ϕ/°	κ	λ	N	Z	χ	m		
	[20, 50]	$[10^{-3}, 10^{-1}]$	[0.01, 1.0]	[0.5, 3.0]	[0.5, 2.0]	[0, 1]	[0, 10]		

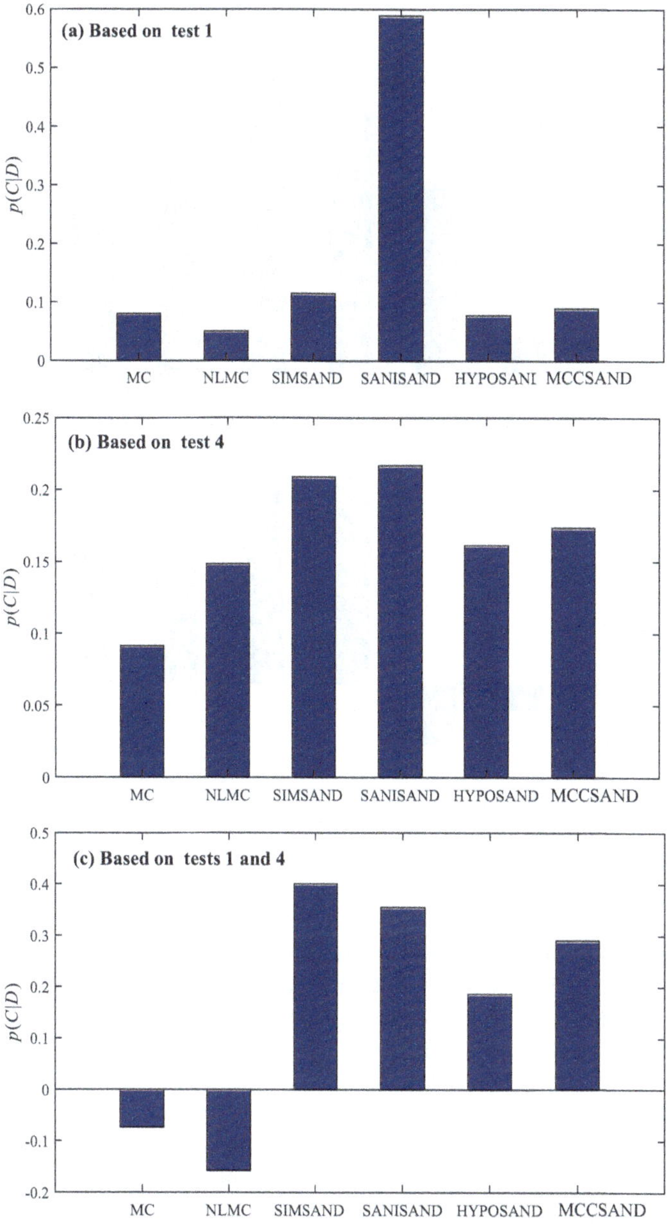

Fig. 6.3 Model class selection results based on: **a** test 1; **b** test 4; **c** tests 1 and 4; **d** tests 4 and 5; **e** tests 1, 4 and 5

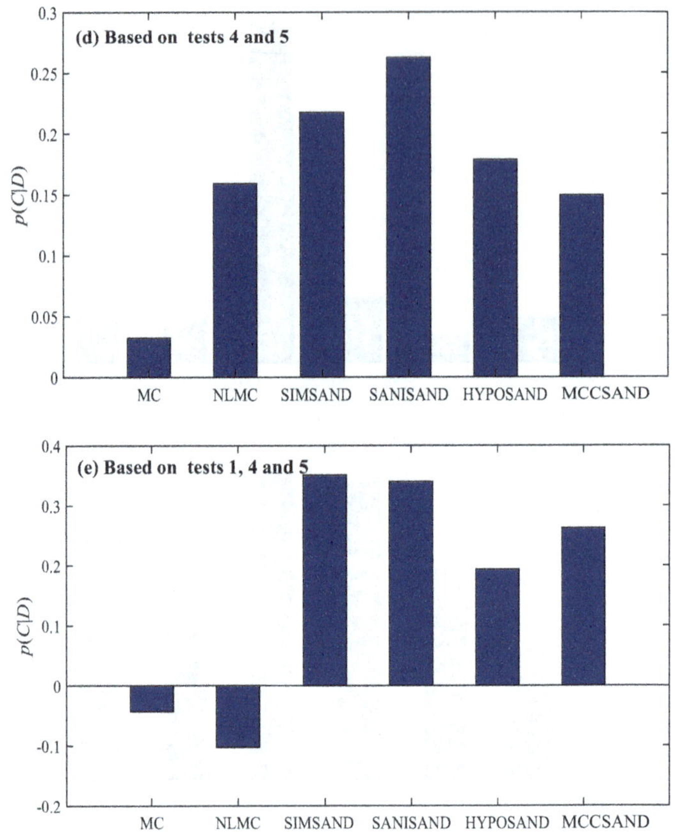

Fig. 6.3 (continued)

HYPOSAND and MCCSAND was inferior to that of SIMSAND and SANISAND, it is also acceptable in practice. Furthermore, compared to the model classes implementing the concept of critical state (i.e., SIMSAND, SANISAND, HYPOSAND and MCCSAND), MC and NLMC failed in simulating sand behaviors.

As mentioned, a reliable soil model should have a reasonable tradeoff between predictive capability and generalization ability. However, for a given soil model, the predictive performance for unseen test data indicating generalization ability relies heavily on the precision of the parameters used. Accordingly, the evaluation of generalization ability can serve as a remedy for Bayesian model class selection. To evaluate the generalization ability of each model using its optimal parameters, the testing tests were simulated and the empirical errors in terms of q and e for drained tests or u for undrained tests were simultaneously calculated according to Eq. (6.18), as shown in Fig. 6.4. The magnitude of empirical error reflects predictive ability along with the accuracy and reliability of optimal parameters obtained by Bayesian model class

Table 6.8 Results of model class selection based on limited number of tests

Tests	Test 1		Test 4		Tests 1 and 4		Tests 4 and 5		Tests 1, 4 and 5	
Model class	Ln(S)	p(C|D)	Ln(S)	p(C|D)	Ln(S)	p(C|D)	Ln(S)	p(C|D)	Ln(S)	p(C|D)
MC	49.16	0.080	143.42	0.091	−71.09	−0.074	100.75	0.032	−86.20	−0.043
NLMC	30.80	0.055	232.76	0.148	−151.87	−0.158	496.46	0.159	−204.74	−0.103
SIMSAND	70.29	0.114	328.02	0.209	385.25	0.401	678.79	0.218	697.85	0.351
SANISAND	361.91	0.589	340.57	0.217	341.22	0.355	819.48	0.263	676.03	0.340
HYPOSAND	47.40	0.077	253.25	0.161	178.79	0.186	557.30	0.179	384.78	0.193
MCCSAND	54.96	0.089	273.10	0.174	279.40	0.291	466.10	0.149	522.10	0.262

Table 6.9 Identified MAP parameters of MC and NLMC with their uncertainties

Model class	MC			NLMC		
Parameters	$\phi/°$	$\psi/°$	E/kPa	$\phi/°$	k_p	$\psi/°$
Based on test 1	30.7 (0.17)	5.3 (0.08)	49,413 (2032)	31.1 (0.23)	0.0010 (3.0E-5)	5.3 (0.09)
Based on test 4	27.7 (0.11)	0.1 (0.04)	3595 (35)	31.8 (0.08)	0.0222 (2.7E-4)	0.3 (0.07)
Based on test 1 and 4	28.8 (0.41)	9.1 (0.34)	3254 (102)	30.8 (0.88)	0.0097 (2.42E-3)	2.8 (0.80)
Based on test 4 and 5	27.2 (0.17)	0.0 (0.02)	5189 (83)	31.5 (0.05)	0.0227 (1.5E-4)	0.1 (0.03)
Based on test 1, 4 and 5	28.3 (0.24)	3.5 (0.32)	5368 (127)	29.7 (0.58)	0.0129 (2.04E-3)	1.3 (0.67)

selection from limited training tests for a given sand model. The large value of empirical error indicates a poor performance in terms of prediction.

In contrast to the results of Bayesian model class selection, some different findings from the standpoint of generalization ability were noted. When only one training test on a dense sample was involved in the Bayesian model class selection, the best model was SANISAND and the worst was NLMC in terms of the generalization ability (see Fig. 6.4a), consistent with the result obtained by Bayesian model class selection. When the involved test in model class selection was performed on a loose sample, the best model was still SANISAND while the worst was MCCSAND (see Fig. 6.4b). It can be shown that the parameters identified from tests on dense sand are more reliable that those identified from tests on loose sand for the MCCSAND, and just the opposite for NLMC model, both of which indicate that the parameters of the NLMC and MCCSAND models are sensitive to the types of tests involved. Therefore, both the NLMC and MCCSAND models are not recommended to be applied to geotechnical practice when only one test on either dense or loose sample is available. Furthermore, MC and HYPOSAND exhibit similar but poor performance for both cases using one test, and thus more attentions should be paid when using them in practice.

For cases with two tests, the most suitable models were SIMSAND and SANISAND (see Fig. 6.4c and d). Due to the lack of the ability to balance the predictions between dilatancy and contractive behaviors, the inaccurate optimum parameters can lead to a weak generalization ability of MC and NLMC, and thus neither is recommended when only one test on a dense sample or only one test on a loose sample is available for parameter identification in practice. The MCCSAND is also not recommended when training tests only on loose sample are available. For the case with three tests, the most appropriate models are SIMSAND and SANISAND, while the worst is still MCCSAND (see Fig. 6.4e).

In summary, the poor generalization ability of MC and NLMC when the involved training tests are performed on both dense and loose sand samples is mainly attributable to poor predictive ability and inaccurate optimum parameters. The

Table 6.10 Identified MAP parameters of SIMSAND and their uncertainties

SIMSAND	e_{ref}	λ	ξ	$\phi/°$	k_p	A_d	n_p	n_d
Based on test 1	0.765 (3.59E-3)	0.0122 (2.05E-3)	0.835 (3.78E-2)	27.8 (1.78E-1)	1.47E-03 (7.84E-5)	0.46 (1.96E-2)	1.8 (7.28E-2)	3.9 (2.16E-1)
Based on test 4	0.718 (8.07E-4)	0.0076 (9.56E-4)	0.313 (2.75E-2)	29.3 (4.54E-2)	7.75E-03 (4.04E-4)	0.73 (2.24E-2)	1.4 (1.06E-1)	4.2 (1.51E-1)
Based on test 1 and 4	0.750 (7.29E-4)	0.0254 (6.21E-4)	0.684 (1.54E-2)	29.3 (3.61E-2)	3.72E-03 (1.17E-4)	0.55 (6.54E-3)	2.8 (3.28E-2)	6.8 (7.45E-2)
Based on test 4 and 5	0.775 (2.47E-3)	0.0478 (2.63E-3)	0.464 (1.53E-2)	29.1 (3.92E-2)	9.36E-03 (3.59E-4)	0.64 (1.84E-2)	1.5 (1.04E-1)	6.0 (2.73E-1)
Based on test 1, 4 and 5	0.751 (1.03E-3)	0.0265 (9.47E-4)	0.628 (1.17E-2)	29.0 (4.11E-2)	4.54E-03 (1.35E-4)	0.56 (2.35E-2)	3.0 (3.66E-2)	7.0 (4.06E-1)

Table 6.11 Identified MAP parameters of SANISAND and their uncertainties

SANISAND	e_{ref}	λ	ξ	ϕ/°	h_0	A_d	n_p	n_d
Based on test 1	0.735 (1.02E-3)	0.0158 (5.80E-4)	0.832 (7.94E-3)	28.6 (0.0358)	19.3 (0.37)	0.54 (7.0E-3)	1.1 (7.5E-2)	6.7 (8.59E-2)
Based on test 4	0.710 (1.19E-3)	0.0052 (8.70E-4)	0.501 (3.13E-2)	28.6 (0.0593)	19.9 (0.38)	0.80 (2.25E-2)	1.1 (8.65E-2)	3.4 (1.48E-1)
Based on test 1 and 4	0.740 (1.23E-3)	0.0169 (8.81E-4)	0.803 (2.57E-2)	28.6 (0.0429)	29.0 (0.85)	0.61 (1.85E-2)	2.2 (5.95E-2)	6.1 (2.07E-1)
Based on test 4 and 5	0.738 (8.13E-4)	0.0225 (6.23E-4)	0.615 (8.66E-3)	28.1 (0.0259)	18.6 (0.12)	0.52 (9.7E-3)	0.6 (4.15E-2)	7.7 (1.92E-1)
Based on test 1, 4 and 5	0.739 (5.87E-4)	0.0156 (4.71E-4)	0.795 (1.15E-2)	28.7 (0.0275)	25.5 (0.46)	0.76 (3.96E-2)	2.3 (3.96E-2)	4.2 (1.89E-1)

Table 6.12 Identified MAP parameters of HYPOSAND and their uncertainties

HYPOSAND	e_{ref}	λ	ξ	E/kPa	ϕ/°	A_d
Based on test 1	0.803 (3.72E-3)	0.056 (3.40E-3)	0.759 (0.031)	14,109 (192)	27.1 (0.092)	1.44 (0.051)
Based on test 4	0.669 (3.74E-3)	0.003 (5.58E-4)	0.919 (0.101)	4132 (84)	31.4 (0.079)	0.91 (0.032)
Based on test 1 and 4	0.734 (1.85E-3)	0.015 (1.23E-3)	0.733 (0.023)	7883 (83)	28.7 (0.093)	2.47 (0.032)
Based on test 4 and 5	0.746 (3.80E-3)	0.035 (3.06E-3)	0.574 (0.030)	4509 (78)	30.7 (0.057)	1.64 (0.048)
Based on test 1, 4 and 5	0.740 (1.65E-3)	0.021 (1.50E-3)	0.682 (0.026)	7860 (71)	28.8 (0.073)	2.53 (0.032)

performance of HYPOSAND is not quite satisfactory compared to SIMSAND and SANISAND regardless of the number and types of involved tests. The poor generalization ability of the MCCSAND is found when the involved tests on a loose sample are dominant, which can be attributed to the inaccurate identified parameters because its predictive ability is acceptable based on the results of Bayesian model class selection. The results also suggest that the tests performed on dense sand are beneficial to identify reliable parameters of the MCCSAND.

As stated by Wood [62], simple yet adequate models are favored on the basis of practicality. In general, the model complexity can be expressed using the number of parameters and the complexity of model equations. Compared to Bayesian model class selection, the Akaike information criterion (AIC) [63] and Bayesian information criterion (BIC) [64] are the two commonly used criteria for model class selection accounting for the effect of the number of parameters. For AIC, the penalty term is taken to be the number of adjustable parameters of model class. When the number of data points N is large, the penalty term will disappear which means that the contribution of the number of model parameters is little. Accordingly, the AIC cannot

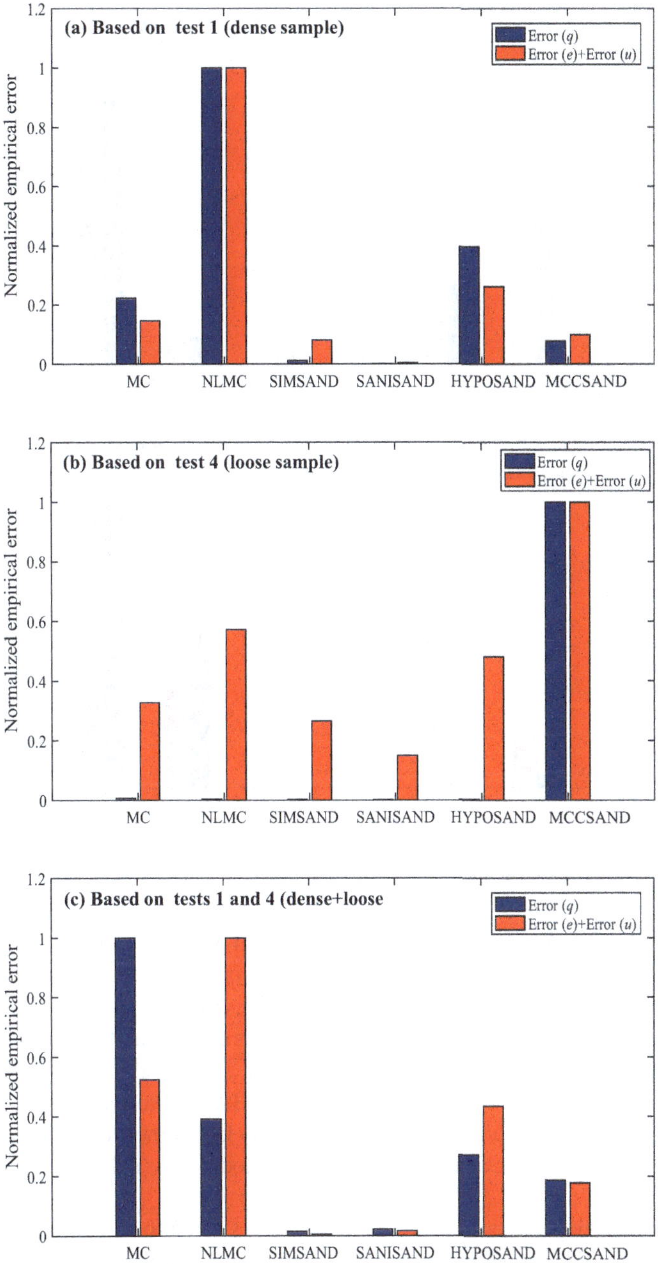

Fig. 6.4 Results of empirical errors based on: **a** test 1; **b** test 4; **c** tests 1 and 4; **d** tests 4 and 5; **e** tests 1, 4 and 5

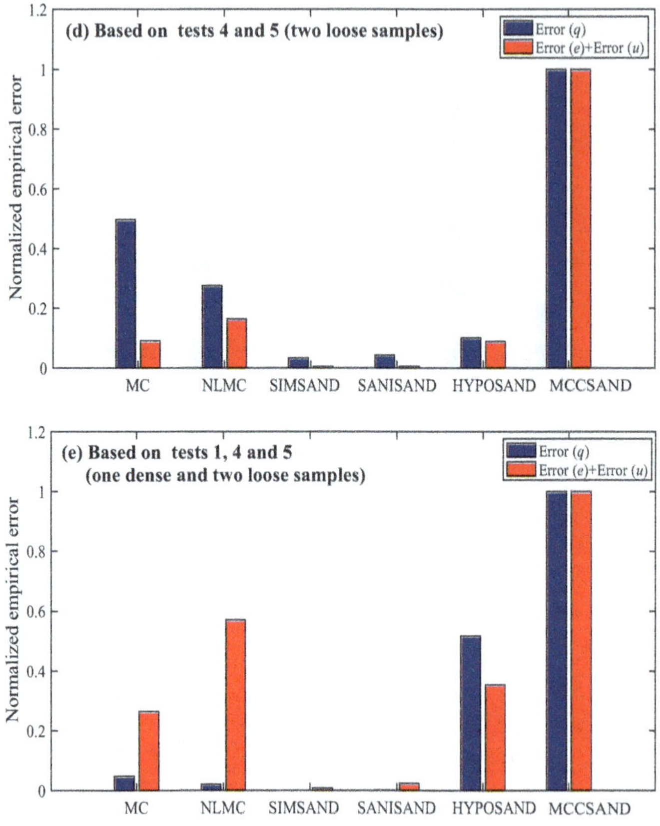

Fig. 6.4 (continued)

replace the Bayesian model class selection method if N is sufficiently large. For BIC, the penalty term increases with the number of data points N. According to Yuen [25], for large N, the BIC is equivalent to the Bayesian approach using equal prior plausibilities for all model candidates. However, in practice, the Bayesian model class selection is especially useful when N is not large, so the selection of model class is difficult by the user's judgment. Therefore, the BIC cannot replace the Bayesian model class selection method if N is not sufficiently large since the residual term has an important contribution. Overall, the Bayesian model class selection is superior to AIC and BIC.

Numerical convergence is easier to obtain when simple formulas are used to deal with complex geotechnical problems. Due to the implementation of the bounding surface concept-based hardening law with a small elastic domain for SANISAND, increasing the model complexity, SIMSAND was finally selected as the most appropriate sand model in terms of the predictive ability and generalization ability for simulating monotonic behaviors, consistent with the results obtained by the optimization method [8]. Therefore, SIMSAND is selected as an outstanding sand model regardless of the methods (deterministic and probabilistic) used.

To show the performance of the selected model class SIMSAND and its MAP parameters, a series of simulations on testing data that have been used in the evaluation of generalization ability were performed, as shown in Fig. 6.5. Satisfactory agreement is found between the numerical simulations and experiments that contractive and dilatancy behaviors, even static liquefaction, can be easily reproduced, demonstrating

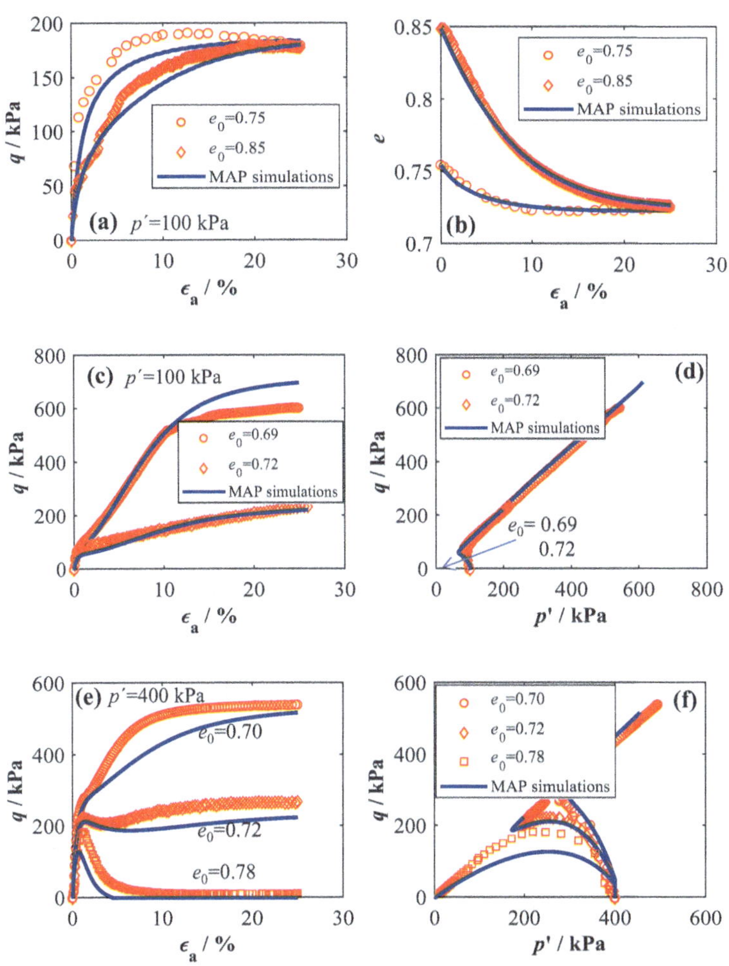

Fig. 6.5 Comparison of triaxial tests between simulations and experiments on Hostun sand

Table 6.13 Identified MAP parameters of MCCSAND and their uncertainties

SANISAND	$\phi/°$	κ	λ	N	Z	χ	m
Based on test 1	33.2 (0.26)	0.044 (0.004)	0.18 (0.009)	2.22 (0.069)	0.94 (0.008)	0.61 (0.024)	1.45 (0.133)
Based on test 4	30.3 (0.3)	0.045 (0.007)	0.12 (0.003)	1.47 (0.016)	1.45 (0.032)	0.52 (0.018)	7.26 (0.469)
Based on test 1 and 4	29.8 (0.3)	0.021 (0.003)	0.25 (0.006)	2.51 (0.049)	0.88 (0.004)	0.48 (0.008)	2.41 (0.031)
Based on test 4 and 5	29.6 (0.5)	0.035 (0.002)	0.22 (0.003)	2.59 (0.021)	0.78 (0.004)	0.96 (0.013)	1.30 (0.112)
Based on test 1, 4 and 5	28.4 (0.3)	0.066 (0.006)	0.17 (0.003)	2.15 (0.031)	0.72 (0.002)	0.22 (0.020)	0.95 (0.043)

an excellent predictive ability for SIMSAND as well as the rationality of identified parameters (Table 6.13).

6.5 Validation by Other Sand

6.5.1 *Model Class Selection Based on Tests of Karlsruhe Sand*

A series of drained triaxial tests performed on Karlsruhe fine sand by Wichtmann and Triantafyllidis [65] was selected for this case. According to Wichtmann and Triantafyllidis [65], the test material "Karlsruhe fine sand" has almost no fines content, a mean grain size $d_{50} = 0.14$ mm and a uniformity coefficient $C_u = d_{60}/d_{10} = 1.5$. The minimum and maximum void ratios were $e_{min} = 0.677$ and $e_{max} = 1.054$. Most of the grains have a subangular shape.

The initial void and confining pressure for each test are summarized in Table 6.14. The experimental results for all triaxial tests are shown in Fig. 6.6. Three drained triaxial tests (tests 3, 4 and 9) with approximate density and different confining pressures following the industrial standard were selected as training data. The remaining tests were used to evaluate generalization ability. $K_0 = 44$ and $n = 0.68$ were determined from the 1-D compression tests following those studies [8, 16, 31].

Table 6.14 Summary of drained triaxial tests on Karlsruhe fine sand

Test number	1	2	3	4	5	6	7	8	9
e_0	0.996	0.840	0.734	0.735	0.706	0.697	0.960	0.840	0.718
p'/kPa	50	50	50	100	200	300	400	400	400

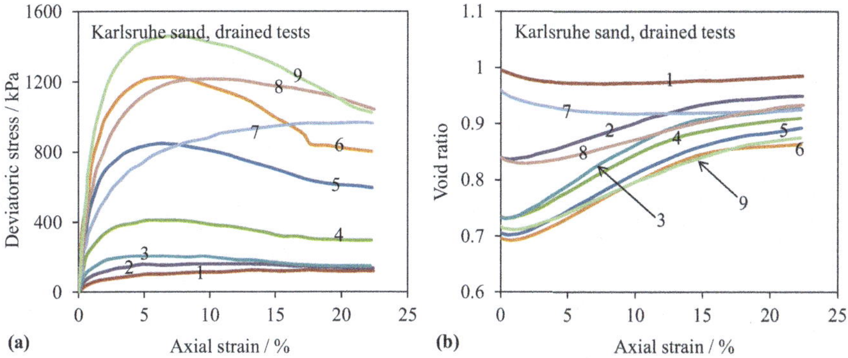

Test number	1	2	3	4	5	6	7	8	9
e_0	0.996	0.840	0.734	0.735	0.706	0.697	0.960	0.840	0.718
p' / kPa	50	50	50	100	200	300	400	400	400

Fig. 6.6 Results of drained triaxial tests on Karlsruhe fine sand

6.5.2 *Results and Discussion*

Figure 6.7 shows the model class selection results based on three drained tests of Karlsruhe fine sand. It was found that the most plausible model class was SIMSAND and the worst was still MC, which is consistent with preliminary results obtained on tests of Hostun sand. SANISAND also had an acceptable performance and was inferior only to SIMSAND. Table 6.15 summarizes the MAP parameters of all sand models for Karlsruhe fine sand. It can be seen that the parameters identified for SIMSAND, SANISAND, HYPOSAND and MCCSAND are in appropriate ranges compared to the experimental investigations ($\phi = 33.1^\circ$ and $e_{\mathrm{ref}} = 1.067$) by Wichtmann and

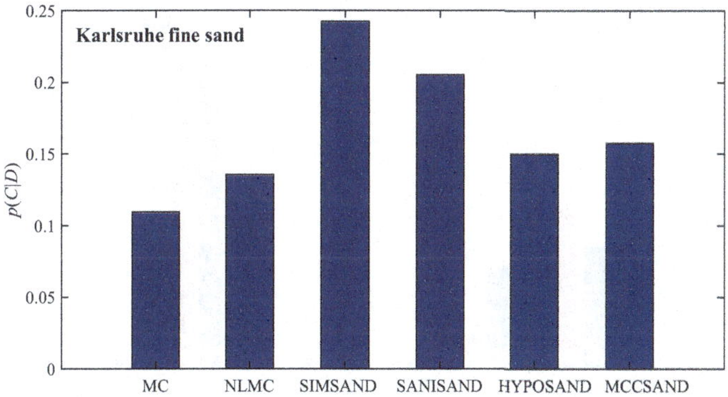

Fig. 6.7 Model class selection results based on training data on Karlsruhe fine sand

Table 6.15 MAP parameters of all sand models for Karlsruhe fine sand

MC	E/kPa	$\phi/°$	$\psi/°$					
	64,755	38.9	13.2					
NLMC	$\phi/°$	$\psi/°$	k_p					
	39	11.3	6.44×10^{-5}					
SIMSAND	e_{ref}	λ	ξ	$\phi/°$	k_p	A_d	n_p	n_d
	1.053	0.0238	0.743	33.6	3.84×10^{-3}	0.75	1.6	1.8
SANISAND	e_{ref}	λ	ξ	$\phi/°$	h_0	A_d	n_p	n_d
	1.046	0.0137	0.881	33.2	35.8	0.60	1.2	2.3
HYPOSAND	e_{ref}	λ	ξ	E/kPa	$\phi/°$	A_d		
	1.183	0.020	0.832	18,940	29.0	0.65		
MCCSAND	$\phi/°$	κ	λ	N	Z	χ	m	
	34.3	0.044	0.18	2.33	0.96	0.05	3.19	

Triantafyllidis [65]. However, for MC and NLMC, the identified parameters are not located in the reasonable ranges due to the defects of these models.

Using the obtained optimal parameters, the evaluation of generalization ability for all model classes was conducted as shown in Fig. 6.8. SIMSAND and SANISAND were found to be the most suitable models, while the HYPOSAND model was worst in terms of generalization ability evaluated using finite testing data. Furthermore, the MCCSAND model also showed good generalization ability due to the parameters identified from tests on dense sand samples, confirming the validity of previous suggestions.

Overall results demonstrate that SIMSAND is the most plausible/suitable sand model in terms of predictive ability and generalization ability for Karlsruhe fine sand. Figure 6.9 shows the comparison of testing tests on Karlsruhe fine sand between experiments and simulations. Acceptable agreement is found between the numerical

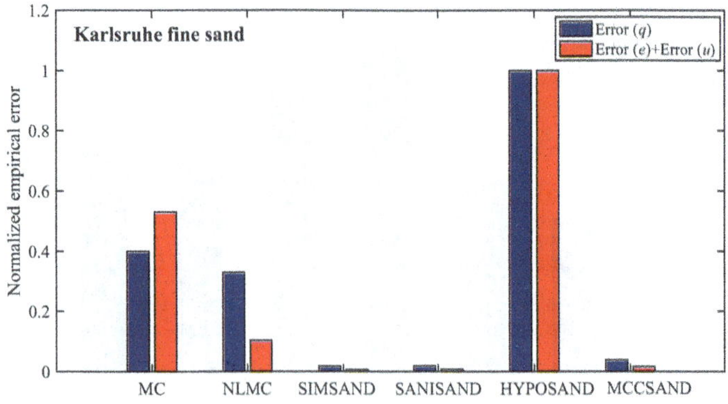

Fig. 6.8 Results of empirical errors based on testing data for Karlsruhe fine sand

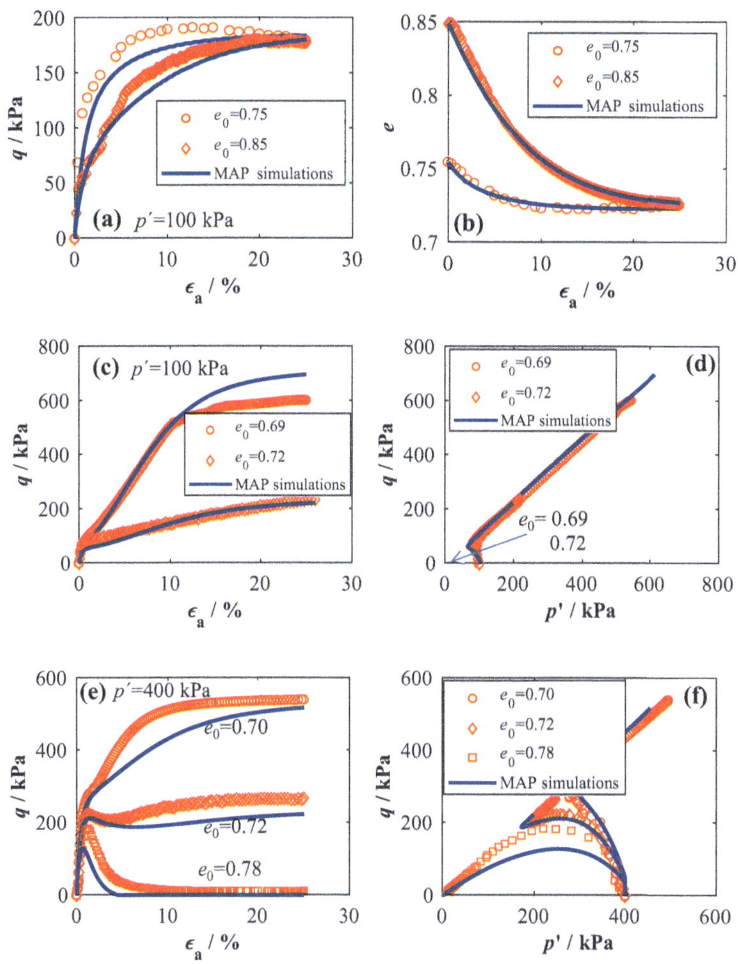

Fig. 6.9 Comparison of triaxial tests between simulations and experiments for Karlsruhe fine sand

simulations and experiments that the contractive and dilatancy behaviors can be adequately reproduced, demonstrating an excellent predictive ability for SIMSAND as well the rationality of identified parameters.

6.6 Discussion

In the above investigation, four advanced constitutive models were adopted to describe the typical mechanical behaviors of sand under monotonic loading. However, the behaviors of granular materials are complex, and an accurate modeling remains an open challenge. One of the main difficulties lies in a clear and efficient definition of the internal state of such materials. Usually two types of internal parameters have to be considered: (a) one or several scalar parameters characterizing the density state with respect to a critical density, such as the evolution of e/ec in considered models of this study, which is a usual way to define the isotropic internal state, and (b) one or several tensorial parameters characterizing the anisotropic internal state, such as the critical state considering fabric anisotropy. In complex loading paths, experimental studies have indicated that the behavior of a granular soil under shear is predominantly anisotropic. Such anisotropic behavior of sand can be effectively modeled by incorporating the fabric tensor and its evolution. Another effective way to simulate such an anisotropic behavior is the use of micromechanics-based models. Note that the four adopted sand models in this study only consider the internal isotropic state variables and the anisotropic behavior of sand cannot be described, which is a limitation of this study. Nevertheless, the approach can also be applied to models considering anisotropy with objective tests of complex loading paths.

The considered tests (training and testing tests) are only monotonous triaxial tests (drained and undrained) in this study. The proposed analysis can give an estimation of the ability of the considered models to give good simulation performance of monotonic loading defined from an initial isotropic internal state. However, the loading paths in practice are more complex, such as (a) a loading with the stress principal directions different from the ones of an anisotropic initial internal state, (b) a cyclic loading, and (c) a loading with significant evolutions of the principal directions. Again, the approach can also be applied to these cases if adopted constitutive models further considering anisotropy with effect of stress reversal.

Furthermore, the test results sometimes are not consistent themselves, which is probably caused by (a) an inherent spatial variability of soil properties, (b) experimental uncertainty (measurement scatter) due to limitations of the experimental techniques, and (c) sampling uncertainty (statistical uncertainty) due to the limited number of soil samples used in the investigation. Such inconsistency would lead to variability of parameters for model of interest, which can be quantified by the proposed Bayesian approach with DE-TMCMC from available experimental data. In practice, such inconsistency (experimental and sampling uncertainties) can be incorporated into probabilistic analyses using random field methods.

The model class selection using Bayesian approach from triaxial tests can be considered as a basic work before extending the approach to the practice. The presented work using triaxial tests to select "best model" can give a comprehensive understanding of different sand models on modeling the mechanical behaviors. Although the appropriateness of various constitutive models in FEM (not at the meso-level of triaxial tests, but at the level of responses of real structures such as settlements, deflections) is well studied, choosing a suitable model for the engineering of interest solved by FEM is still a challenge. Therefore, it is important to engage the proposed approach to shed a different light on past conclusions from these studies. The proposed approach is more applicable to comparable models (same level of numerical sophistication), and an engineer cannot easily distinguish which models to be adopted in FEM.

6.7 Conclusions

This study has presented a selection of sand models along with parameter identification using Bayes' theorem, thus evaluating generalization ability based on test data. First, the principles of Bayesian model class selection and the enhanced TMCMC method were briefly introduced. Then, the procedure for model class selection and evaluation of generalization ability was presented. To conduct the model class selection, six representative sand models (MC, NLMC, SIMSAND, SANISAND, HYPOSAND and MCCSAND) were chosen. Three drained triaxial tests on Hostun sand were selected to make five combinations as five groups of training data. Then, for each combination, Bayesian model class selection was performed using the enhanced TMCMC method. The optimum parameters corresponding to the maximum of posterior PDF with their uncertainties for each sand model were summarized. The plausibilities were compared for six selected sand model classes. The generalization ability of each selected sand model was then evaluated by empirical error computed on new test data. Finally, a validation case based on Karlsruhe fine sand was carried out. Based on the obtained results, some conclusions were drawn, as follows:

(1) The SIMSAND model was demonstrated to be the most plausible/suitable sand model in terms of predictive ability, generalization ability and model complexity regardless of the number and types of training tests involved.
(2) The poor generalization ability of MC and NLMC was mainly attributed to poor predictive ability and inaccurate MAP parameters. Between both models, MC is recommended when the available test data used for parameter identification is limited.
(3) Although the performance of HYPOSAND was not quite satisfactory compared to SIMSAND and SANISAND regardless of the number of tests involved, it was better than the MCCSAND—unsurprisingly better than MC and NLMC—because of its insensitivity to parameters.
(4) The poor generalization ability of the MCCSAND was found when the involved training tests on a loose sample were dominant, which can be attributed to inaccurately identified parameters because its predictive ability was acceptable based on the results of Bayesian model class selection. The results also suggest that the tests performed on dense sand are beneficial to identify reliable parameters of the MCCSAND.

Future studies will focus on model class selection according to boundary value problems.

References

1. Shen SL, Xu YS (2011) Numerical evaluation of land subsidence induced by groundwater pumping in Shanghai. Can Geotech J 48(9):1378–1392
2. Shen S-L, Wu H-N, Cui Y-J, Yin Z-Y (2014) Long-term settlement behaviour of metro tunnels in the soft deposits of Shanghai. Tunnelling Underground Space Technol 40:309–323
3. Karstunen M, Yin ZY (2010) Modelling time-dependent behaviour of Murro test embankment. Geotechnique 60(10):735–749
4. Jiang M, Harris D, Zhu H (2007) Future continuum models for granular materials in penetration analyses. Granular Matter 9(1–2):97–108
5. Chai J-C, Shen JS-L, Liu MD, Yuan D-J (2018) Predicting the performance of embankments on PVD-improved subsoils. Comput Geotech 93:222–231
6. Jin Y-F, Yin Z-Y, Wu Z-X, Daouadji A (2018) Numerical modeling of pile penetration in silica sands considering the effect of grain breakage. Finite Elem Anal Des 144:15–29
7. Jin Y-F, Yin Z-Y, Zhou W-H, Huang H-W (2019) Multi-objective optimization-based updating of predictions during excavation. Eng Appl Artif Intell 78:102–123
8. Jin Y-F, Yin Z-Y, Shen S-L, Hicher P-Y (2016) Selection of sand models and identification of parameters using an enhanced genetic algorithm. Int J Numer Anal Methods Geomech 40(8):1219–1240
9. Jiang M, Harris D, Yu H (2005) Kinematic models for non-coaxial granular materials. Part I: theory. Int J Numer Anal Methods Geomech 29(7):643–661
10. Jiang M, Zhang F, Sun Y (2014) An evaluation on the degradation evolutions in three constitutive models for bonded geomaterials by DEM analyses. Comput Geotech 57:1–16
11. Jin Y-F, Yin Z-Y, Wu Z-X, Zhou W-H (2018) Identifying parameters of easily crushable sand and application to offshore pile driving. Ocean Eng 154:416–429
12. Jiang M, Li T, Chareyre B (2016) Fabric rates applied to kinematic models: evaluating elliptical granular materials under simple shear tests via discrete element method. Granular Matter 18(3):1–15
13. Lyu H-M, Shen JS, Arulrajah A (2018) Assessment of geohazards and preventative countermeasures using AHP incorporated with GIS in Lanzhou, China. Sustainability 10(2):304
14. Ren D-J, Shen S-L, Arulrajah A, Wu H-N (2018) Evaluation of ground loss ratio with moving trajectories induced in DOT tunnelling. Can Geotech J 55(6):894–902
15. Whittle A, Davies R (2006) Nicoll Highway collapse: evaluation of geotechnical factors affecting design of excavation support system. In: International conference on deep excavations, p 30
16. Jin Y-F, Wu Z-X, Yin Z-Y, Shen JS (2017) Estimation of critical state-related formula in advanced constitutive modeling of granular material. Acta Geotech 12(6):1329–1351
17. Akeju OV, Senetakis K, Wang Y (2017) Bayesian parameter identification and model selection for normalized modulus reduction curves of soils. J Earthquake Eng 1–29
18. Beck JL, Yuen K-V (2004) Model selection using response measurements: Bayesian probabilistic approach. J Eng Mech 130(2):192–203
19. Cao Z, Wang Y (2014) Bayesian model comparison and selection of spatial correlation functions for soil parameters. Struct Saf 49:10–17
20. Cheung SH, Beck JL (2009) Bayesian model updating using hybrid Monte Carlo simulation with application to structural dynamic models with many uncertain parameters. J Eng Mech 135(4):243–255
21. Muto M, Beck JL (2008) Bayesian updating and model class selection for hysteretic structural models using stochastic simulation. J Vib Control 14(1–2):7–34
22. Ritto T, Nunes L (2015) Bayesian model selection of hyperelastic models for simple and pure shear at large deformations. Comput Struct 156:101–109
23. Tan F, Zhou W-H, Yuen K-V (2016) Modeling the soil water retention properties of same-textured soils with different initial void ratios. J Hydrol 542:731–743
24. Worden K, Hensman J (2012) Parameter estimation and model selection for a class of hysteretic systems using Bayesian inference. Mech Syst Signal Process 32:153–169

25. Yuen K-V (2010) Recent developments of Bayesian model class selection and applications in civil engineering. Struct Saf 32(5):338–346
26. Yuen K-V (2010) Bayesian methods for structural dynamics and civil engineering: Wiley
27. Zhang L, Li D-Q, Tang X-S, Cao Z-J, Phoon K-K (2017) Bayesian model comparison and characterization of bivariate distribution for shear strength parameters of soil. Comput Geotech 95:110–118
28. Zhang J, Wang H, Huang H, Chen L (2017) System reliability analysis of soil slopes stabilized with piles. Eng Geol 229:45–52
29. Chiu CF, Yan WM, Yuen K-V (2012) Estimation of water retention curve of granular soils from particle-size distribution—a Bayesian probabilistic approach. Can Geotech J 49(9):1024–1035
30. Jin Y-F, Yin Z-Y, Riou Y, Hicher P-Y (2017) Identifying creep and destructuration related soil parameters by optimization methods. KSCE J Civil Eng 21(4):1123–1134
31. Jin Y-F, Yin Z-Y, Shen S-L, Hicher P-Y (2016) Investigation into MOGA for identifying parameters of a critical-state-based sand model and parameters correlation by factor analysis. Acta Geotech 11(5):1131–1145
32. Jin Y-F, Yin Z-Y, Shen S-L, Zhang D-M (2017) A new hybrid real-coded genetic algorithm and its application to parameters identification of soils. Inv Prob Sci Eng 25(9):1343–1366
33. Yin Z-Y, Jin Y-F, Shen JS, Hicher P-Y (2018) Optimization techniques for identifying soil parameters in geotechnical engineering: comparative study and enhancement. Int J Numer Anal Methods Geomech 42(1):70–94
34. Yin Z-Y, Jin Y-F, Shen S-L, Huang H-W (2017) An efficient optimization method for identifying parameters of soft structured clay by an enhanced genetic algorithm and elastic–viscoplastic model. Acta Geotech 12(4):849–867
35. Toni T, Welch D, Strelkowa N, Ipsen A, Stumpf MP (2009) Approximate Bayesian computation scheme for parameter inference and model selection in dynamical systems. J R Soc Interface 6(31):187–202
36. Yuen KV, Mu HQ (2015) Real-time system identification: an algorithm for simultaneous model class selection and parametric identification. Comput-Aided Civil Infrastruct Eng 30(10):785–801
37. Schanz T, Vermeer P, Bonnier P (1999) The hardening soil model: formulation and verification. In: Beyond 2000 in computational geotechnics, pp 281–96
38. Vermeer P (1978) A double hardening model for sand. Geotechnique 28(4):413–433
39. Jefferies M (1993) Nor-Sand: a simple critical state model for sand. Geotechnique 43(1):91–103
40. Yu H (1998) CASM: A unified state parameter model for clay and sand. Int J Numer Anal Methods Geomech 22(8):621–653
41. Gajo A, Wood M (1999) Severn-Trent sand: a kinematic-hardening constitutive model: the q–p formulation. Geotechnique 49(5):595–614
42. Yao Y, Hou W, Zhou A (2009) UH model: three-dimensional unified hardening model for overconsolidated clays. Geotechnique 59(5):451–469
43. Yao Y, Sun D, Luo T (2004) A critical state model for sands dependent on stress and density. Int J Numer Anal Methods Geomech 28(4):323–337
44. Yao Y, Sun D, Matsuoka H (2008) A unified constitutive model for both clay and sand with hardening parameter independent on stress path. Comput Geotech 35(2):210–222
45. Yao Y-P, Kong L-M, Zhou A-N, Yin J-H (2014) Time-dependent unified hardening model: three-dimensional elastoviscoplastic constitutive model for clays. J Eng Mech 141(6):04014162
46. Taiebat M, Dafalias YF (2008) SANISAND: Simple anisotropic sand plasticity model. Int J Numer Anal Methods Geomech 32(8):915–948
47. Wu Z-X, Yin Z-Y, Jin Y-F, Geng X-Y (2017) A straightforward procedure of parameters determination for sand: a bridge from critical state based constitutive modelling to finite element analysis. Eur J Environ Civil Eng 1–23
48. Kolymbas D (1991) An outline of hypoplasticity. Arch Appl Mech 61(3):143–151
49. Mašín D (2005) A hypoplastic constitutive model for clays. Int J Numer Anal Methods Geomech 29(4):311–336

50. Wu W, Bauer E, Kolymbas D (1996) Hypoplastic constitutive model with critical state for granular materials. Mech Mater 23(1):45–69
51. Wu W, Kolymbas D (2000) Hypoplasticity then and now. In: Constitutive modelling of granular materials, pp 57–105
52. Wang S, Wu W, Yin Z-Y, Peng C, He X-Z (2018) Modelling time-dependent behaviour of granular material with hypoplasticity. Int J Numer Anal Methods Geomech 42(12):1331–1345
53. Richart F, Hall J, Woods R (1970) Vibrations of soils and foundations. International series in theoretical and applied mechanics. Prentice-Hall, Englewood Cliffs, NJ
54. Yin Z-Y, Wu Z-X, Hicher P-Y (2018) Modeling monotonic and cyclic behavior of granular materials by exponential constitutive function. J Eng Mech 144(4):04018014
55. He J, Jones JW, Graham WD, Dukes MD (2010) Influence of likelihood function choice for estimating crop model parameters using the generalized likelihood uncertainty estimation method. Agric Syst 103(5):256–264
56. Liu Y-J, Li G, Yin Z-Y, Dano C, Hicher P-Y, Xia X-H et al (2014) Influence of grading on the undrained behavior of granular materials. CR Mec 342(2):85–95
57. Li G, Liu Y-J, Dano C, Hicher P-Y (2014) Grading-dependent behavior of granular materials: from discrete to continuous modeling. J Eng Mech 141(6):04014172
58. Betz W, Papaioannou I, Straub D (2016) Transitional markov chain monte carlo: observations and improvements. J Eng Mech 142(5):04016016
59. Ching J, Chen Y-C (2007) Transitional Markov chain Monte Carlo method for Bayesian model updating, model class selection, and model averaging. J Eng Mech 133(7):816–832
60. Ching J, Wang J-S (2016) Application of the transitional Markov chain Monte Carlo algorithm to probabilistic site characterization. Eng Geol 203:151–167
61. Lee S-H, Song J (2017) System identification of spatial distribution of structural parameters using modified transitional Markov Chain Monte Carlo method. J Eng Mech 143(9):04017099
62. Wood DM (2003) Geotechnical modelling. CRC Press
63. Akaike H (1977) On entropy maximization principle. Appl Stat 27–41
64. Schwarz G (1978) Estimating the dimension of a model. Ann Stat 6(2):461–464
65. Wichtmann T, Triantafyllidis T (2016) An experimental database for the development, calibration and verification of constitutive models for sand with focus to cyclic loading: Part I—tests with monotonic loading and stress cycles. Acta Geotech 11(4):739–761

Chapter 7
Parametric Identification of Advanced Soil Models for Sand

Abstract Parametric identification using Bayesian approach is appealing in geotechnical engineering compared to that using deterministic method, because the soil uncertainty can be taken into account. To date, this approach has been verified only for certain conventional simple constitutive models. This chapter presents an enhanced version of the differential evolution transitional Markov chain Monte Carlo (DE-TMCMC) method and a competitive Bayesian parameter identification approach for use in advanced soil models. To realize the intended computational savings, a parallel computing implementation of DE-TMCMC is achieved using the single program/multiple data (SPMD) technique in MATLAB, with comparison of the proposed DE-TMCMC and the original TMCMC for identification of parameters of a critical-state-based sand model from laboratory tests in terms of robustness and effectiveness based on multiple independent calculations. Results indicate that DE-TMCMC is highly robust and efficient in identifying the parameters of advanced soil models. Finally, Bayesian parameter identification is applied in conjunction with DE-TMCMC to identify parameters of an elasto-viscoplastic model from two in situ pressuremeter tests. All results demonstrate the excellent ability of the enhanced Bayesian parameter identification approach to identify soil model parameters from both laboratory and in situ tests.

7.1 Introduction

Constitutive models play an important role in geotechnical engineering and can be classified as (1) linear-elastic model, (2) elastic perfectly plastic models (such as the Mohr–Coulomb), (3) nonlinear models (such as hardening soil model [1], nonlinear Mohr–Coulomb [2, 3]), (4) critical-state-based advanced models (such as the modified Cam-clay model [4]; Nor-Sand model [5]; CSAM model [6]; Severn–Trent model [7]; UH models [8–11]; SANISAND model [12]; SIMSAND model [2, 13, 14]; ANICREEP model [15, 16]), hypoplasticity models [3, 17–21] and (5) micromechanical models [22–28]. The last two categories are usually called advanced soil models, and such models allow the garnering of increasingly accurate and reliable descriptions of state-dependent mechanical behaviors of soils, resulting

W. Zhou et al., *Practice of Bayesian Probability Theory in Geotechnical Engineering*,
https://doi.org/10.1007/978-981-15-9105-1_7

in complexities as well as some additional parameters. The accuracy of parameters of a model significantly influences its modeling performance and can even result in inaccurate predictions [29–35]. Accordingly, parameter identification is a vital issue surrounding the application of advanced constitutive models in geotechnical engineering.

Yin et al. [36] distinguished three approaches of parameters determination from experimental data: analytical methods, empirical correlations and inverse analysis methods. Among them, inverse analysis produces a relatively objective determination of the parameters for an adopted soil model, even for those without direct physical meaning, and thus has been widely adopted. The key methods of inverse analysis are divided into two categories: (1) deterministic methods and (2) probabilistic methods. Deterministic methods merely focus on finding a set of fixed values for the input parameters of concern [2, 13, 14, 35–41] without taking into account the variability and uncertainty of soils. In contrast, the probabilistic methods are competitive because of their consideration of uncertainty. Of these, the Bayesian approach to parameter identification has been applied in different fields [42–59] in which parameters of concern have been treated as random variables and expressed in terms of posterior distributions and statistics. Up to now, such applications have involved only certain conventional simple soil models (e.g., the linear elastic model [53], the one-dimensional elasto-plastic model [52], the hardening soil model [44]), whereas parameter identification involving advanced constitutive models (e.g., critical state-based models) has rarely been reported. This motivates the method introduced in this chapter.

In Bayesian parameter identification, the Markov chain Monte Carlo (MCMC) simulation is generally known for its ability to efficiently derive a posterior distribution [47, 50, 51, 60–62]. The adopted MCMC highly governs the performance of identified parameters in terms of accuracy and reliability. MCMC sampling (e.g., Metropolis–Hastings algorithm [63]) produces samples that are statistically dependent, possibly reducing the efficiency of statistical estimators, and indeed most of them become inefficient when the number of variables is large [64]. In seeking to overcome such issues, the transitional MCMC (TMCMC) proposed by Ching and Chen [65] has proved to be more efficient for high-dimensional problems [64–69]. Evidently, such efforts have improved the performance of MCMC, but some observations have indicated that the posteriors estimated using the TMCMC tend to fall into local convergence when solving certain problems [64]. Accordingly, further improvement of TMCMC for Bayesian parameter identification is necessary, especially for advanced soil models.

This chapter aims to present an efficient approach to Bayesian parameters identification for advanced soil models through the enhancement of TMCMC. To this end, Bayesian parameter identification is first introduced in principle. To improve the performance of TMCMC, a differential evolution-Markov chain algorithm is implemented in the process of new samples. To save on computational costs, a parallel computing implementation of DE-TMCMC is achieved using the single program/multiple data (SPMD) technique in MATLAB. The enhanced DE-TMCMC-based parameter identification approach is validated by identifying parameters of a critical-state-based sand model from synthetic numerical tests with/without noise first and then from real laboratory tests. Finally, an application of the proposed approach to an elasto-viscoplastic clay model based on two in situ pressuremeter tests is presented.

7.2 Enhanced DE-TMCMC-Based Bayesian Identification with Parallel Computing

7.2.1 *Framework of Bayesian Parameter Identification*

According to Yuen [70], the Bayesian approach can update model parameters and characterize uncertainties using their posterior probability distribution functions (PDFs). Generally, deformation and stress are two extremely important indicators for soil behaviors. The measurement produced by a laboratory test usually contains two curves, such as the curves ε_a–q and ε_a–e for the drained triaxial test or the curves ε_a–q and ε_a–u for the undrained triaxial test (where ε_a is axial strain, q is deviatoric stress, e is void ratio and u is excess pore water pressure). Accordingly, a goodness-of-fit function involving these two important indicators is reasonable. According to Jin et al. [2, 13], a normalized goodness-of-fit function is adopted due to the error independent of the magnitude of different variables (e.g., q and e or u), which is expressed as:

$$J_g(\mathbf{b}; D) = \frac{1}{N_0 N} \sum_{j=1}^{N_0} \left[\sum_{i=1}^{N} \left(\frac{U^i_{\text{obs}} - U^i_{\text{num}}}{U^i_{\text{obs}}} \right)^2 \right]_j \tag{7.1}$$

where N is the number of measured values; N_0 is the number of curves for one test; U^i_{obs} is the value of measurement point i; and U^i_{num} is the value of calculation at point i.

With multiple observations and types of observations, likelihood values for each observation must be combined into an overall value for each candidate parameter set [71]. When the measured data D involve M tests during Bayesian parameter identification, the likelihood function is expressed as:

$$\ln p(D|\boldsymbol{\theta}) = \sum_{i=1}^{M} \mathrm{w}_i \ln p(D_i|\boldsymbol{\theta}) \tag{7.2}$$

where M is the number of involved tests, w_i is weight of $p(D_i|\boldsymbol{\theta})$, and $p(D_i|\boldsymbol{\theta})$ is the likelihood corresponding to the test i. In this study, the weight of each likelihood for all involved tests is considered the same and thus is equal to 1.

Because the soil model involves high-dimensional nonlinear functions, the posterior $p(\boldsymbol{\theta}|D)$ must be evaluated numerically, such as by the TMCMC method. The posterior PDF $p(\boldsymbol{\theta}|D)$ represents the updated belief about the parameter vector $\boldsymbol{\theta}$ after obtaining the evidence of D. An accurate estimator of the parameters $\boldsymbol{\theta}$ for the adopted soil model is the maximum a posteriori (MAP) estimation. The MAP parameter vector $\boldsymbol{\theta}_{\text{MAP}}$ can be computed as follows:

$$\boldsymbol{\theta}_{\text{MAP}} = \arg\max p(\boldsymbol{\theta}|D) = \arg\max \frac{p(\boldsymbol{\theta})p(D|\boldsymbol{\theta})}{p(D)} \tag{7.3}$$

where the arg max is to find the points within a domain for a given function at which the function values are maximized.

7.2.2 Proposition of Enhanced DE-TMCMC

The TMCMC method was originally developed by Ching and Chen [65] as a combination of the sequential particle filter method [72] and MCMC. The method begins with the prior distribution $p(\boldsymbol{\theta})$ and makes a gradual transition to the posterior by optimization at each round of samplings. The key idea of TMCMC is that of proposal density, which corresponds to the jth round of sampling $p(\boldsymbol{\theta})_j$ determined as,

$$p(\boldsymbol{\theta})_j \propto p(\boldsymbol{\theta}) \cdot L(\boldsymbol{\theta}|D)^{q_j} \tag{7.4}$$

where $q_j \in [0, 1]$ is chosen following $q_0 = 0 < q_1 < \cdots < q_m = 1$ with $j = 0, 1, \ldots, m$ denoting the stage level. Consequently, $p(\boldsymbol{\theta})_0$ equals the prior distribution $p(\boldsymbol{\theta})$ for $j = 0$, and $p(\boldsymbol{\theta})_m$ is the posterior distribution $p(\boldsymbol{\theta}|D)$ for $j = m$.

Details about execution of the TMCMC algorithm can be found in Ching and Chen (2007) [65]. In this study, only certain key steps are summarized:

1. Calculate the q_j. If $q_j > 1$, then set $q_j = 1$.
2. For all samples $k = 1, 2, \ldots, N_s$, compute a weighting coefficient $w_{j,k}$:

$$w_{(j,k)}=\left[L\left(\boldsymbol{\theta}_{(j-1,k)}|D\right)\right]^{q_j-q_{j-1}} \tag{7.5}$$

3. Compute the mean of the weighting coefficient

$$S_j = \frac{1}{N_s}\sum_{k=1}^{N_s} w_{(j,k)} \tag{7.6}$$

4. Compute the covariance matrix of the Gaussian proposal distribution

$$\boldsymbol{\Sigma}_j = \beta^2 \cdot \sum_{k=1}^{N_s} \left[\frac{w_{(j,k)}}{S_j N_s} \left(\boldsymbol{\theta}_{(j-1,k)} - \overline{\theta}_j \right) \cdot \left(\boldsymbol{\theta}_{(j-1,k)} - \overline{\theta}_j \right)^T \right] \tag{7.7}$$

with

$$\overline{\theta}_j = \frac{\sum_{l=1}^{N_s} w_{(j,l)} \boldsymbol{\theta}_{(j-1,l)}}{\sum_{l=1}^{N_s} w_{(j,l)}} \tag{7.8}$$

5. For each l in [1, 2, …, N_s], set $\boldsymbol{\theta}^c_{(j,l)} = \boldsymbol{\theta}_{(j-1,l)}$. Then for $k = 1, 2, \ldots, N_s$, do the following MCMC sampling:
   - 5.1 Select the index l from the set [1, 2, …, N_s] using the sequential importance sampling (SIS) method, where each l is assigned probability $w_{(j,l)} \Big/ \sum_{n=1}^{N_s} w_{(j,n)}$.
   - 5.2 Propose a new sample: Draw $\boldsymbol{\theta}^c$ from the normal distribution $N(\boldsymbol{\theta}^c_{(j,l)}, \boldsymbol{\Sigma}_j)$.
   - 5.3 The remaining steps are identical to those in the Metropolis–Hastings algorithm.
6. If $q_j = 1$, then stop the iteration; otherwise set $j = j + 1$ and continue the above process.

The details of the original TMCMC method, with its MATLAB code, can be found in Ching and Chen [65].

To improve the performance of original TMCMC, a differential evolution-Markov chain algorithm proposed by Vrugt [73] was adopted in this study to replace the process that proposes a new sample in step 5.2, which can be generated as:

$$\boldsymbol{\theta}^{\text{new}}_{(j,l)} = \boldsymbol{\theta}^c_{(j,l)} + d\boldsymbol{\theta}_{(j,l)} \tag{7.9}$$

with

$$d\boldsymbol{\theta}_{(j,l)} = (1 + \lambda) \cdot \gamma \cdot \left[\left(\boldsymbol{\theta}^{\text{best}}_j - \boldsymbol{\theta}^c_{(j,l)} \right) + \left(\boldsymbol{\theta}_{(j,a)} - \boldsymbol{\theta}_{(j,b)} \right) \right] + \zeta \tag{7.10}$$

where $\boldsymbol{\theta}^{\text{new}}_{(j,l)}$ is the new sample; $\boldsymbol{\theta}^c_{(j,l)}$ is the current sample; $\boldsymbol{\theta}^{\text{best}}_j$ is the sample corresponding to the maximum weight in the current iteration; d is the dimension of $\boldsymbol{\theta}$; $\boldsymbol{\theta}_{(j,a)}$ and $\boldsymbol{\theta}_{(j,b)}$ are two vectors consisting of d variables, where the indices a and b are two integers drawn from [1,…, N_s]; $\gamma = 2.38 \Big/ \sqrt{2\delta d}$ is the jump rate; δ denotes the number of chain pairs used to generate the jump with a default value of $\delta = 3$ according to Vrugt [73]. The values of λ and ζ are sampled independently from the uniform distribution [$-c$, c] and the normal distribution $N(0, c^*)$, respectively. In this study, the $c = 0.1$ and $c^* = 10^{-12}$ were employed, as recommended by Vrugt [73].

After differential evolution, a binomial crossover operation forms the final sample,

$$\boldsymbol{\theta}_{(j,l)}^{\text{new}} = \begin{cases} \boldsymbol{\theta}_{(j,l)}^{\text{new}}, & \text{if rand}(0,\,1) \leq \text{CR or } l = l_{\text{rand}} \\ \boldsymbol{\theta}_{(j,l)}^{c}, & \text{otherwise} \end{cases} \tag{7.11}$$

where rand(0, 1) is a uniform random number within [0, 1]; $l_{\text{rand}} = \text{randint}\ (1,\, d)$ is an integer randomly chosen from 1 to d and is newly generated for each l; the crossover probability CR $\in$ [0, 1] corresponds roughly to the average fraction of the vector components that are inherited from the mutation vector, with CR = 0.9 taken in this study.

For simplicity's sake, the original TMCMC is referred as "O-TMCMC," and the enhanced TMCMC as "DE-TMCMC" in the following sections.

7.2.3 Parallel Computing DE-TMCMC

Stochastic simulation algorithms, such as the TMCMC algorithm, used in Bayesian inverse modeling require numerous simulation runs when used for geotechnical engineering problems. For large-order computational models, the computational demands involved in the TMCMC sampling algorithm are excessive, sometimes to the point of being unacceptable. Accordingly, high-performance computing (HPC) techniques are vital for their ability to reduce computation time. Most MCMC algorithms involve a single Markov chain and are thus not parallelizable. Hence parallel computing algorithms should be implemented for use with the enhanced DE-TMCMC algorithm, efficiently distributing computations on multicore CPUs [74, 75].

In this study, the single program/multiple data (SPMD) technique in MATLAB was adopted to achieve a parallel computing implementation of DE-TMCMC. Typical applications appropriate for SPMD are those that require simultaneous execution of a program on multiple data sets when communication or synchronization is required between the workers. The SPMD efficiently distributes the computations involved in the DE-TMCMC algorithm to available heterogeneous GPUs and multicore CPUs so that the number of log-likelihood evaluations is the same for each computer worker. The SPMD technique is computationally efficient when the computational time for a log-likelihood evaluation is the same independent of the location of sample in the parameter space. More information about the SPMD technique can be found in the MATLAB manual.

Figure 7.1 shows the parallel computing strategy in DE-TMCMC using the SPMD technique, where log-likelihood represents the user-defined function for computing the log-likelihood; $N_$workers is the total number of computer workers (e.g., 8 for a 4-core CPU) and "labindex" is the index of computer workers, automatically identified in MATLAB. The entire calculation will be distributed among different computer

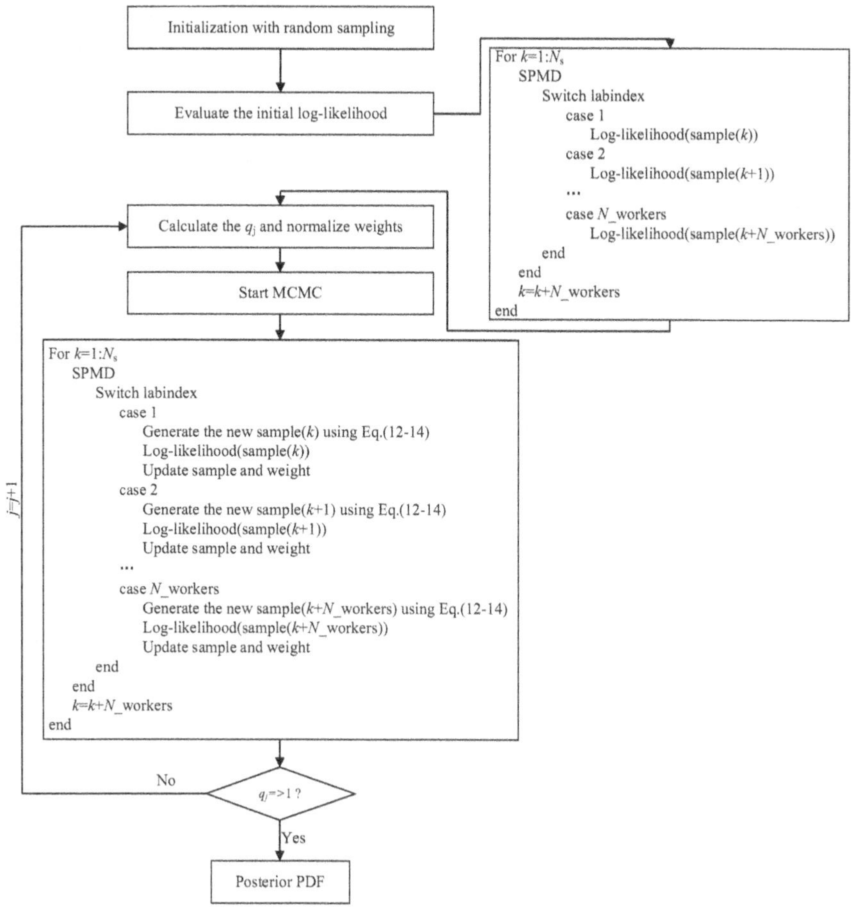

Fig. 7.1 Parallel computing strategy in DE-TMCMC using SPMD technique

workers according to the value of labindex. Note that the number of samples N_s should be an integer multiple of the number of computer workers, *N*_workers, in parallel computation.

Not surprisingly, drastic reductions of time can be achieved using this parallel implementation of the DE-TMCMC algorithm. Furthermore, the time savings differ for the same case on different computers with different settings: powerful computers with many multicore CPUs can save more computational time. The corresponding MATLAB code is shown in Appendix E.

7.3 Performance of Parallel Computing DE-TMCMC

In order to assess the efficiency of the parallel computing DE-TMCMC methodology for the identification of parameters of advanced constitutive models, several MCMC simulations were carried out using the O-TMCMC and DE-TMCMC, respectively, for a comparison on synthetic data and real experimental data. The following cases were considered:

(1) Identifying parameters of SIMSAND model (see Appendix F) from synthetic test data for the numerical validation;
(2) Identifying parameters of SIMSAND model from real experimental data for the applicability.

7.3.1 Numerical Validation on Synthetic Laboratory Tests

To generate the synthetic tests, a set of typical parameters of SIMSAND model were adopted, as summarized in Table 7.1. Using these values, three synthetic drained triaxial tests ($e_0 = 0.6$ and $p_0^{'} = 800$ kPa, $e_0 = 0.7$ and $p_0^{'} = 200$ kPa, $e_0 = 0.8$ and $p_0^{'} = 25$ kPa) were computed. Note that the strain measurement (e.g., linearized strain suitable for small strain, logarithmic strain suitable for large strain) may influence the obtained parameters such as stiffness or critical-state-related parameters, depending on their contributions to the material model. In these synthetic tests only the linearized strain measurement was used as usually adopted in analyzing laboratory tests [73]. According to Jin et al. [2], tests with wider range of confining stress involved in parameter identification would be helpful on finding the accurate CSL. To increase the difficulty of evaluation, the computed stress and void ratio of three tests were contaminated by additive Gaussian noise with standard deviation 0.1. Adding Gaussian noise to synthetic data is a relatively intuitive and quantitative approach for investigating this problem. The purpose of adding Gaussian noise is to show the anti-interference ability of the proposed algorithm in parameter identification. The results of three synthetic drained tests without/with noise are shown in Fig. 7.2.

Table 7.1 Identification results of SIMSAND from synthetic data and the bounds used in the calculation

Parameters	υ	K_0	n	e_{ref}	λ	ξ	ϕ_μ	k_p	A_d	n_p	n_d
Upper bound	–	–	–	1.5	0.1	1.0	50	0.01	2.0	0.1	0.1
Lower bound				0.5	0.001	0.01	20	0.0001	0.5	10.0	10.0
True value	0.2	100	0.6	0.750	0.03	0.60	30.0	0.003	1.0	2.0	2.0
Average $\boldsymbol{\theta}_{MAP}$ (without noise)				0.750	0.03	0.60	30.0	0.003	1.0	2.0	2.0
Average $\boldsymbol{\theta}_{MAP}$ (with noise)				0.753	0.034	0.57	30.05	0.0031	0.98	2.04	2.15

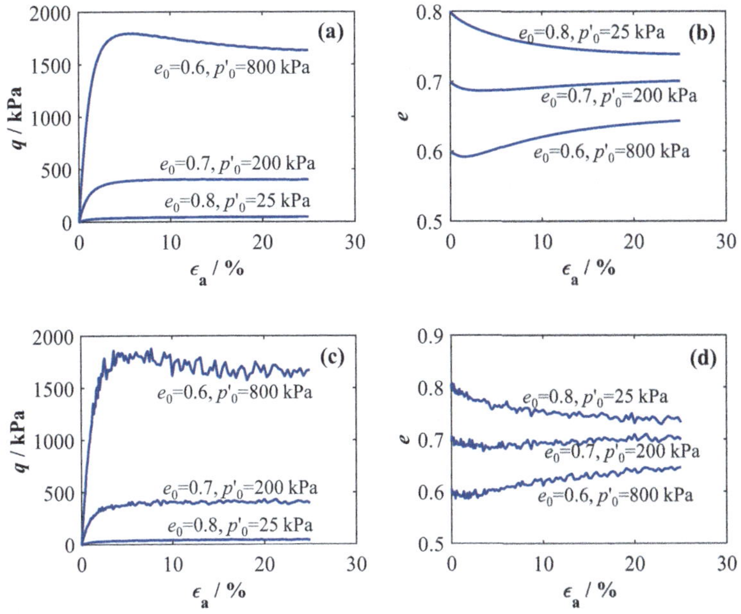

Fig. 7.2 Results of synthetic drained triaxial tests computed by SIMSAND using true parameters: **a**, **b** without noise; **c**, **d** with noise

For both cases, 2000 samples per stage was implemented, and thus 2000m (m representing total number of stages in TMCMC) samples were totally generated in TMCMC simulation. According to previous studies [64, 65, 68, 69, 76], the total number of stages for different cases were not always the same. The prior PDF of all uncertain parameters was assumed to follow a multivariable uniform distribution. These were independently uniformly distributed within the lower and upper bounds. The bounds of each prior PDF for all parameters are summarized in Table 7.1. In this case, the Poisson's ratio υ, bulk modulus K_0 and nonlinear elasticity parameter n were fixed. The remaining parameters of SIMSAND along with the uncertainty σ_ε were identified using the Bayesian parameter identification, adopting O-TMCMC and DE-TMCMC, respectively. Note that both O-TMCMC and DE-TMCMC were enhanced with the parallel computing technique to save the computational costs.

For both TMCMC simulations, 10 independent runs were performed to avoid the randomness in order to show the performance of the DE-TMCMC. For Bayesian analysis, the generated samples should be represented by the "entire" PDF to evaluate the performance of adopted approaches. However, the multidimensional distributions over all the space of unknown parameters cannot be directly illustrated due to its high dimension. Thus, the obtained posterior PDF of each parameter was presented by mean and standard deviation values for a convenient comparison between O-TMCMC and DE-TMCMC. Figures 7.3 and 7.4 show the mean values of all parameters and the uncertainties for DE-TMCMC and O-TMCMC on

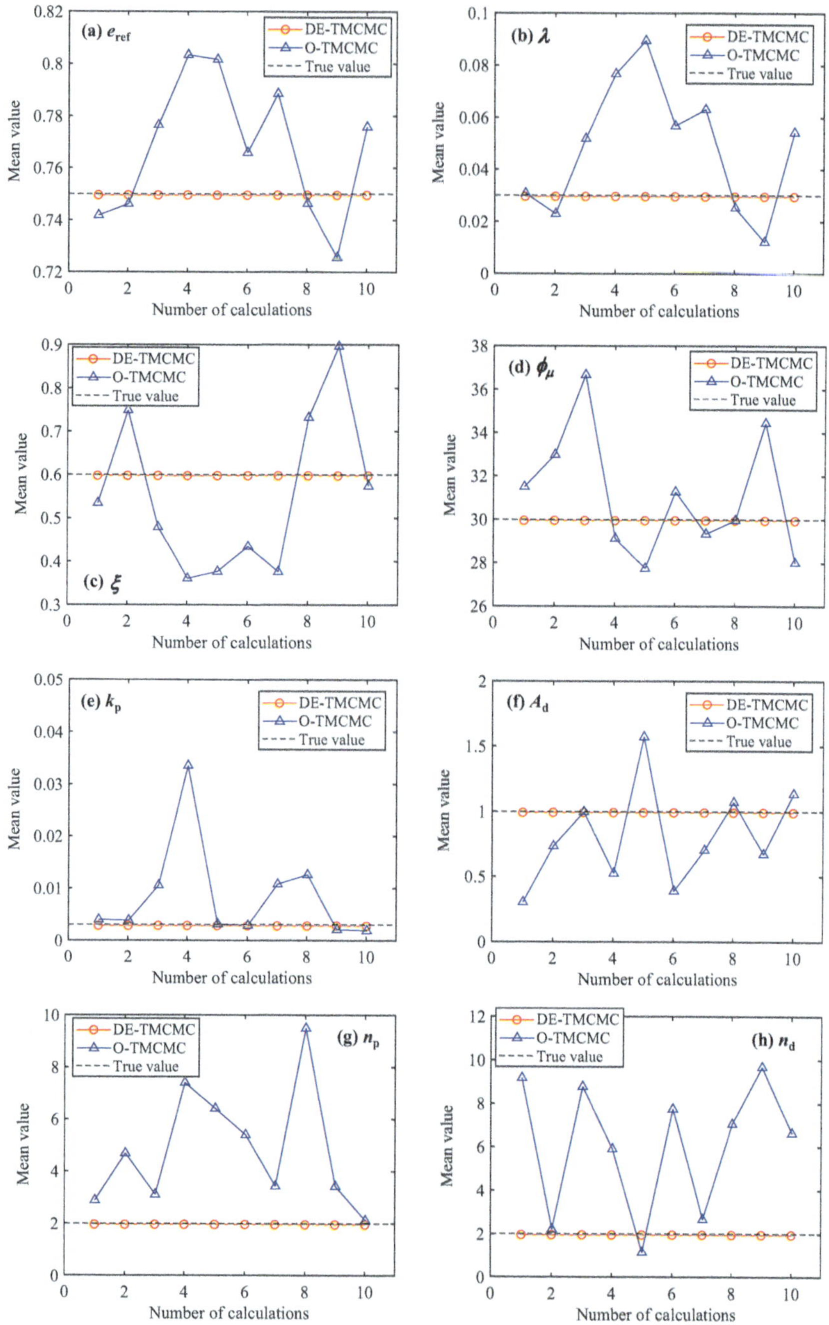

Fig. 7.3 Comparison of MAP value of each parameter and the uncertainty for O-TMCMC and DE-TMCMC on synthetic data without noise

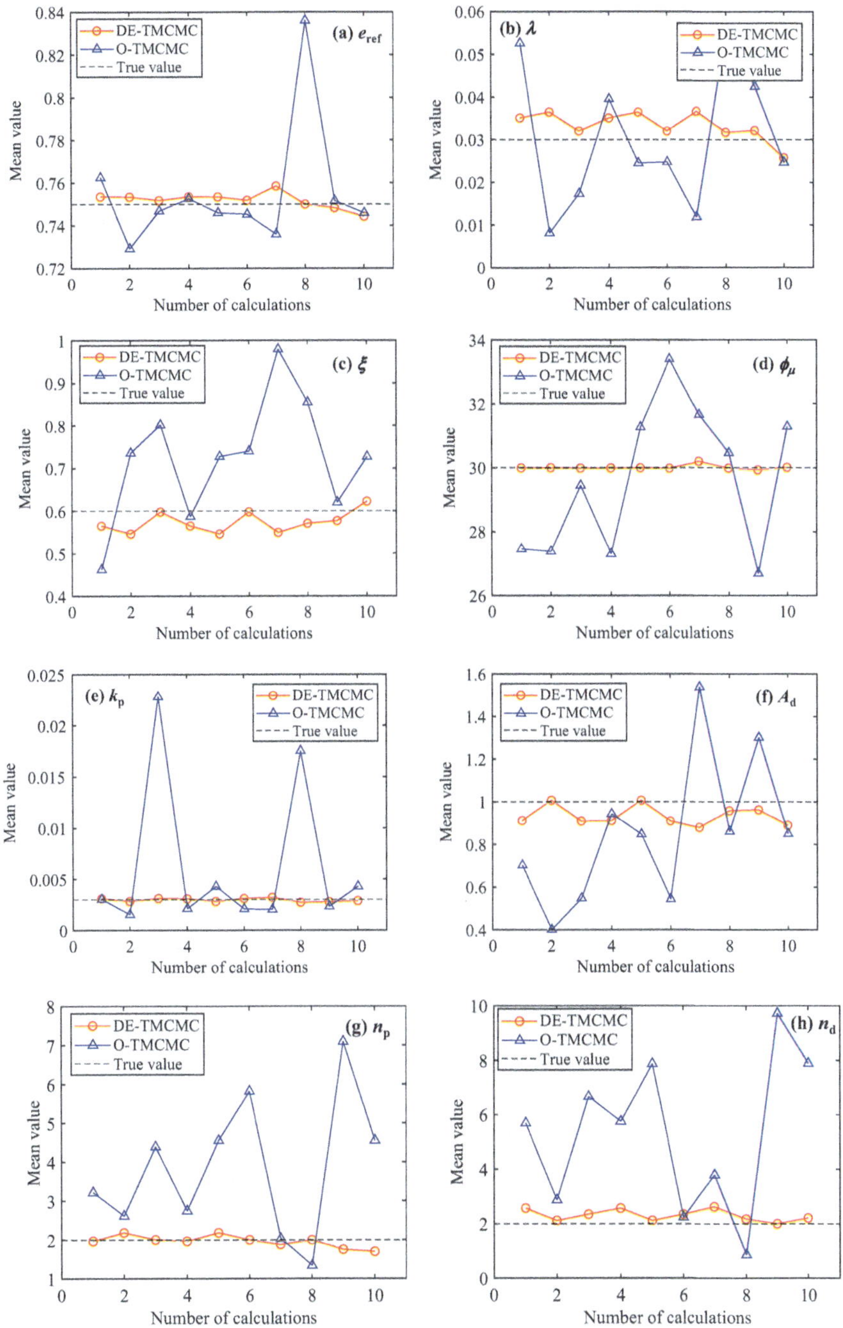

Fig. 7.4 Comparison of MAP value of each parameter and the uncertainty for O-TMCMC and DE-TMCMC on synthetic data with noise

synthetic data without/with noise, respectively. Regardless of whether the used data is noisy, the parameters identified by DE-TMCMC in comparison to O-TMCMC fluctuate slightly, and basically stabilize to a constant value among multiple calculations, demonstrating the excellent performance of the proposed DE-TMCMC on attaining the robustness.

A good MCMC method is not only robust but also effective that the concerned parameters should be identified. The presented results in Fig. 7.3 also show that the value of each parameter identified by DE-TMCMC is closer to the true value than the one identified by O-TMCMC on synthetic data without noise, which indicates the effectiveness of the DE-TMCMC. Moreover, as shown in Fig. 7.4 the similar exact parameters can also be identified by DE-TMCMC even though the used data is noisy, which indicates the outstanding anti-interference capability of the DE-TMCMC. Thus, the DE-TMCMC is more suitable and reliable than O-TMCMC when identifying the parameters for advanced sand models from noisy data.

Table 7.1 summarizes the average MAP values of SIMSAND parameters identified by DE-TMCMC. Except three CSL-related parameters (e_{ref}, λ and ξ), others are approximately equal to the true values. Figure 7.5 shows the comparison of CSL computed by the parameters identified by DE-TMCMC from the data without/with noise. Slight difference between three CSLs is found at the stress level ranging from 100 to 1000 kPa. Using the MAP parameters, three drained triaxial tests were simulated by SIMSAND, as shown in Fig. 7.6. Good agreement between simulations and synthetic data shows that the identified parameters are accurate. Secondly, although the parameters of CSL are inaccurate, the impact on the simulation results is not significant.

Overall, the proposed DE-TMCMC is superior to O-TMCMC in identifying parameters of advanced sand model in terms of robustness, effectiveness and reliability regardless of whether the used data is noisy or not.

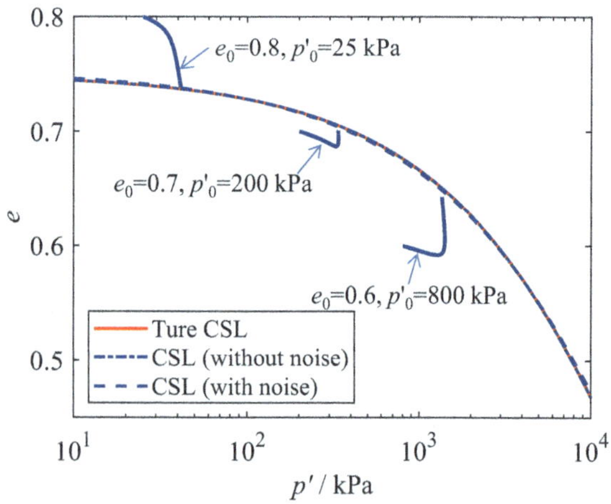

Fig. 7.5 Comparison of CSL computed by the identified parameters from the data without/with noise by DE-TMCMC

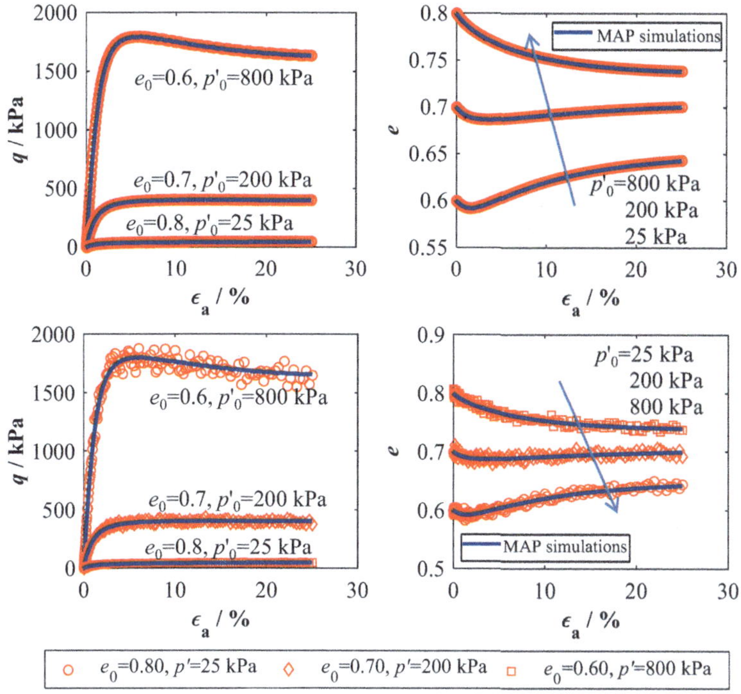

Fig. 7.6 Comparison between experimental and MAP simulations of the synthetic drained tests

7.3.2 Performance on Real Laboratory Tests

A series of drained triaxial tests performed on Toyoura sand by Verdugo and Ishihara [77] was selected for this case. Toyoura sand is a uniform fine quartzitic sand consisting of sub-rounded to sub-angular particles having a mean grain size d_{50} = 0.17 mm and a uniformity coefficient $C_u = d_{60}/d_{10} = 1.7$. The minimum and maximum void ratios are $e_{\min} = 0.597$ and $e_{\max} = 0.977$.

The initial void and confining pressure for each test is summarized in Table 7.2. The experimental results for all triaxial tests are shown in Fig. 7.7. To form a comprehensive group of experiments that can well reflect the common sand behaviors, following the recommendation of Jin et al. [2], three triaxial tests (one drained (test 2) and two undrained (tests 7 and 12)) were selected as objectives for parameter identification. Other remaining tests were used to assess the accuracy of obtained parameters.

In this study, the Poisson's ratio was assumed to be 0.2, a common value for most sands. Except for the Poisson's ratio (υ), the remaining parameters of SIMSAND along with the uncertainty were identified using DE-TMCMC combining with Bayesian method.

Table 7.2 Summary of triaxial tests on Toyoura sand

Test type	Drained tests						Undrained tests							
Test number	1	2	3	4	5	6	7	8	9	10	11	12	13	14
e_0	0.996	0.916	0.831	0.959	0.885	0.809	0.735	0.833						
p'/MPa	0.1	0.1	0.1	0.5	0.5	0.5	0.1	1	2	3	0.1	1	2	3

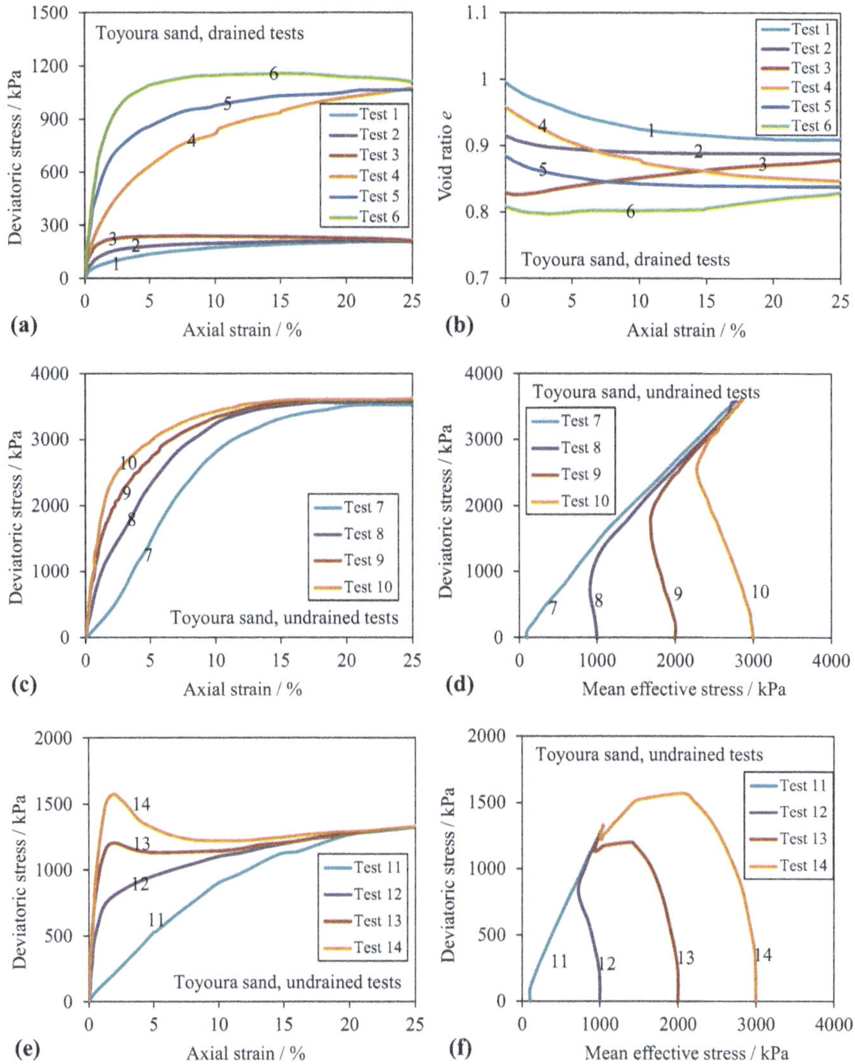

Fig. 7.7 Results of triaxial tests of Toyoura sand

7.3.3 Results and Discussions

Similarly, this case was also run independently for 10 times. To show the effectiveness of the proposed DE-TMCMC in identifying parameters of sand model, the identification of parameters was also conducted by optimization techniques [13, 36] for comparison. The obtained parameters are summarized in Table 7.3. Although the optimization-based identification does not take the soil uncertainty into account,

Table 7.3 Bounds in prior PDF, reference values, mean MAP values and standard deviations in posterior PDF for all parameters

Parameters and bounds	K_0	n	e_{ref}	λ□	ξ□	ϕ□	k_p	A_d	n_p	n_d	□σ_ε□
	[50, 300]	[0.1, 0.9]	[0.5, 1.5]	$[10^{-3}, 10^{-1}]$	[0.1, 1.0]	[20, 50]	$[10^{-3}, 10^{-1}]$	[0.1, 2]	[0, 10]	[0, 10]	[0, 1]
□$_{MAP}$ O-TMCMC	172.6 (13.2)	0.449 (0.029)	0.928 (0.004)	0.0212 (0.0016)	0.81 (0.028)	30.74 (0.67)	0.0010 (0.001)	0.68 (0.065)	5.35 (0.98)	5.00 (0.49)	0.28 (0.008)
□$_{MAP}$ DE-TMCMC	144.6 (6.96)	0.463 (0.018)	0.904 (0.002)	0.0097 (6.62e−4)	0.933 (0.019)	32.18 (0.26)	0.0043 (4.42e−4)	0.42 (0.023)	3.157 (0.52)	4.24 (0.43)	0.057 (0.0055)
Reference values	130	0.5	0.920	0.0131	0.78	32.40	0.0030	0.50	1.60	4.10	–

the results can be deemed a set of reference values to assess the effectiveness of the proposed DE-TMCMC: a closer distance between the solution of optimization and generated samples of MCMC means more reliable parameters achieved, indicating a more effective MCMC method adopted or developed.

Figure 7.8 compares the mean values of each parameter and the uncertainty for two approaches under multiple calculations. Similar phenomenon that the parameters identified by DE-TMCMC fluctuate slightly is found. In contrast, the results of O-TMCMC are unsatisfactory with large fluctuations, which obviously demonstrate the bad robustness of O-TMCMC, especially for the friction angle ϕ which should be almost constant for a given granular material. Except three CSL-related parameters, other parameters identified by DE-TMCMC are very close to the reference values compared to O-TMCMC. The reason leading to the inaccurate CSL is also the small stress range of test involved in the parameter identification. Without considering this, the comparison between DE-TMCMC and O-TMCMC demonstrates that the DE-TMCMC is robust and effective in identifying parameters of SIMSAND from real laboratory tests.

Figure 7.9 shows the values of standard deviation of each parameter and the uncertainty for two approaches. Large fluctuation in data of standard deviation indicates the instability of posterior PDF. Furthermore, the large value of standard deviation means the generated samples distributed in a large range, implying the existence of local convergence. For granular material, the friction angle ϕ is a key parameter relating to material strength. In SIMSAND, the plastic modulus k_{p} and dilatancy-related parameters A_{d} and n_{d} are key parameters relating to material deformation. Thus, the accuracy of ϕ, k_{p}, A_{d} and n_{d} can be an indicator of the effectiveness of DE-TMCMC. The evolutions of four selected parameters corresponding to the larger standard deviation at final stage with $q_j = 1$ for O-TMCMC and DE-TMCMC were plotted. For ϕ□ and k_{p}, the comparisons at the calculations 2nd, 3rd, 6th and 10th are shown in Fig. 7.10. Figure 7.11 shows the comparison of evolutions at final stage for A_{d} and n_{d}. For DE-TMCMC, the parameters gradually converge to a narrow range from the prior uniform distribution. However, for O-TMCMC, the identification falls into local convergence and thus results in unreasonable values of ϕ, k_{p}, A_{d} and n_{d} for Toyoura sand [12, 13]. In contrast, this situation is overcome by DE-TMCMC with a domain of more reliable parameters. All comparisons demonstrate that the proposed DE-TMCMC is robust and can overcome the problem of local convergence for parameter identification of advanced soil models.

The average MAP parameters and the standard deviation of posterior PDF—identified by O-TMCMC and DE-TMCMC—are also summarized in Table 7.3. The MAP parameters identified using DE-TMCMC are more close to those obtained by Taiebat and Dafalias [12] and Jin et al. [13], which indicates that the parameters obtained by DE-TMCMC are more reasonable and reliable than those by O-TMCMC. Furthermore, finding the uncertainty of parameters is also important compared to the MAP values, which is the characteristic of Bayesian approach unlike the deterministic

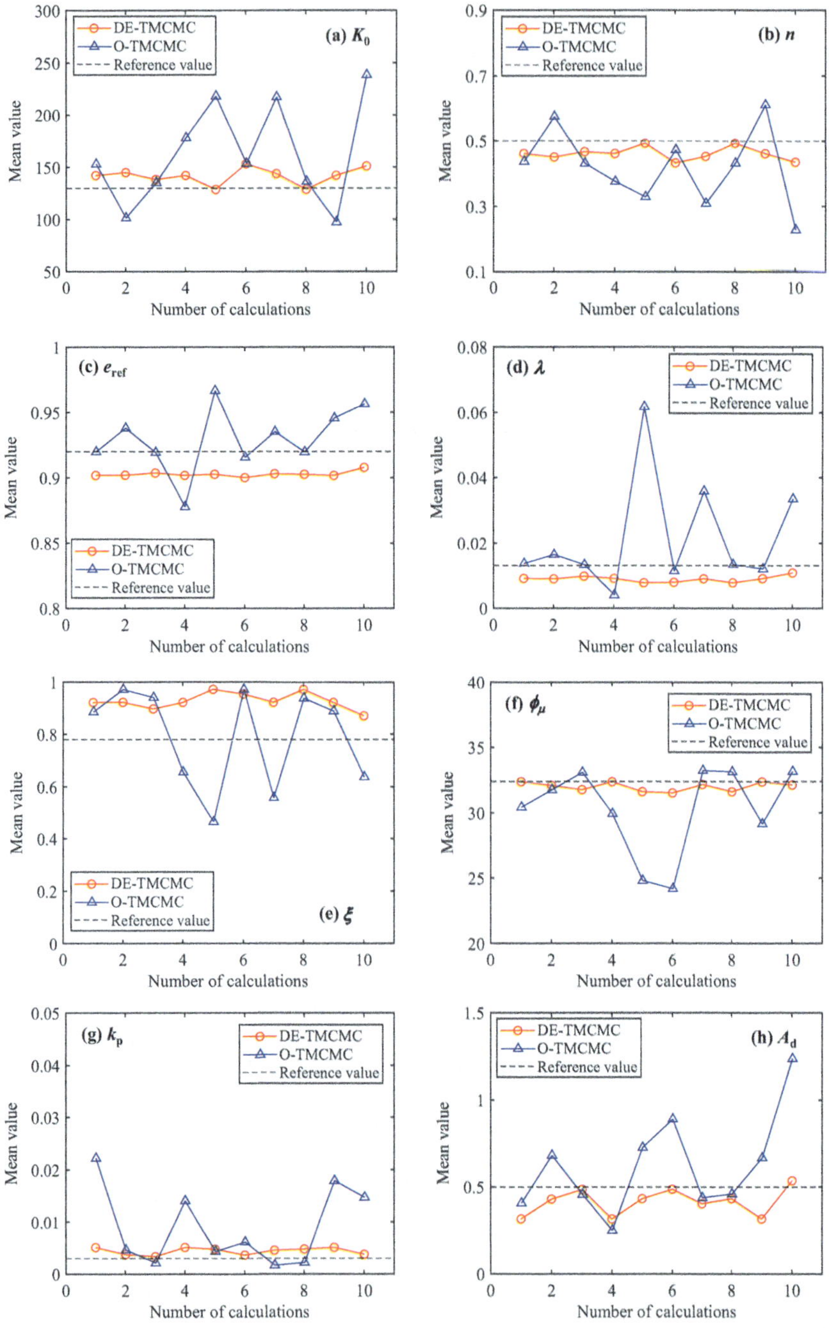

Fig. 7.8 Comparison of mean value of each parameter and the uncertainty for O-TMCMC and DE-TMCMC

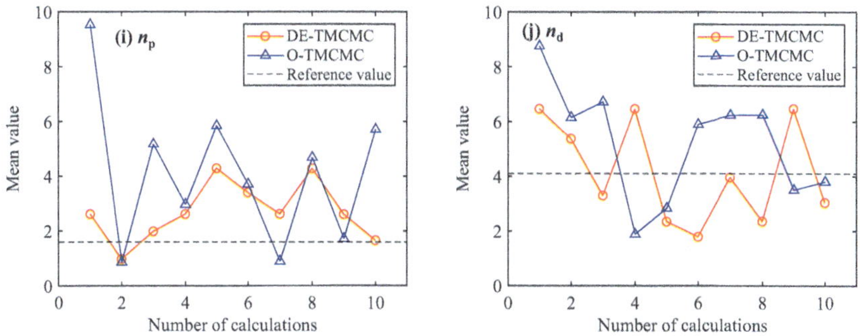

Fig. 7.8 (continued)

methods. The uncertainty of obtained parameters can be used to assess the uncertainty of predictions. Therefore, using all the obtained 2000 samples, the uncertainty of predictions for two used approaches was compared by error bar plot, as shown in Fig. 7.12 on objective tests 2, 7 and 12. In fact, the outstanding predictive ability of SIMSAND model that can well capture the behavior of Toyoura sand is known before conducting this case [13, 78]. Thus, the O-TMCMC overstates the uncertainty of predictions compared to the proposed DE-TMCMC, indicating that the proposed DE-TMCMC is more effective than O-TMCMC in terms of the reliability of identified parameters. To demonstrate the performance of the MAP parameters identified by DE-TMCMC, a series of predictions were conducted, as shown in Fig. 7.13. Good agreement can be found between the numerical predictions and experiments, demonstrating the rationality of the parameters obtained by the Bayesian parameter identification with DE-TMCMC.

Furthermore, some of the test results are not consistent themselves, which is probably caused by: (a) an inherent spatial variability of soil properties, (b) experimental uncertainty (measurement scatter) due to limitations of the experimental techniques and (c) sampling uncertainty (statistical uncertainty) due to the limited number of soil samples used in the investigation. Such inconsistency would lead to variability of parameters for model of interest, which can be quantified by the proposed Bayesian approach with DE-TMCMC from available experimental data. In practice, such inconsistency (experimental and sampling uncertainties) can be incorporated into probabilistic analyses using random field methods, such as the approach developed by Mašín [79].

Overall, through the above comparisons, the enhanced DE-TMCMC is highly robust and efficient in identifying the parameters of advanced soil models.

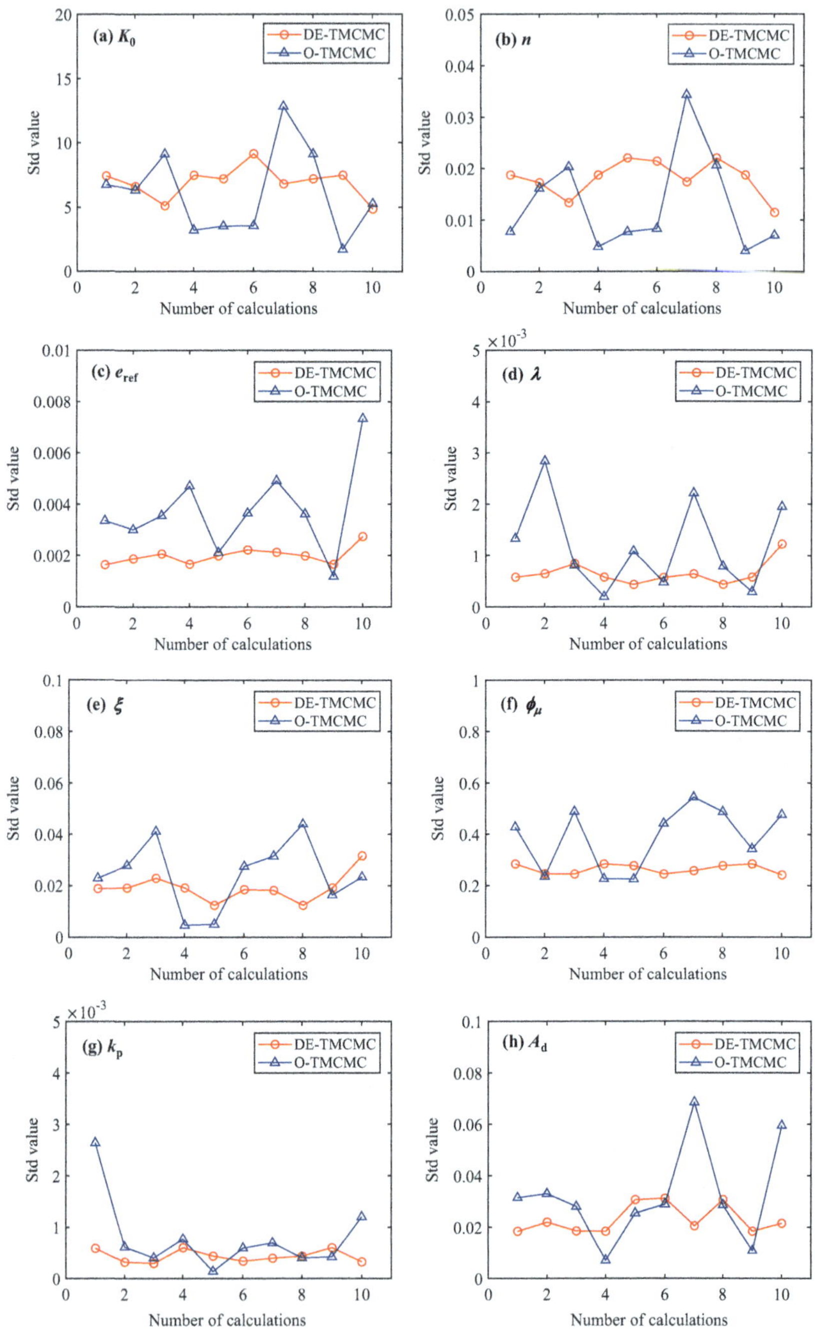

Fig. 7.9 Comparison of standard deviation of each parameter and the uncertainty for O-TMCMC and DE-TMCMC

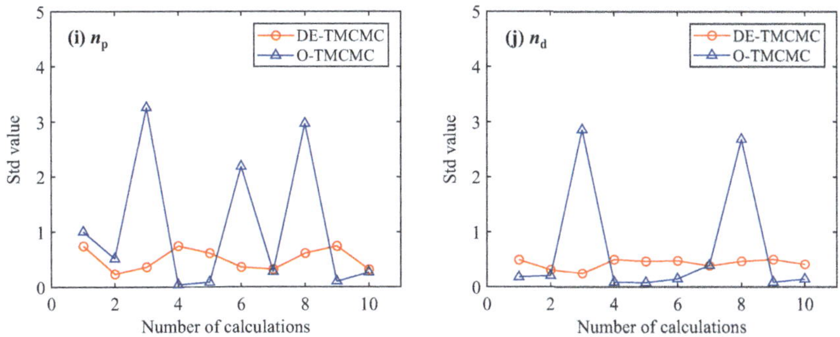

Fig. 7.9 (continued)

7.4 Application to In Situ Testing

7.4.1 Parametric Identification Parameters from Pressuremeter Test

To evaluate the performance of Bayesian parameter identification in conjunction with the DE-TMCMC through the use of in situ testing, two self-boring pressuremeter (SBP) tests performed in the Burswood Peninsula site at a depth of 5.25 m by Lee and Fahey [80] were selected as the objectives. A two-dimensional finite element model, with boundaries in an axisymmetric condition to simulate the pressuremeter, was created in a commercial finite element code ABAQUS, as shown in Fig. 7.14. The upper and bottom sides were fixed only for vertical displacement, whereas the right side was fixed for horizontal displacement. The loading could then be generated by applying horizontal displacement at the left side. Accordingly, the horizontal displacement had its biggest value at the left side and gradually decreased to zero as it reached the right side. Altogether, 240 four-node reduced-integration elements (CAX4RP) were used to simulate the soil. The same displacement seen in a typical field test was applied, and at each step, the same displacement increment was applied.

For the two tests, the expansion phase was conducted at two different rates up to around 10% of the cavity strain: 0.167%/s and 0.0185%/s, respectively. The initial effective vertical stress σ'_{v0} was 31 kPa, and the lateral earth coefficient K_0 was 0.9, where the initial value of vertical stress is estimated from the effective unit weight of soils with the depth, and the horizontal stress relating to the earth pressure coefficient is measured as the initial value of lateral stress from test results according to the site investigation and two SBP tests conducted by Lee and Fahey [80]. The water pore pressure was 38.8 kPa. In this case, the elasto-viscoplastic model ANICREEP was adopted to simulate the rate-dependency behavior of soft clay. More details of ANICREEP with its finite element implementation can be found in Yin et al. [16, 36]. To help identify the rate-dependent parameters, the initial void ratio e_0, the swelling index κ, and the compression index λ, with the permeability ($k = 3.3 \times 10^{-9}$ m/s)

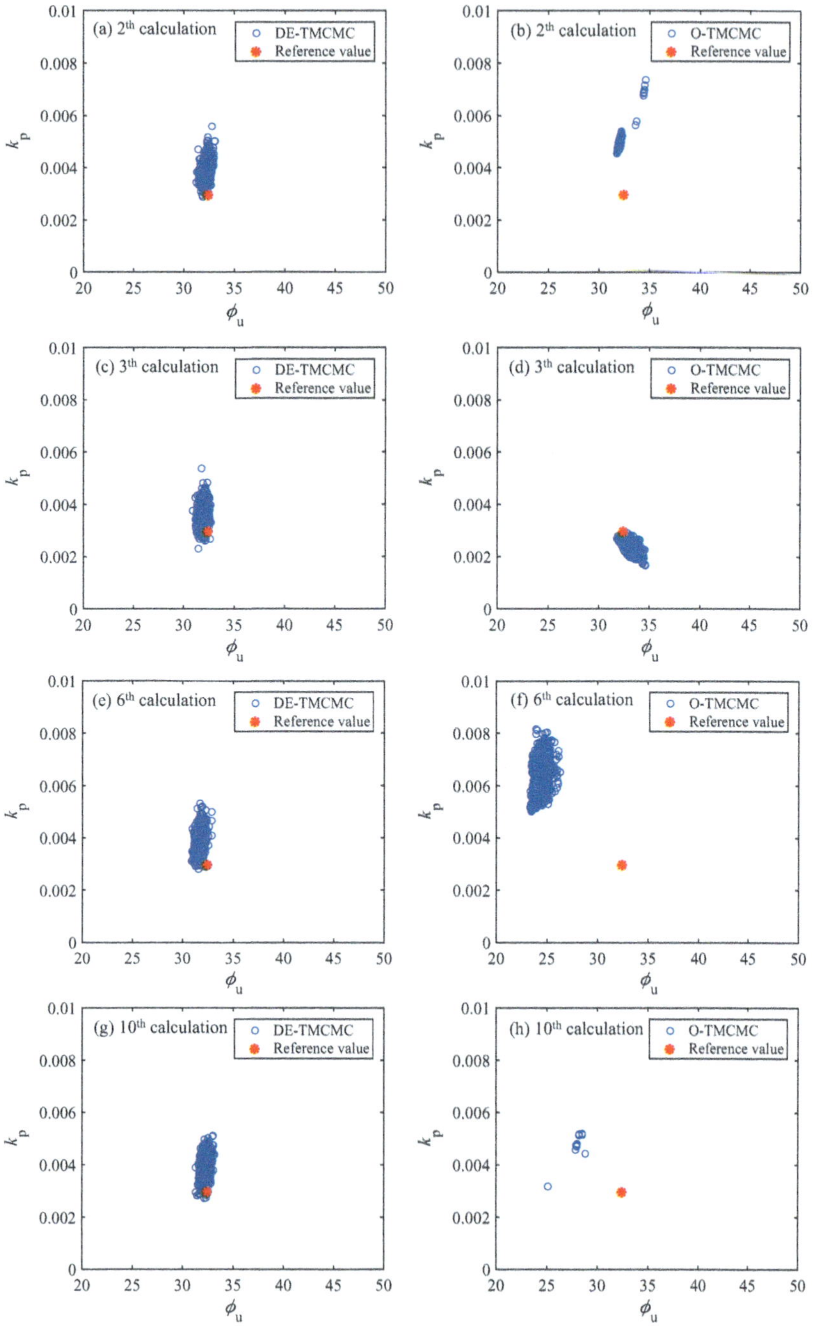

Fig. 7.10 Comparison of evolution of samples between O-TMCMC and DE-TMCMC at final stage for ϕ and k_p

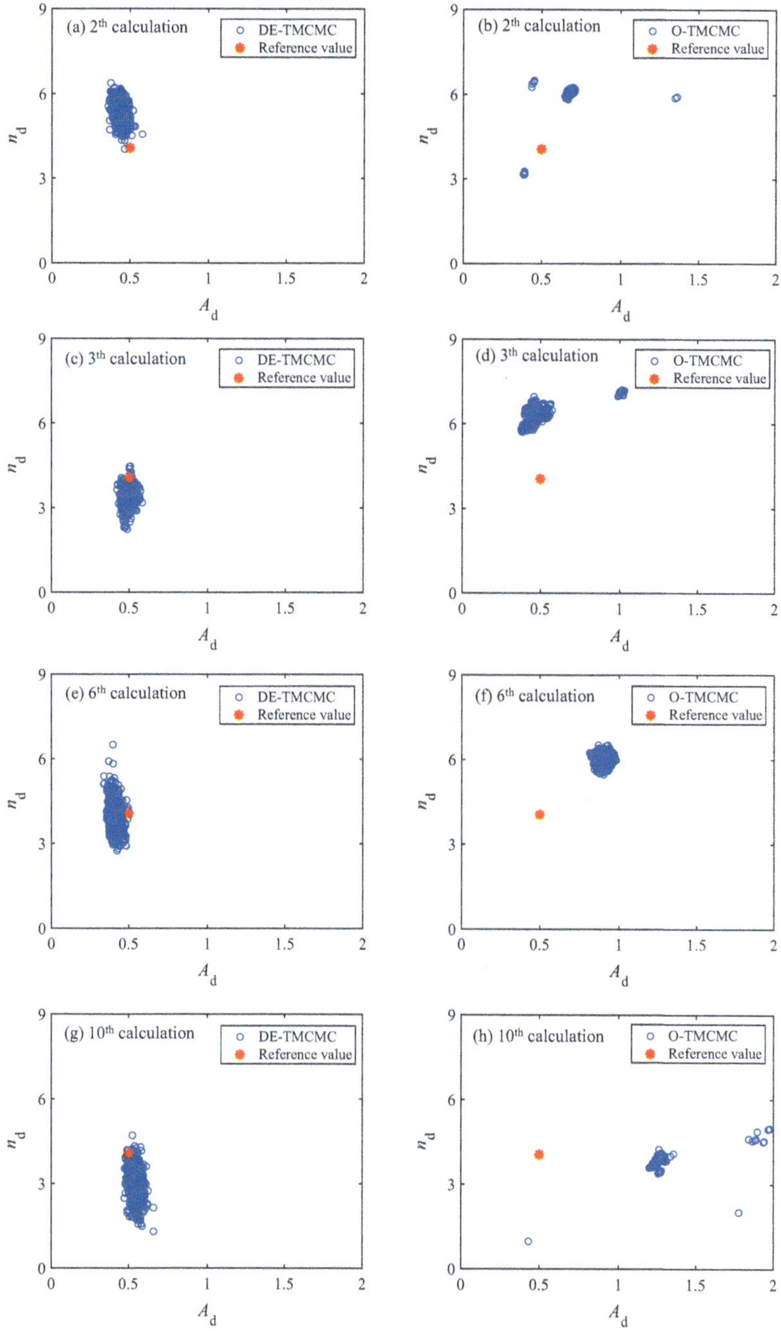

Fig. 7.11 Comparison of evolution of samples between O-TMCMC and DE-TMCMC at final stage for A_d and n_d

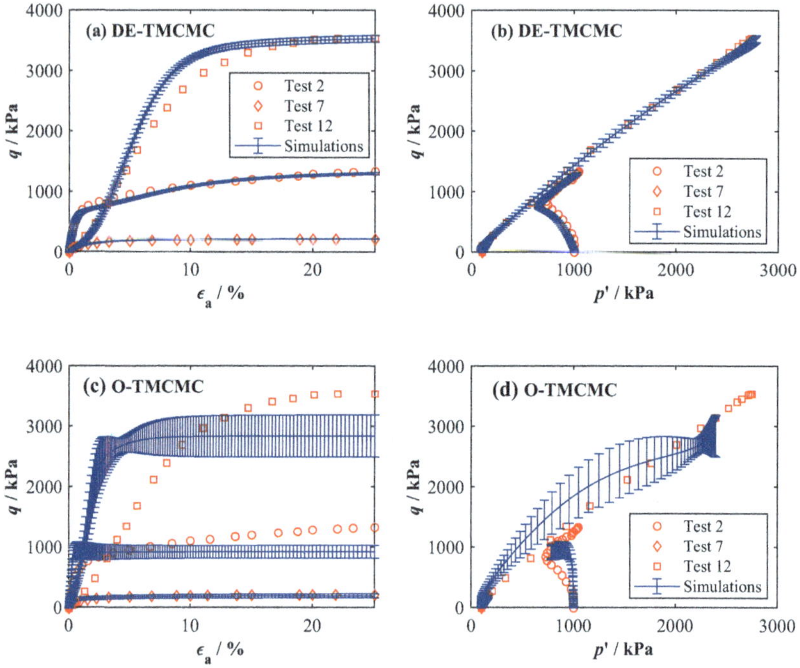

Fig. 7.12 Simulations using parameters of final state for objective tests 2, 7 and 12

provided by Lee and Fahey [80] were fixed in the parameter identification. Note that the swelling index κ, the compression index λ and the permeability k could not be put into the Bayesian identification process since the test results of oedometer tests and excess pore pressure measurements in SBP testing are not available. The Poisson's ratio υ was assumed to be 0.3 for soft clays [38]. The other parameters (the slope of critical state line M, the initial preconsolidation pressure σ'_{p0}, and the secondary compression coefficient $C_{\alpha\mathrm{e}}$) were determined by Bayesian identification with DE-TMCMC. The bounds of uniform distribution of parameters are given in Table 7.4. The settings of DE-TMCMC were the same as those given previously. For the purpose of comparison, the case was also carried out using O-TMCMC.

7.4.2 Results and Discussions

Figure 7.15 shows the posterior PDFs of three identified parameters of ANICREEP model using DE-TMCMC and O-TMCMC. The value of M distributes primarily in a large range within [1.5, 1.8] for DE-TMCMC, with MAP value of 1.71, but for a narrow range [0.95, 1.05] for O-TMCMC, with MAP value of 0.98. Similarly, the identified σ'_{p0} and $C_{\alpha\mathrm{e}}$ also distribute in a large range for DE-TMCMC, with MAP

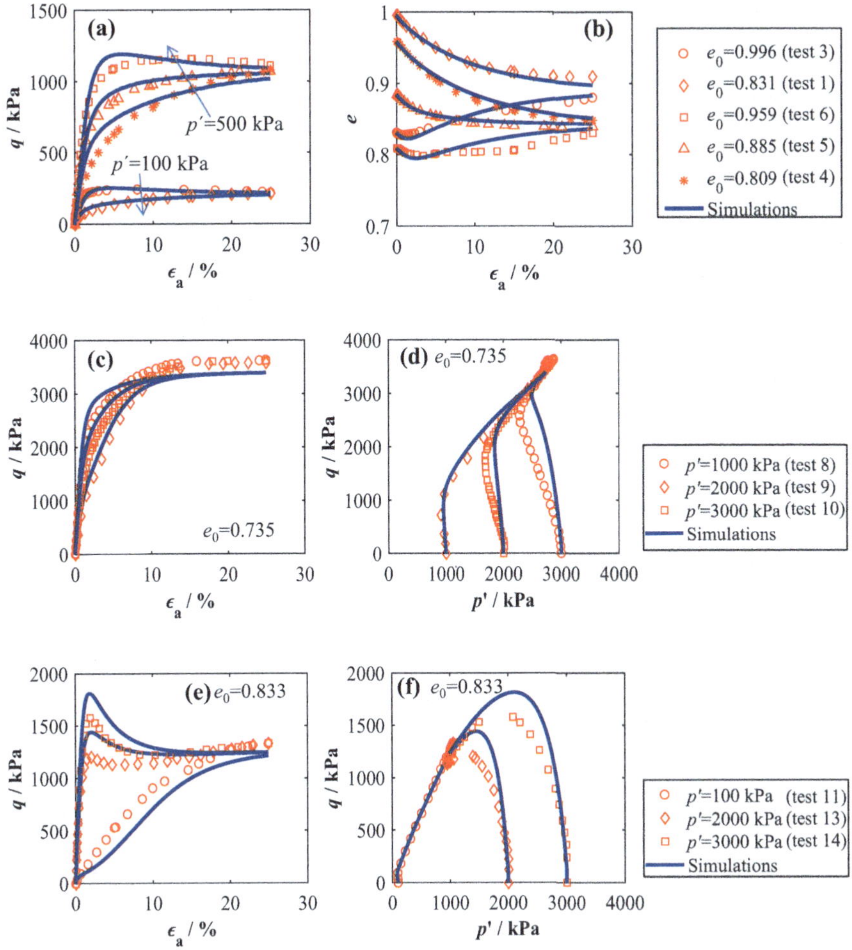

Fig. 7.13 Comparison between simulations and experiments of triaxial tests on Toyoura sand

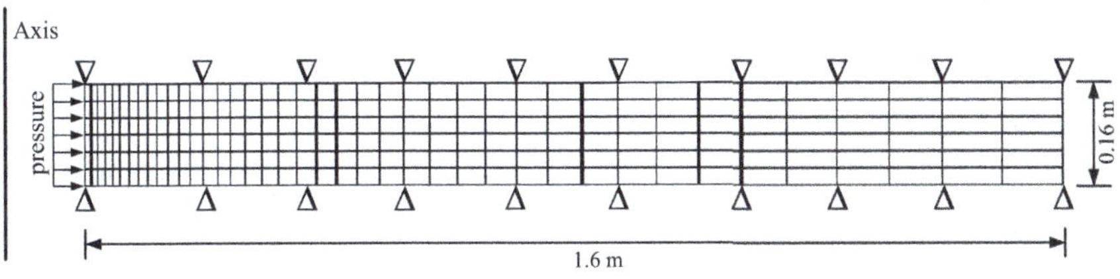

Fig. 7.14 Finite element model of pressuremeter test

Table 7.4 Search domain for three parameters of ANICREEP model

Parameters	M	σ'_{p0}/kPa	$C_{\alpha e}$
Lower bound	0.5	10	0.001
Upper bound	2.0	200	0.05

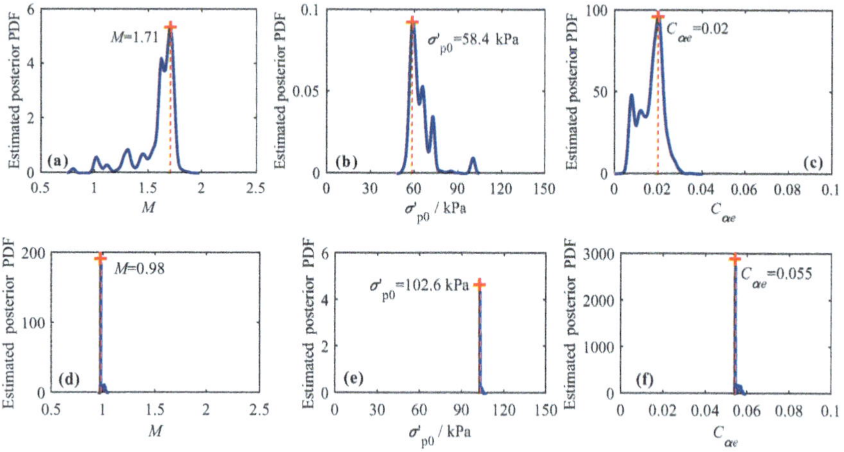

Fig. 7.15 Posterior PDF of parameters of ANICREEP model for Burswood clay: **a–c** by DE-TMCMC; **d–f** by O-TMCMC

values of 58.4 kPa and 0.020, compared to a narrow range for O-TMCMC, with MAP values of 102.6 kPa and 0.055. According to Low [81], the identified M (= 1.71) and σ'_{p0} (= 58.4 kPa) were close to measured values, and the identified parameter $C_{\alpha e}$ = 0.020 is identical to the $C_{\alpha e} = 0.020$ experimentally estimated by Low [81] based on the strain-rate dependency parameter $\beta = 0.055$ of Burswood clay using the unified formulation of Yin et al. [82, 83]. These comparisons demonstrate that much more reasonable parameters can be achieved through Bayesian parameter identification with DE-TMCMC than those with O-TMCMC.

All parameters of ANICREEP for Burswood clay are summarized in Table 7.5. Using all solutions obtained by DE-TMCMC, numerous simulations of two pressuremeter tests were performed. Figure 7.16 compares the predictions with uncertainties and measurements for the SBP tests in terms of total pressure at the cavity

Table 7.5 MAP values of ANICREEP model for Burswood clay

MAP Parameters	υ	e_0	κ	λ	M	σ'_{p0}/kPa	$C_{\alpha e}$
DE-TMCMC	0.3	2.8	0.036	0.40	1.71	58.4	0.020
O-TMCMC					0.98	102.6	0.055

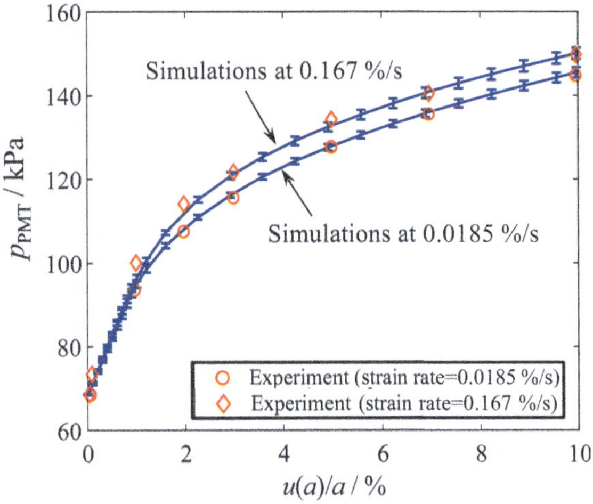

Fig. 7.16 Comparison between simulated and experimental results for SBP tests on Burswood clay

wall versus cavity strain. The measurements are mostly covered by the simulations. Accordingly, all findings demonstrate that the Bayesian parameter identification using the DE-TMCMC is efficient and applicable for in situ testing.

7.5 Discussions

Using statistical methods in calibrating has a long tradition in soil mechanics, which has been reviewed in the introduction. The calibrating procedure is an inverse problem and the solution is generally not unique, that is, many parameter combinations provide similar acceptable results in the sense that a user-defined error measurement is small enough [14, 41, 84]. Moreover, the solutions depend on user-defined error measurement, for which no general rule exists. However, the performance of parameters identification using the proposed error function in the likelihood defined in the proposed method has been highlighted in various cases [2, 13, 14, 34–38, 84]. The defined error function considers both the effect of quantities for different test curves (such as deviatoric stress–axial strain curve for different confining pressures) and the effect of dimension/unit for different variables (such as deviatoric stress and volumetric strain in the same test). Thus, a set of reliable parameters can be obtained by the proposed method to represent the material behavior.

In traditional calibration, one uses experiments suitable for calibrating subsets of the material parameters. The critical state friction angle, for example, can be calibrated from three triaxial tests, without considering other parameters such as the elastic stiffness. Although one has better sense of how the specific parameter influences a specific behavior of the soil by doing such calibration works, such way can be successful only for few conventional laboratory tests, such as 1D compression and 3D triaxial tests. Identification of parameters from in situ or field tests cannot be completed by traditional calibration methods. Moreover, the number of parameters directly determined by such way is limited. With the development of constitutive modeling, numerous advanced soil models with a large number of parameters have been proposed [36]. The traditional calibration method becomes ineffective for a large number of parameters, in particular becomes invalid for parameters without clear physical meaning. For this situation, the trial-and-error method is a popular choice. In practice, such calibrations often depend on the user's preferences and experiences. In that sense, calibrating of the material parameters by the proposed method is more favorable than other traditional calibration methods.

Furthermore, the local convergence problem inevitably exists for most inverse methods. However, the proposed Bayesian identification with DE-TMCMC has the ability to prevent the identification process from the local convergence, which has been demonstrated by aforementioned results and the comparisons to original TMCMC (see Figs. 7.10 and 7.11).

Overall, the proposed Bayesian parametric identification has advantages in dealing with different parameter calibration works, even when some parameters do not have physical meaning.

7.6 Conclusions

To identify parameters and quantify the associated uncertainty, a Bayesian parametric identification method using an enhanced DE-TMCMC with parallel computing technique has been presented in this chapter. The DE-TMCMC, enhanced through implementing a differential evolution into TMCMC to replace the process of proposing a new sample, has been presented. To save on computational costs, a parallel computing DE-TMCMC was achieved by implementing the SPMD technique.

The performance of DE-TMCMC for Bayesian parameter identification was first validated through comparison to the original TMCMC for the case of identifying parameters of a critical state-based model from triaxial tests in terms of robustness and effectiveness: three synthetic drained triaxial tests with/without noise for numerical validation, and three real drained triaxial tests on Toyoura sand for applicability. To accommodate the randomness, the calculations were independently conducted 10 times. The results demonstrate that the DE-TMCMC is highly robust and efficient, producing reasonably ranged posterior PDFs as well as a set of accurate parameters regardless of whether the used data is noisy.

Finally, the enhanced DE-TMCMC-based parameter identification approach was applied to two in situ pressuremeter tests on soft clays. The elastic viscoplastic model ANICREEP was adopted to accommodate the time effect for clays. The identified MAP parameters agree well with values measured in oedometer and triaxial tests, indicating a high level of efficiency for the proposed approach and the applicability to the in situ testing.

In addition to the parametric identification by DE-TMCMC, this approach can also be used to estimate the evidence of the chosen model class conditioning on the measured data, a key component for Bayesian model class selection and model averaging [64, 65].

References

1. Vermeer P (1978) A double hardening model for sand. Geotechnique 28(4):413–433
2. Jin Y-F, Yin Z-Y, Shen S-L, Hicher P-Y (2016) Selection of sand models and identification of parameters using an enhanced genetic algorithm. Int J Numer Anal Methods Geomech 40(8):1219–1240
3. Kolymbas D (1991) An outline of hypoplasticity. Arch Appl Mech 61(3):143–151
4. Roscoe KH, Burland J (1968) On the generalized stress-strain behaviour of wet clay. Engineering Plasticity. Cambridge University Press, Cambridge, UK, pp 535–609
5. Jefferies M (1993) Nor-Sand: a simle critical state model for sand. Geotechnique 43(1):91–103
6. Yu H (1998) CASM: A unified state parameter model for clay and sand. Int J Numer Anal Methods Geomech 22(8):621–653
7. Gajo A, Wood M (1999) Severn-Trent sand: a kinematic-hardening constitutive model: the q–p formulation. Geotechnique 49(5):595–614
8. Yao Y, Hou W, Zhou A (2009) UH model: three-dimensional unified hardening model for overconsolidated clays. Geotechnique 59(5):451–469
9. Yao Y, Sun D, Luo T (2004) A critical state model for sands dependent on stress and density. Int J Numer Anal Methods Geomech 28(4):323–337
10. Yao Y, Sun D, Matsuoka H (2008) A unified constitutive model for both clay and sand with hardening parameter independent on stress path. Comput Geotech 35(2):210–222
11. Yao Y-P, Kong L-M, Zhou A-N, Yin J-H (2014) Time-dependent unified hardening model: Three-dimensional elastoviscoplastic constitutive model for clays. J Eng Mech 141(6):04014162
12. Taiebat M, Dafalias YF (2008) SANISAND: Simple anisotropic sand plasticity model. Int J Numer Anal Methods Geomech 32(8):915–948
13. Jin Y-F, Wu Z-X, Yin Z-Y, Shen JS (2017) Estimation of critical state-related formula in advanced constitutive modeling of granular material. Acta Geotech 12(6):1329–1351
14. Jin Y-F, Yin Z-Y, Shen S-L, Hicher P-Y (2016) Investigation into MOGA for identifying parameters of a critical-state-based sand model and parameters correlation by factor analysis. Acta Geotech 11(5):1131–1145
15. Yin ZY, Chang CS, Karstunen M, Hicher PY (2010) An anisotropic elastic-viscoplastic model for soft clays. Int J Solids Struct 47(5):665–677
16. Yin ZY, Karstunen M, Chang CS, Koskinen M, Lojander M (2011) Modeling time-dependent behavior of soft sensitive clay. J Geotechn Geoenviron Eng 137(11):1103–1113
17. Mašín D (2005) A hypoplastic constitutive model for clays. Int J Numer Anal Methods Geomech 29(4):311–336
18. Wu W, Bauer E, Kolymbas D (1996) Hypoplastic constitutive model with critical state for granular materials. Mech Mater 23(1):45–69
19. Wang S, Wu W, Yin Z-Y, Peng C, He X-Z (2018) Modelling time-dependent behaviour of granular material with hypoplasticity. Int J Numer Anal Methods Geomech 42(12):1331–1345
20. Von Wolffersdorff PA (1996) A hypoplastic relation for granular materials with a predefined limit state surface. Mec Cohes Frict Mater Int J Exp Model Comput Mater Struct 1(3):251–271
21. Kolymbas D (1985) A generalized hypoelastic constitutive law. In: Proceedings of the XI International Conference on Soil Mechanics and Foundation Engineering. Balkema, Rotterdam, San Francisco, p 2626

22. Chang CS, Hicher PY (2005) An elasto-plastic model for granular materials with microstructural consideration. Int J Solids Struct 42(14):4258–4277
23. Yin ZY, Chang CS, Hicher PY, Karstunen M (2009) Micromechanical analysis of kinematic hardening in natural clay. Int J Plast 25(8):1413–1435
24. Yin ZY, Chang CS, Hicher PY (2010) Micromechanical modelling for effect of inherent anisotropy on cyclic behaviour of sand. Int J Solids Struct 47(14–15):1933–1951
25. Yin Z-Y, Zhao J, Hicher P-Y (2014) A micromechanics-based model for sand-silt mixtures. Int J Solids Struct 51(6):1350–1363
26. Yin Z, Chang C, Hicher P, Karstunen M (2008) Microstructural modeling of rate-dependent behavior of soft soil
27. Yin ZY, Chang CS (2009) Microstructural modelling of stress-dependent behaviour of clay. Int J Solids Struct 46(6):1373–1388
28. Xiong H, Nicot F, Yin Z (2017) A three-dimensional micromechanically based model. Int J Numer Anal Methods Geomech 41(17):1669–1686
29. Shen SL, Xu YS (2011) Numerical evaluation of land subsidence induced by groundwater pumping in Shanghai. Can Geotech J 48(9):1378–1392
30. Wu H-N, Shen S-L, Yang J (2017) Identification of tunnel settlement caused by land subsidence in soft deposit of Shanghai. J Perform Constr Facil 31(6):04017092
31. Xu Y-S, Ma L, Shen S-L, Sun W-J (2012) Evaluation of land subsidence by considering underground structures that penetrate the aquifers of Shanghai. China. Hydrol J 20(8):1623–1634
32. Shen S-L, Wu Y-X, Misra A (2017) Calculation of head difference at two sides of a cut-off barrier during excavation dewatering. Comput Geotech 91:192–202
33. Ren D-J, Shen S-L, Arulrajah A, Wu H-N (2018) Evaluation of ground loss ratio with moving trajectories induced in DOT tunnelling. Can Geotech J 55(6):894–902
34. Jin Y-F, Yin Z-Y, Wu Z-X, Daouadji A (2018) Numerical modeling of pile penetration in silica sands considering the effect of grain breakage. Finite Elem Anal Des 144:15–29
35. Jin Y-F, Yin Z-Y, Wu Z-X, Zhou W-H (2018) Identifying parameters of easily crushable sand and application to offshore pile driving. Ocean Eng 154:416–429
36. Yin Z-Y, Jin Y-F, Shen JS, Hicher P-Y (2018) Optimization techniques for identifying soil parameters in geotechnical engineering: comparative study and enhancement. Int J Numer Anal Methods Geomech 42(1):70–94
37. Jin Y-F, Yin Z-Y, Shen S-L, Zhang D-M (2017) A new hybrid real-coded genetic algorithm and its application to parameters identification of soils. Inverse Probl Sci Eng 25(9):1343–1366
38. Yin Z-Y, Jin Y-F, Shen S-L, Huang H-W (2017) An efficient optimization method for identifying parameters of soft structured clay by an enhanced genetic algorithm and elastic–viscoplastic model. Acta Geotech 12(4):849–867
39. Knabe T, Datcheva M, Lahmer T, Cotecchia F, Schanz T (2013) Identification of constitutive parameters of soil using an optimization strategy and statistical analysis. Comput Geotech 49:143–157
40. Levasseur S, Malécot Y, Boulon M, Flavigny E (2008) Soil parameter identification using a genetic algorithm. Int J Numer Anal Methods Geomech 32(2):189–213
41. Papon A, Riou Y, Dano C, Hicher PY (2012) Single-and multi-objective genetic algorithm optimization for identifying soil parameters. Int J Numer Anal Methods Geomech 36(5):597–618
42. Zhang X, Srinivasan R, Bosch D (2009) Calibration and uncertainty analysis of the SWAT model using Genetic Algorithms and Bayesian Model Averaging. J Hydrol 374(3):307–317
43. Cao Z, Wang Y (2014) Bayesian model comparison and selection of spatial correlation functions for soil parameters. Struct Safety 49:10–17
44. Miro S, König M, Hartmann D, Schanz T (2015) A probabilistic analysis of subsoil parameters uncertainty impacts on tunnel-induced ground movements with a back-analysis study. Comput Geotech 68:38–53
45. Ritto T, Nunes L (2015) Bayesian model selection of hyperelastic models for simple and pure shear at large deformations. Comput Struct 156:101–109

46. Akeju OV, Senetakis K, Wang Y (2017) Bayesian parameter identification and model selection for normalized modulus reduction curves of soils. J Earthq Eng, 1–29
47. Zhang L, Li D-Q, Tang X-S, Cao Z-J, Phoon K-K (2017) Bayesian model comparison and characterization of bivariate distribution for shear strength parameters of soil. Comput Geotech 95:110–118
48. Cividini A, Maier G, Nappi A (1983) Parameter estimation of a static geotechnical model using a Bayes' approach. Int J Rock Mech Min Sci Geomech Abstr, 215–226
49. Murakami A, Shinmura H, Ohno S, Fujisawa K (2017) Model identification and parameter estimation of elastoplastic constitutive model by data assimilation using the particle filter. Int J Numer Anal Methods Geomech
50. Hsiao EC, Schuster M, Juang CH, Kung GT (2008) Reliability analysis and updating of excavation-induced ground settlement for building serviceability assessment. J Geotech Geoenviron Eng 134(10):1448–1458
51. Hsein Juang C, Luo Z, Atamturktur S, Huang H (2012) Bayesian updating of soil parameters for braced excavations using field observations. J Geotech Geoenviron Eng 139(3):395–406
52. Most T (2010) Identification of the parameters of complex constitutive models: least squares minimization vs. Bayesian updating. Reliab Optim Struct Syst, 119
53. Honjo Y, Wen-Tsung L, Guha S (1994) Inverse analysis of an embankment on soft clay by extended Bayesian method. Int J Numer Anal Methods Geomech 18(10):709–734
54. Qi X-H, Zhou W-H (2017) An efficient probabilistic back-analysis method for braced excavations using wall deflection data at multiple points. Comput Geotech 85:186–198
55. Ancey C (2005) Monte Carlo calibration of avalanches described as Coulomb fluid flows. Philos Trans R Soc Math Phys Eng Sci 363(1832):1529–1550
56. Eckert N, Parent E, Richard D (2007) Revisiting statistical–topographical methods for avalanche predetermination: Bayesian modelling for runout distance predictive distribution. Cold Reg Sci Technol 49(1):88–107
57. Fischer J-T, Kofler A, Fellin W, Granig M, Kleemayr K (2015) Multivariate parameter optimization for computational snow avalanche simulation. J Glaciol 61(229):875–888
58. Gauer P, Medina-Cetina Z, Lied K, Kristensen K (2009) Optimization and probabilistic calibration of avalanche block models. Cold Reg Sci Technol 59(2–3):251–258
59. Hellweger V, Fischer J-T, Kofler A, Huber A, Fellin W, Oberguggenberger M (2016) Stochastic methods in operational avalanche simulation-from back calculation to prediction. In: International snow science workshop 2016 proceedings. Montana State University Library, Colorado, USA, pp 1357–1381
60. Zhang J, Zhang LM, Tang WH (2009) Bayesian framework for characterizing geotechnical model uncertainty. J Geotech Geoenviron Eng 135(7):932–940
61. Zhang R, Mahadevan S (2000) Model uncertainty and Bayesian updating in reliability-based inspection. Struct Saf 22(2):145–160
62. Zhang L, Zuo Z, Ye G, Jeng D, Wang J (2013) Probabilistic parameter estimation and predictive uncertainty based on field measurements for unsaturated soil slope. Comput Geotech 48:72–81
63. Hastings WK (1970) Monte Carlo sampling methods using Markov chains and their applications. Biometrika 57(1):97–109
64. Lee S-H, Song J (2017) System identification of spatial distribution of structural parameters using modified transitional Markov chain Monte Carlo method. J Eng Mech 143(9):04017099
65. Ching J, Chen Y-C (2007) Transitional Markov chain Monte Carlo method for Bayesian model updating, model class selection, and model averaging. J Eng Mech 133(7):816–832
66. Angelikopoulos P, Papadimitriou C, Koumoutsakos P (2015) X-TMCMC: adaptive kriging for Bayesian inverse modeling. Compute Methods Appl Mech Eng 289:409–428
67. Ortiz GA, Alvarez DA, Bedoya-Ruíz D (2015) Identification of Bouc–Wen type models using the transitional Markov chain Monte Carlo method. Comput Struct 146:252–269
68. Betz W, Papaioannou I, Straub D (2016) Transitional markov chain Monte Carlo: observations and improvements. J Eng Mech 142(5):04016016
69. Ching J, Wang J-S (2016) Application of the transitional Markov chain Monte Carlo algorithm to probabilistic site characterization. Eng Geol 203:151–167

70. Yuen K-V (2010) Bayesian methods for structural dynamics and civil engineering. Wiley
71. He J, Jones JW, Graham WD, Dukes MD (2010) Influence of likelihood function choice for estimating crop model parameters using the generalized likelihood uncertainty estimation method. Agric Syst 103(5):256–264
72. Chopin N (2002) A sequential particle filter method for static models. Biometrika 89(3):539–552
73. Vrugt JA (2016) Markov chain Monte Carlo simulation using the DREAM software package: theory, concepts, and MATLAB implementation. Environ Model Softw 75:273–316
74. Angelikopoulos P, Papadimitriou C, Koumoutsakos P (2012) Bayesian uncertainty quantification and propagation in molecular dynamics simulations: a high performance computing framework. J Chem Phys 137(14):144103
75. Hadjidoukas PE, Angelikopoulos P, Papadimitriou C, Koumoutsakos P (2015) Π4U: a high performance computing framework for Bayesian uncertainty quantification of complex models. J Comput Phys 284:1–21
76. Tan F, Zhou W-H, Yuen K-V (2016) Modeling the soil water retention properties of same-textured soils with different initial void ratios. J Hydrol 542:731–743
77. Verdugo R, Ishihara K (1996) The steady state of sandy soils. Soils Found 36(2):81–91
78. Wu Z-X, Yin Z-Y, Jin Y-F, Geng X-Y (2017) A straightforward procedure of parameters determination for sand: a bridge from critical state based constitutive modelling to finite element analysis. Eur J Environ Civ Eng, 1–23
79. Mašín D (2015) The influence of experimental and sampling uncertainties on the probability of unsatisfactory performance in geotechnical applications. Géotechnique 65(11):897–910
80. Lee Goh A, Fahey M (1991) Application of a 1-dimensional cavity expansion model to pressuremeter and piezocone tests in clay. In: Proceeding of the seventh international conference on computer methods and advances in geomechanics, Cairns, pp 255–260
81. Low HE (2009) Performance of penetrometers in deepwater soft soil characterisation. University of Western Australia
82. Yin Z-Y, Zhu Q-Y, Yin J-H, Ni Q (2014) Stress relaxation coefficient and formulation for soft soils. Géotechn Lett, 45–51
83. Yin Z-Y, Yin J-H, Huang H-W (2015) Rate-dependent and long-term yield stress and strength of soft Wenzhou marine clay: experiments and modeling. Mar Georesour Geotechnol 33(1):79–91
84. Jin Y-F, Yin Z-Y, Zhou W-H, Huang H-W (2019) Multi-objective optimization-based updating of predictions during excavation. Eng Appl Artif Intell 78:102–123

Chapter 8
Estimation of Pullout Shear Strength of Grouted Soil Nails

Abstract Pullout shear resistance of grouted soil nails is affected by many factors, such as overburden pressure, grouting pressure, soil dilation and degree of saturation of soil. Due to the complexity of pullout mechanism, these factors have not been well incorporated in current design methods. In this chapter, Bayesian analysis is performed to investigate the relative importance of several key factors and to build a new design formula to estimate maximum pullout shear stress of grouted soil nails. By using a series of laboratory soil nail pullout test data, Bayesian inference is performed to select the predictive formula with suitable complexity and to identify its parameters. It has been found that the most important factors are the degree of saturation and the product of grouting pressure and overburden pressure. It is shown that the proposed optimal model exhibits significantly stronger correlation with the measurements than the existing effective stress method for the laboratory and field tests.

8.1 Introduction

Soil nailing is an effective technique to stabilize new steep cuts or to upgrade the existing soil slopes. In the past few decades, soil nailing has been widely used owing to its technical and economical advantages over conventional stabilization methods. Drilling and grouting is the most popular construction method in Hong Kong and in many other places. A grouted soil nail is normally constructed by placing a steel bar in the middle of a drillhole in a slope and then grouting with cement slurry. At present, many design methods for soil nailed system are available in different countries and regions. Most of them were derived from classical slope stability theory with the consideration of additional resisting forces provided by soil nails. The most commonly used methods include the French method [1], German method [2], Davis method [3], UK method [4], FHWA design method [5, 6] and Hong Kong design method [7, 8]. These methods vary in the safety factor, the assumed slope failure surface, the determination method of ultimate pullout resistance, and so on. Methods for determining soil nail ultimate pullout resistance include the effective stress method used in Hong Kong and UK, empirical correlation with SPT-N values

W. Zhou et al., *Practice of Bayesian Probability Theory in Geotechnical Engineering*,
https://doi.org/10.1007/978-981-15-9105-1_8

used in Japan, correlation with pressuremeter tests used in France and correlation with soil types adopted in USA [9]. Except the effective stress method, all the other methods are empirical and thus only suitable under specific conditions. Pun and Shiu [9] compared hundreds of pullout test results with the estimated pullout resistance and found that the actual pullout resistance was generally several times higher than the estimation by the Hong Kong design method, that is, the effective stress method.

It is well known that soil nail pullout resistance is influenced by a number of factors, such as the method of installation, normal stress, overburden pressure (OP), degree of saturation (Sr), grouting pressure (GP) and soil properties. Laboratory and field pullout tests have been conducted by many researchers to study the soil nail pullout resistance. Schlosser [10] suggested that the normal pressure is dependent on the overburden pressure for driven nails but its value may be very low for grouted nails. Pradhan et al. [11] conducted laboratory study on soil nails in loose sandy fills and found that pullout resistance of the soil nails increases with overburden pressure. Su [12] and Su et al. [13] observed significant stress release in the soil around the drillhole due to drilling process and the pullout resistance is independent of the overburden pressure for normally grouted soil nails. Furthermore, the pullout resistance of soil nails is strongly influenced by the degree of saturation and dilation of the shearing zone [14, 15]. Yin et al. [16] analyzed a few laboratory pullout test records in unsaturated soil and concluded that pullout resistance increases with grouting pressure and the relationship is approximately linear. Recently, Yin and Zhou [17] indicated that the shear stress on the soil nail interface is influenced by the multiple effects between the overburden pressure and grouting pressure.

However, the aforementioned research provided only qualitative description of the relationship between the soil nail pullout resistance and one or two of its many influencing factors. Due to the complexity of the mechanism of soil nail pullout resistance, the general formula considering multiple influencing factors is rare in the literature according to the authors' knowledge. Luo et al. [18] proposed a pullout model of a soil nail considering the effects of soil dilation and overburden pressure on shear stress of nail–soil interface. Zhang et al. [19] studied the uncertainties of soil nail pullout resistance with consideration of overburden pressure, grout length, soil suction and soil dilation. Yin et al. [20] proposed a simple method to calculate the maximum shear stress of nail–soil interface, with consideration of the following influencing factors: soil dilation, overburden pressure and grouting pressure. In addition, Xue et al. [21] developed a finite element program to study the interaction mechanism between soil mass and nails based on the modified Goodman element model.

In order to construct a reliable relationship for soil nail pullout shear strength, systematic analysis is performed using Bayesian model class selection in conjunction with multivariate linear regression. A series of laboratory soil nail pullout test data on the completely decomposed granite (CDG) soil are used to analyze the model class candidates by Bayesian method. The most plausible formula for the maximum pullout shear stress is proposed. It is shown that the proposed model provides significantly better estimation than the well-known effective stress method.

8.2 Construction of Predictive Formula Candidates

Su et al. [14, 13] reported a laboratory soil nail pullout test study and found that the overburden pressure has no significant influence on soil nail pullout resistance under unsaturated condition without grouting pressure. For the influence of degree of saturation, it was concluded that the maximum pullout shear stress reached the highest values when the degree of saturation was between 50 and 75% [14]. Yin et al. [16] studied the influence of grouting pressure on soil nail pullout resistance based on limited laboratory pullout tests under unsaturated condition ($S_r = 50\%$) and low grouting pressure (130 kPa or below). In addition, Yin and Zhou [17] carried out a series of laboratory pullout tests under nearly saturated condition with different combinations of grouting pressure and overburden pressure. They observed that there were multiple effects by the grouting pressure and overburden pressure on soil nail pullout resistance [22].

By incorporating the observations of different influencing factors discussed above, the "full model class" for the estimation of the maximum pullout shear stress ($\bar{\tau}_{\max}$) is proposed as follows:

$$\bar{\tau}_b = b_1 \times GP + b_2 \times S_r^2 + b_3 \times S_r + b_4 \times OP + b_5 \times GP \times OP + b_6 \quad (8.1)$$

where $\bar{\tau}_b$ is the estimation value of $\bar{\tau}_{\max}$. The influence of *GP* is different in unsaturated soil and saturated soil conditions as mentioned in the above section. Here the contribution from the grouting pressure *GP* alone is represented by a linear relationship in the first term. The fifth term, as discussed later, also includes the effect of *GP*. For the degree of saturation S_r, a quadratic function is used to represent the relationship between $\bar{\tau}_b$ and S_r, that is, the second and third terms in Eq. (8.1). This is necessary to model the behavior that the maximum occurred in the mid range of the degree of saturation observed in Su et al. [14]. The effect of overburden pressure *OP* is considered in the fourth term. Although it was discussed in Su et al. [14] and [13] that the overburden pressure alone has no significant effect on the maximum pullout shear stress in the cases without pressure grouting, this term is included to validate whether the Bayesian model class selection method is successful in selecting a predictive formula without this term. Yin and Zhou [17] reported that the soil nail pullout resistance is hardly or slightly dependent on *OP* when *GP* is low and for $S_r = 98\%$. However, the resistance increases with *OP* when *GP* is considerable. They suggested that there are multiple effects of *GP* and *OP* on the soil nail pullout resistance under saturated soil condition. Therefore, the product of the overburden pressure and grouting pressure: $GP \times OP$ is taken as fifth influencing factor candidate to reconfirm from a Bayesian perspective if this observation is significant. Finally, the constant term b_6 is used to account for any possible offset.

In order to propose a reliable (neither over-fitted nor under-fitted) predictive model class, other model class candidates are constructed based on the model in Eq. (8.1). These model class candidates are generated by including different combinations of terms on the right hand side of Eq. (8.1), for example, $\bar{\tau}_b = b_1 \times GP + b_5 \times GP \times OP$.

There are $2^6 - 1 = 63$ model class candidates to be analyzed by Bayesian method. The selected model class possesses rational balance between data fitting capability and robustness to measurement noise and modeling error.

8.3 Bayesian Approach

8.3.1 Bayesian Model Class Selection

As stated in Sect. 1.2.2, the probability/plausibility of a model class conditional on a database D can be obtained by using the Bayes' theorem as follows [23]:

$$P(C_j|D) = \frac{p(D|C_j) \cdot P(C_j)}{p(D)}, j = 1, 2, \ldots, N_C \tag{8.2}$$

The quantity $p(D|C_j)$ is called the evidence of the model class C_j provided by the data D and it can be obtained using the theorem of total probability:

$$p(D|C_j) = \int_{\Theta} p(D|\boldsymbol{\theta}, C_j)p(\boldsymbol{\theta}|C_j)d\boldsymbol{\theta}, j = 1, 2, \ldots, N_C \tag{8.3}$$

where $\boldsymbol{\theta}$ is the parameter vector in the parameter space $\Theta \subset \Re^{N_j}$ and it is defined in each model class (C_j); $p(\boldsymbol{\theta}|C_j)$ is the prior probability density function (*PDF*) of the uncertain parameter vector $\boldsymbol{\theta}$; $p(D|\boldsymbol{\theta}, C_j)$ is the likelihood function expressing the level of data fitting. It should be noted that one of the main advantages of Bayesian method is that it does not require any ad hoc penalty in the selection of the best model class, such as the Akaike's information criterion. Another advantage is that it allows for the uncertainty quantification even with one set of measurement only.

8.3.2 Multivariate Linear Model

A linear regression problem can be represented by the following relationship:

$$y = y(X; b, C_j) = b_1 x_1 + b_2 x_2 + \cdots + b_{N_b} x_{N_b} \tag{8.4}$$

where y is the quantity of concern (e.g., the maximum pullout shear stress of soil nail in this study); N_b is the number of uncertain coefficients; the variables $x_1, x_2, \ldots, x_{N_b}$ in the measured variable vector X corresponds to the measurements and constant 1 on the right of Eq. (8.1). Use Y to denote the measurement of y and it is assumed that the difference can be adequately modeled as Gaussian random variable:

$$Y = y + \varepsilon \tag{8.5}$$

where ε is a Gaussian random variable with zero mean and variance σ_ε^2 referred hereafter as prediction error variance. It is used to represent the measurement noise and modeling error. The uncertain parameter vector $\boldsymbol{\theta} = \left[b^T, \sigma_\varepsilon^2\right]^T$ includes the coefficients $b_1, b_2, \ldots, b_{N_b}$ and the prediction-error variance σ_ε^2.

Referring to Eqs. (8.1), (8.4) and (8.5), the measurement Y is $\bar{\tau}_{\max}$, and the measured variable vector X includes the corresponding quantities of *GP*, S_r^2, S_r, *OP*, $GP \times OP$ and 1 for the model class. By assuming statistically independent prediction errors in different records, the likelihood function is readily obtained:

$$p(D|\boldsymbol{\theta}, C_j) = (2\pi)^{-\frac{N}{2}}\sigma_\varepsilon{}^{-N}\exp[-\frac{N}{2\sigma_\varepsilon^2}J_g(b; D, C_j)] \tag{8.6}$$

where N is the number of measured records. The goodness-of-fit function $J_g(b; D, C_j)$ represents the degree of data fitting and it is given by:

$$J_g(b; D, C_j) = \frac{1}{N}\sum_{n=1}^{N}[Y(n) - \sum_{l=1}^{N_b} b_l x_l(n)]^2 \tag{8.7}$$

Next, the prior *PDF* of the uncertain parameters is assumed separable:

$$p(\boldsymbol{\theta}|C_j) = p(b, \sigma_\varepsilon^2|C_j) = p(b|C_j)p(\sigma_\varepsilon^2|C_j) \tag{8.8}$$

For the uncertain coefficient vector *b*, Gaussian prior *PDF* is taken with mean b^η and covariance matrix V:

$$p(b|C_j) = (2\pi)^{-\frac{N_b}{2}}|V|^{-\frac{1}{2}}\exp[-\frac{1}{2}(b - b^\eta)^T V^{-1}(b - b^\eta)] \tag{8.9}$$

For the prediction-error variance, the conjugate prior distribution (the inverse Gamma distribution in this case) is used:

$$p(\sigma_\varepsilon^2|C_j) = \frac{\beta_0^{\alpha_0}}{\sigma_\varepsilon{}^{2(\alpha_0+1)}\Gamma(\alpha_0)}\exp(-\frac{\beta_0}{\sigma_\varepsilon^2}) \tag{8.10}$$

where α_0 and β_0 are the shape and scale parameters. Therefore, the product of the prior *PDF* and likelihood function in the integrand of Eq. (8.3) is given by:

$$\begin{aligned} p(\boldsymbol{\theta}|C_j)p(D|\boldsymbol{\theta}, C_j) &= (2\pi)^{-\frac{N_b+N}{2}}|V|^{-\frac{1}{2}}\frac{\beta_0^{\alpha_0}}{\Gamma(\alpha_0)}\sigma_\varepsilon{}^{-2(\alpha_0+1)-N}\exp\left(-\frac{\beta_0}{\sigma_\varepsilon^2}\right) \\ &\times \exp\left(-\tfrac{1}{2}(b - b^\eta)^T V^{-1}(b - b^\eta) - \tfrac{1}{2\sigma_\varepsilon^2}\sum_{n=1}^{N}\left[Y(n) - \sum_{l=1}^{N_b} b_l x_l(n)\right]^2\right) \end{aligned} \tag{8.11}$$

The negative logarithm of this product is defined as the objective function, that is, $J(\boldsymbol{\theta}) = -\ln[p(\boldsymbol{\theta}|C_j)p(D|\boldsymbol{\theta}, C_j)]$. Obtaining the most probable values of the parameters involves a nonlinear optimization problem of the objective function. However, it can be solved iteratively in a series of two linear optimizations as follows. Given any positive value of σ_ε^2, the conditional optimal coefficient vector b^* can be obtained by minimizing the objective function $J(\boldsymbol{\theta})$. This can be done by solving the linear algebraic equation $\partial J(\boldsymbol{\theta})/\partial b = 0$ and the solution is given by:

$$b^*\left(\sigma_\varepsilon^2\right) = \left(N\sigma_\varepsilon^{-2}A + V^{-1}\right)^{-1} \cdot \left[N\sigma_\varepsilon^{-2} \begin{bmatrix} \frac{1}{N}\sum_{n=1}^{N} x_1(n)Y(n) \\ \frac{1}{N}\sum_{n=1}^{N} x_2(n)Y(n) \\ \vdots \\ \frac{1}{N}\sum_{n=1}^{N} x_{N_b}(n)Y(n) \end{bmatrix} + V^{-1}b^{\eta} \right] \tag{8.12}$$

where A is an $N_b \times N_b$ symmetric matrix and its $\left(l, l'\right)$ element is given by:

$$A_{ll'} = \frac{1}{N}\sum_{n=1}^{N} x_l(n)x_{l'}(n) \tag{8.13}$$

Similarly, for a given coefficient vector b, the conditional optimal prediction-error variance σ_ε^2 can be obtained by solving $\partial J(\boldsymbol{\theta})/\partial\sigma_\varepsilon^2 = 0$:

$$\sigma_\varepsilon^{2*}(b) = \frac{NJ_g\left(b; D, C_j\right) + 2\beta_0}{N + 2(\alpha_0 + 1)} \tag{8.14}$$

Therefore, the optimal parameters can be obtained by starting with an arbitrary positive value for σ_ε^2 and then iterate between Eqs. (8.12) and (8.14) until convergence.

For uncertainty quantification, the covariance matrix of uncertain parameter vector $\boldsymbol{\theta}^*$ can be determined by taking the inverse of the Hessian matrix, that is, $\sum_{\boldsymbol{\theta}^*} = \mathbf{H}\left(\boldsymbol{\theta}^*\right)^{-1}$, and it represents the uncertainty of identified parameters. The Hessian matrix $\mathbf{H}\left(\boldsymbol{\theta}^*\right)$ consists of the second derivatives of objective function $J(\boldsymbol{\theta})$:

$$\mathbf{H}_{ll'}\left(\boldsymbol{\theta}^*\right) = -\frac{\partial^2}{\partial\boldsymbol{\theta}_l\partial\boldsymbol{\theta}_{l'}}\ln\left[p\left(\boldsymbol{\theta}|C_j\right)p\left(D|\boldsymbol{\theta}, C_j\right)\right]|_{\boldsymbol{\theta}=\boldsymbol{\theta}^*} \tag{8.15}$$

It can also be written in the following form:

$$\mathbf{H}(\boldsymbol{\theta}^*) = \left[\begin{array}{ccccc} \frac{\partial^2 J}{\partial b_1^2} & -\frac{\partial^2 J}{\partial b_1 \partial b_2} & \cdots & \frac{\partial^2 J}{\partial b_1 \partial b_{N_b}} & \frac{\partial^2 J}{\partial b_1 \partial \sigma_\varepsilon^2} \\ \frac{\partial^2 J}{\partial b_2 \partial b_1} & -\frac{\partial^2 J}{\partial b_2^2} & \cdots & \frac{\partial^2 J}{\partial b_2 \partial b_{N_b}} & \frac{\partial^2 J}{\partial b_2 \partial \sigma_\varepsilon^2} \\ \vdots & \vdots & \ddots & \vdots & \vdots \\ \frac{\partial^2 J}{\partial b_{N_b} \partial b_1} & -\frac{\partial^2 J}{\partial b_{N_b} \partial b_2} & \cdots & \frac{\partial^2 J}{\partial b_{N_b}^2} & \frac{\partial^2 J}{\partial b_{N_b} \partial \sigma_\varepsilon^2} \\ \frac{\partial^2 J}{\partial \sigma_\varepsilon^2 \partial b_1} & -\frac{\partial^2 J}{\partial \sigma_\varepsilon^2 \partial b_2} & \cdots & \frac{\partial^2 J}{\partial \sigma_\varepsilon^2 \partial b_{N_b}} & \frac{\partial^2 J}{\partial (\sigma_\varepsilon^2)^2} \end{array}\right]\Bigg|_{\boldsymbol{\theta}=\boldsymbol{\theta}^*} \tag{8.16}$$

Finally, the evidence integral in Eq. (8.3) can be computed approximately using the following formula [24, 25]:

$$p(D|C_j) \approx \frac{(\beta_0)^{\alpha_0} \Gamma(\alpha^*)}{\sqrt{(2\pi)^N N^{N_b} |A| \cdot |V| (\beta^*)^{\alpha*} \Gamma(\alpha_0)}} \exp\left[-\frac{1}{2}(b^* - b^\eta)^T V^{-1} (b^* - b^\eta)\right] \tag{8.17}$$

where α^*, β^* are the identified shape and scale parameters, respectively.

On the other hand, based on the Bayes' theorem, the exact expression for $p(D|C_j)$ is given by:

$$p(D|C_j) = \frac{p(D|\boldsymbol{\theta}^*, C_j) p(\boldsymbol{\theta}^*|C_j)}{p(\boldsymbol{\theta}^*|D, C_j)} \tag{8.18}$$

where the ratio $p(\boldsymbol{\theta}^*|C_j)/p(\boldsymbol{\theta}^*|D, C_j)$ can be considered as the Ockham factor. This factor is interpreted as a measure of robustness of the model class and its value decreases exponentially with the number of uncertain parameters in the model class.

Equation (8.18) will be used to rank different predictive formulae with different terms and the one with the highest plausibility will be selected as the proposed design formula for maximum pullout shear stress estimation. The proposed formula by this method has a good tradeoff between the data-fitting capability and the robustness to model error. Readers may refer to Appendix A for the MATLAB code for the linear model class selection.

8.4 Analysis of Laboratory Test Data and Proposed Design Formula

The laboratory soil nail pullout tests used in this study were conducted in the pullout box designed by Yin and Su [26] on CDG soils. The soil nails (including grout) are 100 mm in diameter and approximately 1.2 m in length. Thirty tests were performed under different combinations of overburden pressure ($OP = 40$, 80, 120, 200, 240, 300 and 350 kPa), grouting pressure ($GP = 0$, 80, 130, 250 and 300 kPa) and degree of saturation ($S_r = 38$, 50, 75 and 98%). In these tests, all soil nails were pulled out at the end of a five-day curing period after the grouting process. The measured maximum pullout shear stress ($\bar{\tau}_{max}$) during pullout was considered as the average maximum shear stress of the soil nail interface during pullout. The relationship between $\bar{\tau}_{max}$ and the corresponding pullout resistance per length $\hat{T}$ can be written as:

$$\hat{T} = \bar{\tau}_{max} \pi D_m \tag{8.19}$$

where D_m is the diameter of grouted soil nail which is the average of the diameter at the head, middle and end of each soil nail after pullout.

8.4.1 Influence of Different Prior PDFs

In this study, uniform prior plausibility $P(C_j)$ in Eq. (8.2) is used for all model class candidates: $P(C_j) = 1/63$ ($j = 1, 2,...,63$). The uncertain parameters in $\boldsymbol{\theta} = \left[b^T, \sigma_\varepsilon^2\right]^T$ include the corresponding coefficients $b_1, b_2, \ldots, b_6$ and the prediction-error variance σ_ε^2. Large prior variances of $b_1, b_2, \ldots, b_6$ are taken so that they do not affect the identification results significantly. The distribution parameters α_0 and β_0 of prediction-error variance σ_ε^2 are taken as 1 and 1000, respectively.

In order to examine the influence of the choice of prior *PDFs*, two prior *PDFs* are adopted to analyze the laboratory data and they are shown in Table 8.1. The notation $N(\mu, \sigma)$ denotes the Gaussian distribution with mean μ and standard deviation σ. It can be seen that the standard deviations of prior *PDF* 1 are two times of those in prior *PDF* 2. The marginal prior and posterior *PDFs* of parameter b_5 using these two prior *PDFs* are depicted in Fig. 8.1. It can be seen that the two posterior distributions are on top of each other. Although the two prior distributions are substantially different, the results turn out that the corresponding posterior mean values and standard deviations are virtually the same (i.e., 0.0011 and 0.0001, respectively). In other words, the choice of the prior distribution has virtually no effect on the parametric identification results and its distribution interval is sufficiently wide. On the other hand, compared with the prior uncertainty, it is clearly observed that the posterior uncertainty of the parameter is significantly reduced due to the information gained from the data. Furthermore, it is also found that the most plausible model classes for both prior *PDFs* are the same, that is, $\bar{\tau}_b = b_2 \times S_r^2 + b_3 \times S_r + b_5 \times GP \times OP$.

Finally, the maximum pullout shear stresses predicted by the optimal models of the most plausible model classes with the two prior *PDFs* are compared in Fig. 8.2. The coefficient of determination R^2 between these two sets of values is virtually unity,

Table 8.1 Prior PDFs of uncertain coefficients

	b1	b2	b3	b4	b5	b6
Prior PDF 1 $N(\mu, \sigma)$	N(0,1)	N(0,1000)	N(0,1000)	N(0,1)	N(0,0.004)	N(0,300)
Prior PDF 2 $N(\mu, \sigma)$	N(0,0.5)	N(0,500)	N(0,500)	N(0,0.5)	N(0,0.002)	N(0,150)

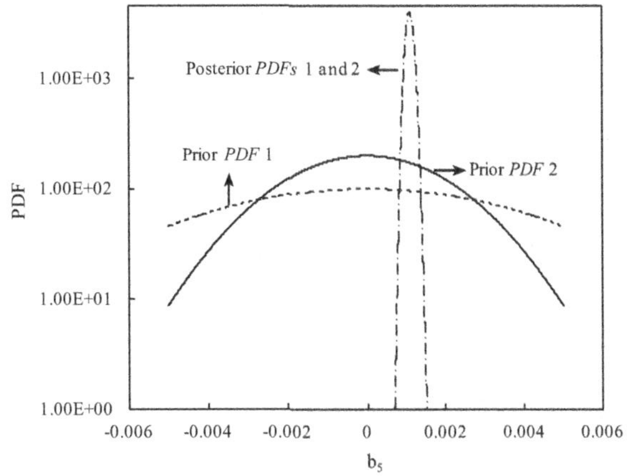

Fig. 8.1 Posterior PDFs of parameter b5 with two different prior PDFs

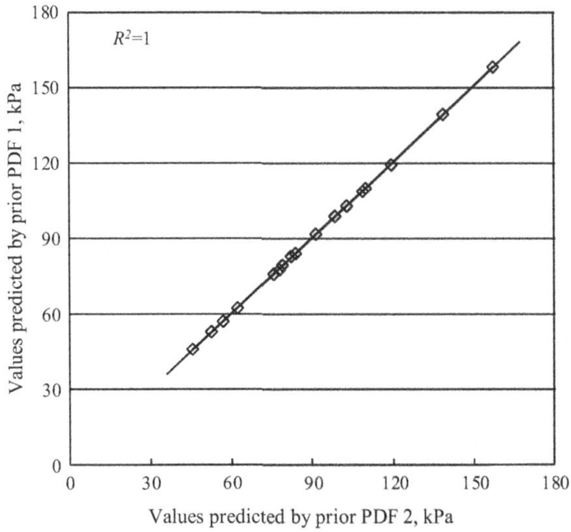

Fig. 8.2 Comparison of maximum pullout shear stress predicted using different prior *PDF*s

implying that the predicted results using these two prior *PDFs* are almost identical. Therefore, it can be concluded that prediction of maximum pullout shear stress is insensitive to the choice of prior distribution.

8.4.2 Proposed Model

Since the identification results are virtually identical using the two prior distributions, the results presented below correspond to prior *PDF* 1 only. Table 8.2 shows the model class selection results of the top 20 model classes. The first column of the

Table 8.2 Model class selection results

Number	Parameters	σ_ε^2	$\ln p(D\|\boldsymbol{\theta}^*, C_j)$	$\ln o_j$	$P(C_j\|D)$
1	2 3 5	277	−125.33	−15.49	0.384
2	1 2 3	295	−126.46	−15.12	0.180
3	1 2 3 5	246	−123.10	−18.55	0.168
4	2 3 5 6	269	−124.79	−17.53	0.086
5	1 2 3 6	284	−125.80	−17.16	0.045
6	1 2 3 5 6	236	−122.26	−20.76	0.043
7	1 2 3 4	266	−124.57	−18.89	0.028
8	2 3 4 5	277	−125.30	−18.73	0.015
9	1 2 3 4 6	249	−123.30	−21.18	0.010
10	2 5 6	361	−130.14	−14.42	0.009
11	1 2 3 4 5	244	−122.89	−21.80	0.008
12	1 2 6	379	−130.98	−14.14	0.005
13	3 5 6	393	−131.65	−13.83	3.65E-03
14	2 3 4 5 6	269	−124.79	−20.74	3.46E-03
15	1 2 5 6	334	−128.73	−17.08	2.63E-03
16	1 2 3 4 5 6	230	−121.81	−24.09	2.42E-03
17	1 3 6	412	−132.46	−13.56	2.13E-03
18	1 3 5 6	369	−130.50	−16.38	8.98E-04
19	1 2 4 6	350	−129.60	−17.60	6.55E-04
20	2 4 5 6	361	−130.12	−17.51	4.22E-04

table shows the ranking of each model class candidate according to their posterior plausibility. In particular, model class 1 is the most plausible model class and it is selected as our proposed model. The second column shows the terms being included in each model class. For example, model class "2 3 5" refers to the predictive formula: $\bar{\tau}_b = b_2 \times S_r^2 + b_3 \times S_r + b_5 \times GP \times OP$. The values in the third and fourth column are the variance of the fitting error and the corresponding maximum log-likelihood value, respectively. It is not surprising to observe that the data-fitting capability of a model class can be enhanced by adding extra adjustable parameters, resulting in larger maximum log-likelihood value and smaller prediction-error variance value (e.g., model class "1 2 3" vs. "1 2 3 5"). The fifth column shows the log-Ockham factor, which serves as the penalty for the complexity of a model class. This factor will decrease by including additional adjustable parameter (e.g., model class "1 2 3" vs. "1 2 3 5"). As a result, the full model class (model class 16) with six adjustable parameters possesses the largest value of the maximum log-likelihood but the smallest log-Ockham factor among all 63 model class candidates. Furthermore, its posterior plausibility ranks only 16th and it is significantly smaller than other much simpler model classes (e.g., model class "2 3 5" with only three parameters). This confirms that the Bayesian model class selection approach penalizes complicated model classes without giving ad hoc penalty. In general, it is inappropriate to select the model class for future prediction according solely to the maximum log-likelihood value.

From Table 8.2, the most plausible model class is "2 3 5", that is, $\bar{\tau}_b = b_2 \times S_r^2 + b_3 \times S_r + b_5 \times GP \times OP$. The degree of saturation and the product of grouting pressure and overburden pressure are important for the estimation of maximum pullout shear stress. This result fully agrees with our early discussion in the second part of this article. Furthermore, both the first- and second-order terms involving the degree of saturation appear in all top 9 model classes, implying that S_r is the major factor. These reconfirm that Bayesian model class selection methodology is effective for geotechnical problems. For the effects of overburden pressure and grouting pressure, it is found that the most plausible model class selected by the Bayesian model class approach does not include the linear terms of *OP* and *GP*. Instead, the product of *OP* and *GP* is included which implies the complex effects of *OP* and *GP*. As observed from laboratory study, the *OP* effect is reduced by the drilling procedure [13]. The *GP* influence on the soil nail pullout resistance is due to one or a combination of three reasons [16]: (a) compaction and densification of soil surrounding the drilling hole, (b) bond strength increase and (c) great roughness of the nail surface. The Bayesian model class selection results not only present consistency with the previous observations [13, 14, 16, 17], but also reveal more insights on the relative importance and relationship of different factors on the soil nail pullout resistance.

Using the proposed iteration procedure with Eqs. (8.12) and (8.14), the optimal parameter vector $\boldsymbol{\theta}^*$ of each model class candidate can be obtained and the results are shown in Table 8.3. Meanwhile, the associated uncertainty of the parameters can be represented by the covariance matrix, and the marginal standard deviation of each parameter is shown in parentheses.

The most plausible model class is selected as the proposed design formula for the maximum pullout shear stress of soil nails:

$$\bar{\tau}_b = -284.5 \times S_r^2 + 325.5 \times S_r + 0.0011 \times GP \times OP \tag{8.20}$$

where $\bar{\tau}_b$ (in kPa) represents the estimation of maximum pullout shear stress. The values of *GP* and *OP* are also in kPa. The degree of saturation S_r uses the digital format. For example, if S_r is equal to 50%, 0.5 will be used in Eq. (8.20). Regarding the degree of saturation, the maximum occurs when $S_r = 0.57$. This is consistent with Su et al. [14] which indicated that the pullout shear stress reached its maximum when the degree of saturation is between 0.5 and 0.75. Furthermore, it can be seen from Table 8.3 that when both the first- and second-order terms of S_r are included in a model class, b_2 are always negative with b_3 positive. Therefore, the polynomial of $b_2 \times S_r^2 + b_3 \times S_r$ is convex with respect to the degree of saturation S_r.

Figure 8.3 shows the comparison between the experimental maximum pullout shear stress and the corresponding predictions by using model class "2 3 5". The large coefficient of determination R^2 (0.752) indicates a good level of data-fitting. Although it is common to use directly the most plausible model class as the predictive formula, that is, Eq. (8.20), there exists a possible more robust/reliable choice to

Table 8.3 Optimal parameters of selected predictive model class and their uncertainties

Number	b_1	b_2	b_3	b_4	b_5	b_6
1		−284.5 (22.6)	325.5 (19.8)		0.0011 (0.0001)	
2	0.30 (0.038)	−277.1 (23.1)	313.1 (20.3)			
3	0.14 (0.069)	−284.2 (21.3)	320.9 (18.8)		0.0006 (0.0002)	
4		−367.1 (93.34)	444.5 (132.0)		0.0011 (0.0001)	−38.42 (42.17)
5	0.31 (0.037)	−372.3 (96.0)	450.0 (135.6)			−44.29 (43.39)
6	0.15 (0.068)	−382.1 (87.8)	461.6 (124.0)		0.0006 (0.0002)	−45.52 (39.67)
7	0.28 (0.037)	−259.9 (23.7)	287.3 (23.5)	0.059 (0.031)		
8		−287.1 (25.6)	329.1 (25.8)	−0.0076 (0.036)	0.0011 (0.0002)	
9	0.28 (0.036)	−386.4 (90.2)	469.2 (127.5)	0.067 (0.030)		−59.85 (41.28)
10		−54.4 (11.07)			0.0010 (0.0001)	102.00 (7.21)
11	0.16 (0.075)	−276.5 (24.5)	309.7 (25.8)	0.023 (0.036)	0.0005 (0.0003)	
12	0.29 (0.043)	−55.5 (11.4)				98.12 (7.36)
13			−71.9 (16.3)		0.0010 (0.0002)	122.59 (12.14)
14		−367.0 (93.4)	444.2 (132.0)	−0.0027 (0.035)	0.0011 (0.0001)	−37.92 (42.67)
15	0.14 (0.081)	−57.1 (10.8)			0.0006 (0.0003)	100.48 (6.99)
16	0.18 (0.073)	−386.6 (86.9)	468.1 (122.8)	0.032 (0.036)	0.0005 (0.0003)	−52.72 (40.02)
17	0.28 (0.044)		−73.2 (16.9)			119.21 (12.33)
18	0.13 (0.085)		−75.3 (16.0)		0.0006 (0.0003)	122.16 (11.75)
19	0.27 (0.043)	−56.1 (11.0)		0.060 (3.59)		89.59 (8.74)
20		−54.6 (11.1)		−0.006 (0.041)	0.0010 (0.0002)	102.91 (9.51)

Note The values in the parentheses denote the uncertainty of parameters

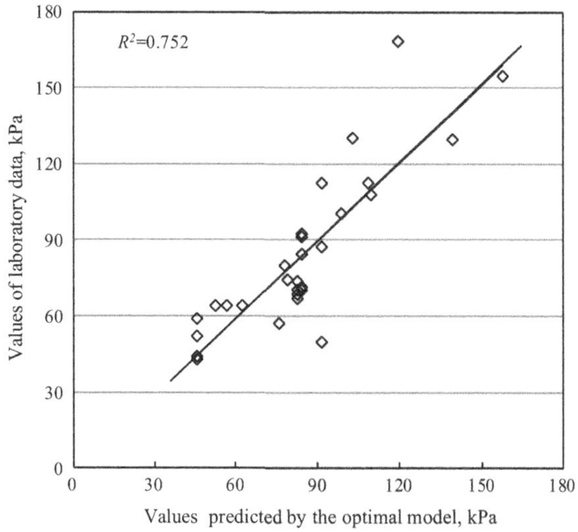

Fig. 8.3 Comparison of maximum pullout shear stress between the laboratory data and the predictions by the optimal model

utilize all model classes with the weightings being the posterior plausibility. This multimodel predictive formula is given by:

$$E[y|X, D, C_1, \ldots, C_{N_C}] = \sum_{j=1}^{N_C} P(C_j|D) \cdot y\big(X; b^*, C_j\big) \tag{8.21}$$

It uses all model classes to provide the "hyper-robust prediction" instead of using only the most plausible model class [27]. To compare the performance of this multi-model formula with the optimal model in Eq. (8.21), their predicted values are compared in Fig. 8.4. The high R^2 value (0.969) indicates that the optimal model class "2 3 5", that is, Eq. (8.21), is almost as robust as the hyper-robust model for fitting the maximum pullout shear stress. Therefore, it is sufficient to use the model in Eq. (8.19) as the proposed design formula.

Figure 8.5 presents the relationship between maximum pullout shear stress and degree of saturation under different overburden pressures and grouting pressures. The shade area represents the prediction by Eq. (8.21) for 40 kPa $\leq$ OP $\leq$ 350 kPa and 0 $\leq$ GP $\leq$ 300 kPa. The scattered points are the measured data from laboratory tests under different S_r values. It is noted that only the test data at $S_r = 50\%$, $OP =$ 80 kPa and $GP = 0$ is a little far away from the prediction area while the other test data are close or within the shaded zone. It should be noted that the laboratory test data is only available for certain values of S_r, OP and GP as it is very expensive and time-consuming to conduct tests under all conditions. From the prediction of Eq. (8.21), one can easily determine the maximum pullout shear stress at any combination of S_r, OP and GP conditions.

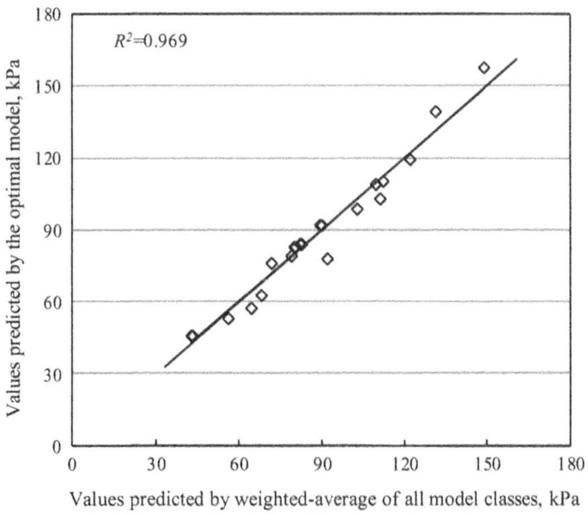

Fig. 8.4 Comparison of maximum pullout shear stress between the predictions by the optimal model and the weighted-average prediction from all model classes

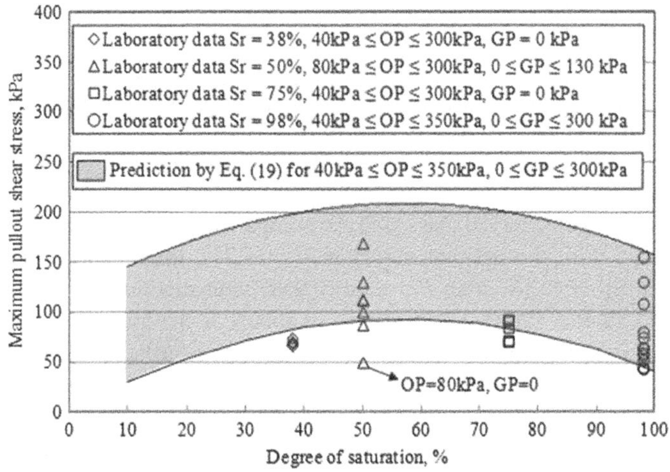

Fig. 8.5 Maximum pullout shear stress versus degree of saturation under different overburden pressure and grouting pressure

8.5 Comparison with the Effective Stress Method

According to the effective stress method, the ultimate soil nail pullout resistance per length T_{ult} (kN/m) can be computed from [7, 28]:

$$T_{\text{ult}} = c' P_c + 2 D_m \sigma_v' \mu^* \tag{8.22}$$

where c' (in kPa) is the effective cohesion of the soil obtained from saturated soil tests; P_c is the outer perimeter of the cement grout sleeve; D_m is the outer diameter of the cement grout sleeve ($D_m = 100$ mm in this case); σ_v' (in kPa) is the effective overburden stress in the soil calculated at mid-depth of the soil nail reinforcement; μ^* is the apparent friction coefficient of soil (μ^* may be taken as $\tan\varphi'$ where φ' is the angle of shearing resistance of the soil under effective stress condition). In the laboratory test, the values of the soil sample, c' and φ', are 3.7 kPa and 34.9°, respectively. Finally, the maximum pullout shear stress of soil nails $\bar{\tau}_d$ can be computed by the following formula:

$$\bar{\tau}_d = \frac{T_{\text{ult}}}{\pi D_m} \tag{8.23}$$

The values obtained from the experimental data are plotted against the corresponding predictions by Eq. (8.23) in Fig. 8.6. The value of R^2 is only 0.114, which is significantly smaller than that of the optimal model ($R^2 = 0.752$). Comparing Figs. 8.3 and 8.6, it can be clearly seen that the optimal model substantially overperforms the effective stress method. The poor predictions by the latter are due to the fact that only the soil effective shear strength and overburden pressure have been taken into account. However, other important factors, such as the degree of saturation, are not incorporated.

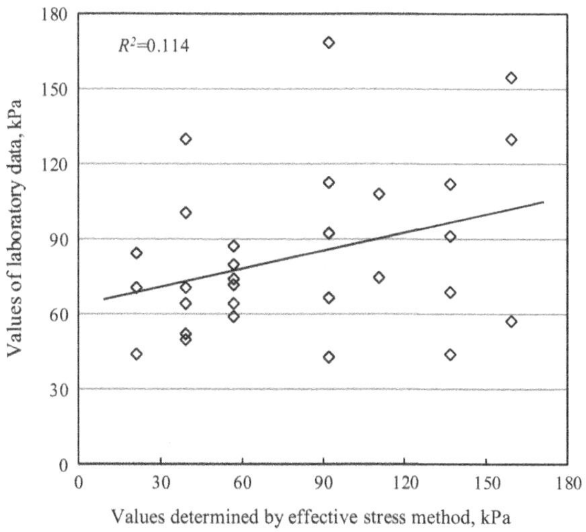

Fig. 8.6 Comparison of maximum pullout shear stress between the laboratory data and the estimations by the effective stress method

8.6 Estimation of Field Pullout Test Data

In order to verify the engineering applicability, eight field pullout test data on CDG soils are analyzed by the optimal model. The field tests were carried out in a uniform natural slope, namely feature No.11SW-C/F220 under the Landslip Preventive Measure (LPM) Program of Geotechnical Engineering Office (GEO) of Hong Kong Government at South of Pok Fu Lam Kennels, Pok Fu Lam, Hong Kong, China. The averagc degree saturation of site soil is about 55%. The average soil nail diameter measured after pulled out of ground is nearly equal to 121.9 mm. The effective cohesion and the friction angle obtained from the site saturated soil tests are 7.7 kPa and 29.9°, respectively. The tests were conducted under different grouting pressures (0, 40, 80, 140 kPa) and overburden pressures (43, 128 kPa). The field soil nail maximum pullout shear stress is estimated by the optimal model and effective stress method, respectively. The coefficient of determination (R^2), mean absolute percentage error (MAPE) and root mean square error (RMSE) for the estimation are presented for comparison. The statistics index MAPE and RMSE are computed as follows:

$$\text{MAPE} = \frac{1}{N}\sum_{i=1}^{N}\frac{|Y_i - y_i|}{Y_i} \times 100\% \tag{8.24}$$

$$RMSE = \sqrt{\frac{\sum_{i=1}^{N}(Y_i - y_i)^2}{N}} \tag{8.25}$$

where Y_i is the measured maximum pullout shear stress; y_i is the predicted maximum pullout shear stress. In general, good predictive models associate with small values of MAPE and RMSE, as well as R^2 close to unity.

Based on the eight field test data, the analysis results about R^2, MAPE and RMSE for the two methods are given in Table 8.4. It is found that the error statistics of the optimal model is better than those of the effective stress method. The values of MAPE and RMSE of the optimal model are 11.34% and 15.44, respectively, which are much less than those of the effective stress method. In addition, the R^2 value of the optimal model is 0.501, which is about two times larger than that of the effective stress method, 0.266. The scatter plots between the estimations by two methods and the field data are shown in Fig. 8.7. It can be seen that predictions by the effective stress method depend only on the value of *OP* but not the grouting pressure *GP*. On the other hand, the proposed optimal model can reflect the changing trend of the test

Table 8.4 The comparison of the proposed optimal model and effective stress method from the field pullout test data

Method	R^2	MAPE (%)	RMSE
Proposed optimal model	0.501	11.34	15.44
Effective stress method	0.266	62.91	67.49

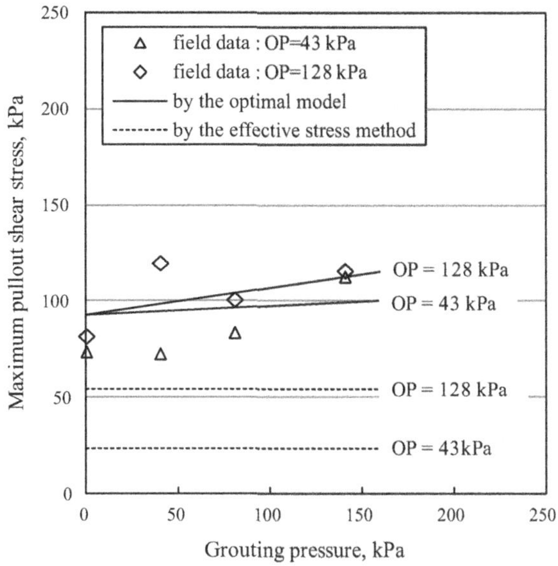

Fig. 8.7 Maximum pullout shear stress versus grouting pressure from field data and the estimations by both two methods

data by the term $GP \times OP$. It is noted that the correlation of the optimal model with the field test results is smaller than that in the previous laboratory data analysis. This is due to the different test conditions in the field and laboratory. The model parameters (b_2, b_3, b_5) may not be the optimal values for the field test condition. Nevertheless, the analysis of field data confirms that the optimal model provides significantly better field maximum pullout shear stress estimation than the traditional effective stress method. Optimal values of model parameters for field conditions can be obtained when sufficient field data are available.

8.7 Conclusions

In this chapter, the design formula candidates were constructed based on observations in the literature and systematic generation with the consideration of multiple influencing factors for soil nail pullout resistance. By using the Bayesian model class selection approach and a series of laboratory soil nail pullout test data, a reliable design formula was proposed. It is found that the most important factors in the estimation of soil nail pullout resistance are the degree of saturation and the product of grouting pressure and overburden pressure. The predictions of maximum pullout shear stress by the proposed optimal formula and the effective stress method were compared using the laboratory and field pullout test data. It is found that the proposed optimal formula possesses far stronger correlation with measurements than the effective stress method.

It should be noted that the presented estimation formula is obtained from limited laboratory pullout tests on CDG soils in Hong Kong and verified by field test data for the same soil. The findings are useful for better understanding of the significant factors in the estimation of soil nail pullout resistance. However, one should be cautious for any extension of the findings in this chapter for other soils and under other test conditions.

References

1. FHWA (1993) French national research project clouterre 1991—recommendations Clouterre 1991. English translation. Soil nailing recommendations, Publication No FHWA-SA-93-026, Federal Highway Administration, Washington DC, USA
2. Stocker MF, Korber GW, Gassler G, Gudehus G (1979) Soil nailing. In: Proceedings of the international conference on soil reinforcement, 20–22 Mar 1979, Paris, France, vol 2, pp 469–474
3. Shen CK, Bang S, Hermann LR (1981) Ground movement analysis of an earth support system. Journal of Geotechnical Engineering Division, ASCE 107(12):1609–1624
4. HA68/94 (1994) Design manual for the reinforcement of highways slopes by reinforced soil and soil nailing techniques. Design manual for roads and bridges, vol 4, No 1. HMSO, London
5. FHWA (1996) Manual for design and construction of soil nail walls. Soil nail walls-demonstration project 103, Publication No FHWA-SA-96-096, Federal Highway Administration, Washington DC, USA
6. FHWA (2003) Soil nail walls. Geotechnical Engineering Circular No 7, Publication No. FHWA0-IF-03-017, Federal Highway Administration, Washington DC, USA
7. GEO (2006) Good practice in design of steel soil nails for soil cut slopes. GEO technical guidance note No 23 (TGN 23), Geotechnical Engineering Office, Civil Engineering and Development Department, Hong Kong Government, Hong Kong
8. GEO (2008) Geoguide 7-guide to soil nail design and construction. Geotechnical Engineering Office Civil Engineering and Development Department, Hong Kong Government, Hong Kong. ISBN: 978-962-02-0375-6, GEO REPORT No 197
9. Pun WK, Shiu YK (2007) Design practice and technical developments of soil nailing in Hong Kong. In: Geotechnical advancements in Hong Kong since 1970s, the HKIE geotechnical division 27th annual seminar, The Hong Kong Institution of Engineers, May 15, 2007, Hong Kong, pp 197–212
10. Schlosser F (1982) Behaviour and design of soil nailing. In: International symposium on recent development in ground improvement techniques, Asian Institute of Technology, 29 Nov–3 Dec 1982, Bangkok, Thailand, pp 399–413
11. Pradhan B, Tham LG, Yue ZQ, Junaideen SM, Lee CF (2006) Soil-nail pullout interaction in loose fill materials. Int J Geomech 6(4):238–247
12. Su LJ (2006) Laboratory pull-out testing study on soil nails in compacted completely decomposed granite fill. PhD thesis, The Hong Kong Polytechnic University
13. Su LJ, Chan CF, Yin JH, Shiu YK, Chiu SL (2008) Influence of overburden pressure on soil-nail pullout resistance in a compacted fill. J Geotech Geoenviron Eng ASCE 134(9):1339–1347
14. Su LJ, Chan TCF, Shiu YK, Cheung T, Yin JH (2007) Influence of degree of saturation on soil nail pull-out resistance in compacted completely decomposed granite fill. Can Geotech J 44(11):1314–1328
15. Su LJ, Yin JH, Zhou WH (2010) Influences of overburden pressure and soil dilation on soil nail pull-out resistance. Comput Geotech 37(4):555–564
16. Yin JH, Su LJ, Cheung WM, Shiu YK, Tang C (2008) The influence of grouting pressure on the pullout resistance of soil nails in compacted completely decomposed granite fill. Geotechnique 59(2):103–113

17. Yin JH, Zhou WH (2009) Influence of grouting pressure and overburden stress on the interface resistance of a soil nail. J Geotech Geoenviron Eng ASCE 135(9):1198–1208
18. Luo SQ, Tan SA, Yong KY (2000) Pull-out resistance mechanism of a soil nail reinforcement in dilative soils. Soils Found 40(1):47–56
19. Zhang LL, Zhang LM, Tang WH (2009) Uncertainties of field pullout resistance of soil nails. J Geotech Geoenviron Eng ASCE 135(7):966–972
20. Yin JH, Hong CY, Zhou WH (2012) Simplified analytical method for calculating the maximum shear stress of nail-soil interface. Int J Geomech 12(3):309–317
21. Xue XH, Yang XG, Liu EL (2011) Application of modified good man model in soil nailing. Int J Geomech. https://doi.org/10.1061/(ASCE)GM.1943-5622.0000171
22. Zhou WH (2008) Experimental and theoretical study on pullout resistance of grouted soil nail. PhD thesis, The Hong Kong Polytechnic University
23. Yuen KV (2010) Recent developments of Bayesian model class selection and applications in civil engineering. Struct Saf 32(5):338–346
24. Yuen KV (2010) Bayesian methods for structural dynamics and civil engineering. Wiley(Asia) Pte Ltd, Singapore
25. Yuen KV, Mu HQ (2011) Peak ground acceleration estimation by linear and nonlinear models with reduced order Monte Carlo simulation. Comput-Aided Civil Infrastruct Eng 26(1):30–47
26. Yin JH, Su LJ (2006) An innovative laboratory box for testing nail pullout resistance in soil. ASTM Geotech Test J 29(6):451–461
27. Beck JL, Yuen KV (2004) Model selection using response measurements: Bayesian probabilistic approach. J Eng Mech ASCE 130(2):192–203
28. Watkins AT, Powell GE (1992) Soil nailing to existing slopes as landslip preventive works. Hong Kong Eng 20(3):20–27

Chapter 9
Selection of Physical and Chemical Properties of Natural Fibers for Predicting Soil Reinforcement

Abstract In this chapter, the physical and chemical properties of natural fibers (i.e., natural moisture content, specific gravity, breaking tensile strength, breaking strain, cellulose, hemicellulose, lignin and ash) were examined for their influences in the soil reinforcement. A total of 11 factors including soil moisture content, soil density and fiber percentage were evaluated by using the Bayesian nonparametric general regression (BNGR) method. The robustness of the BNGR algorithm was validated using k-fold cross-validation. A parametric study was carried out to unveil the effects of these input variables. The results indicated that a higher fiber percentage or larger soil density induces higher reinforcement effect. Besides, the cellulose in a natural fiber is found to have a significant effect on the unconfined compressive strength (UCS) followed by the ash. With respect to the physical properties, the breaking strain is found to be highly correlated with the UCS, while the effects of fiber natural moisture content, specific gravity and breaking tensile strength are less significant parameters.

9.1 Introduction

Fiber reinforcement in soil improvement has become popular in recent decades [1–12] because a large amount of fiber from industry products [13–16], waste fabrics [17], and plants [18, 19] are available. These types of fiber can be used to improve the foundation soil if they are properly treated. However, the rubber fibers or the polypropylene fibers are synthetic and nondegraded. The natural fiber as an alternative material in soil improvement is an economic and environment-friendly approach that attracted the attention of researchers and engineers [20–25]. Different types of plant fiber, such as coir, jute, bamboo, water hyacinth and so on, have been investigated by researchers from the view point of geotechnical engineering [26–28]. The schematic diagram of soil–fiber possible surface interaction is shown in Fig. 9.1. The surface friction between soil particles and the adjacent fibers can lead to stress transfer and interlocking among soil particles [29–34]. The tensile strength of fiber can be mobilized, which affects the stress–strain behavior of fiber-reinforced soil. Hence, the fiber content in the soil is one of the key concerned influencing factors

W. Zhou et al., *Practice of Bayesian Probability Theory in Geotechnical Engineering*,
https://doi.org/10.1007/978-981-15-9105-1_9

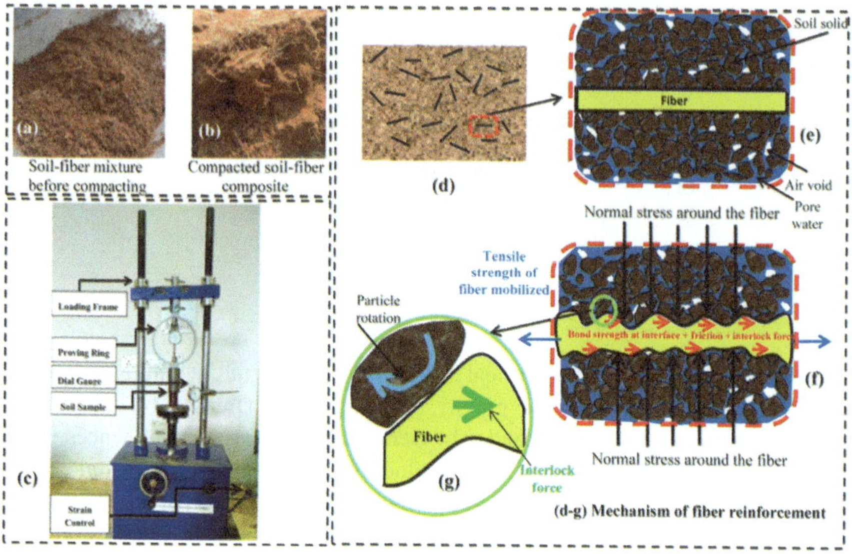

Fig. 9.1 **a** Soil–fiber mixture before compacting; **b** compacted soil–fiber composite; **c** test setup; and **d**–**g** mechanism of fiber reinforcement

on the soil strength. Besides, the intrinsic properties of soil, such as the particle-size distribution, the soil density and the soil moisture content, affect the strength of soil as well. Some researchers [35–41] summarized these influencing factors and established regression models based on experimental data to predict the shear strength of fiber-reinforced soil. Nevertheless, most of the scholars concentrated on these influencing factors, which are mainly based on the properties of soil and fiber content. However, the understanding of the chemical components and physical properties related to the types of natural fiber is not clear. Especially, seldom researchers paid attention on these inherent characteristics of natural fiber and investigated the relationship between these chemical and physical factors and the strength of fiber-reinforced soil. Bordoloi et al. [38] found that the soil samples with different types of natural fiber exhibited different unconfined compression strengths. The reason behind this phenomenon is the distinct chemical and physical properties of these fibers. The four main chemical components in a natural fiber are the cellulose, hemicellulose, lignin and ash. Different proportions of these chemical components result in diverse physical properties, such as the specific gravity, the natural moisture content, the breaking tensile strength and the breaking strain. It is no doubt that these eight parameters (four chemical components and four physical properties) are related to the UCS of fiber-reinforced soil. However, when it comes to a regression model to predict the UCS, inclusion of all these eight parameters is not necessary to improve the predictive power of the regression model, because these eight parameters are correlative and should have intrinsic links among them. It is a challenge task to distinguish the significant influencing factors from the possible input candidates.

Many existing mathematical models [39–47] can be used to bridge the design variables with the interested outputs. Zhou et al. [48] obtained a parametric formula to predict the maximum pullout shear stress of grouted soil nails using the Bayesian model updating. However, a predictive formula had to be constructed based on the design variables, such as the grouting pressure, the overburden pressure and the degree of saturation. It is subjective and the predictive formula designer should have sufficient engineering judgment. Tan et al. [49] adopted the Bayesian method to predict the soil water retention curve considering the influencing factors, such as the water content, matric suction and initial void ratio. Similarly, the predictive formula had to be judged subjectively. Artificial neural network (ANN) is an alternate approach which does not have to define a predictive formula. Meulenkamp and Grima [50] applied a neural network to predict the UCS of rock samples. However, the backpropagation is required in an ANN, which is slow and inefficient. Specht [51] proposed a single-pass learning algorithm, namely general regression neural network (GRNN). This method was adopted by Ceryan et al. [52] to establish a predictive model for the UCS of rock samples and the GRNN is a better choice than other neural networks. However, if plenty of design variables are summarized, it will be difficult to decide which variables are crucial and certain redundant input variables may be included into the generated regression model. Yuen and Ortiz [53] obtained a novel algorithm, namely the Bayesian nonparametric general regression (BNGR), which can be used to effectively select the significant input variables from the available candidates and establish a regression model between inputs and output(s). This BNGR method was applied in the seismic attenuation modeling in Yuen et al.'s study [54]. Based on the selected design variables, the authors found that the site properties are very important to examine the ground motion.

In this study, the chemical and physical properties of natural fibers were discussed and 11 influencing factors were summarized as the input candidates to predict the UCS of natural fiber-reinforced soil. The BNGR method was applied to select a proper set of explanatory input variables from the candidates and build the regression model between the significant input variables and the UCS. The robustness of the BNGR method was examined and the generated regression model was verified using the measured data. The sensitivity analysis and a parametric study were performed to evaluate and discuss the effects of these selected input variables on the UCS.

9.2 Basic Index of Materials

In this study, three types of natural fibers were investigated, that is, coir, jute and water hyacinth. They are environment-friendly and cost-effective. The coir is a natural fiber extracted from the husk of coconut. Jute is a long, soft, shiny vegetable fiber. Water hyacinth (also known as crassipes) is an invasive species which influences the growth of other plants. The properties of these natural fibers depend on the chemical components and physical properties of them. The chemical components of these natural fibers can be generally classified as cellulose, hemicellulose, lignin and ash.

The recommended methods [55–58] were followed to determine the percentage of each component. The physical properties contain specific gravity, natural moisture content, breaking tensile strength and breaking strain.

9.2.1 Chemical Components of Natural Fibers

Cellulose

Cellulose is an organic polymer and the formula is (C6H10O5)n. This is an important structural component of the primary cell wall of natural plants. The biodegradable cellulose is hydrophilic with the contact angle of 20–30°. Plant-derived cellulose is usually found in a mixture with hemicellulose, lignin, and other substances. The function of the cellulose in the nature of fiber is the structural strength and this factor is related to the UCS of fiber-reinforced soil.

Hemicellulose

Hemicellulose is also known as polyose and presents along with cellulose in almost all plant cell walls. It is any of several heteropolymers. Compared to the cellulose which is crystalline, strong and resistant to hydrolysis, the hemicellulose has a random, amorphous structure with little strength and is easily hydrolyzed by dilute acid. In contrast to the cellulose with 7,000–5,000 glucose molecules per polymer, hemicellulose consists of shorter chains of 500–3,000 sugar units. Hemicellulose is often associated with cellulose, but it has different composition. This component mainly contributes to second strength of a fiber, as well as facilitates to absorb moisture from the surrounding environment, so this factor affects the physical properties of a natural fiber.

Lignin

Lignin is a class of complex organic biopolymers and is necessary to form the cell walls because they lend rigidity and do not rot easily. Lignin is lack of a defined primary structure. Its most commonly noted function is the support through strengthening in plant fiber. Lignin fills the spaces in the cell wall between cellulose and hemicellulose components. It is conferring mechanical strength to the cell wall and by extension the plant as a whole. Lignin plays a crucial part in conducting water in plant stems and is related to the natural moisture content. The lignin can show resistant against the microbial erosion and keep the fiber survive longer in the nature.

Ash

The component of ash is related to the texture of a natural fiber and is a measure of the mineral content and other inorganic matter in plants. It contains potassium or potash (K_2CO_3). The percentage of ash is determined based on ASTM E1755-01 [58]. Ash content is found to be highest in water hyacinth, thus indicating its suitability in metal retention.

9.2.2 Physical Properties of Natural Fibers

Specific gravity

The specific gravity is related to the percentage of chemical components and is correlative with the strength of a natural fiber. The proportions of cellulose, hemicellulose, lignin and ash vary with different plants, which result in different specific gravities for different plants. In this study, the specific gravity of natural fiber was determined using the method in IS-2720-Part 3 [59].

Natural moisture content

Based on the properties and functions of hemicellulose and lignin, the natural moisture content could be related to the proportion of these components and would have a significant effect on the breaking strain and breaking tensile strength. Oven drying method is adopted to determine the natural moisture content. The temperature is kept at 105 °C lasting 24 h to conduct this experiment.

Breaking tensile strength

The application of fiber-reinforced soil in geotechnical engineering is commonly randomly distributed fiber-reinforced soil (RDFS). The breaking tensile strength happens when the soil particles have large deformation and this can enhance the shear strength of a soil sample. This property can be related to the cellulose. The breaking tensile strength and breaking strain of fiber were measured based on the method in IS-1670 [60].

Breaking strain

The breaking strain of fiber can enhance the ductility of fiber-reinforced soil after reaching the peak strength. This physical property can be related to the chemical component of cellulose because of the long chain of sugar units in the polymer of cellulose.

From the above discussion, there may exist inner relationships between these chemical components and the physical properties. For instance, a large percentage in cellulose may result in a large value in the breaking tensile strength or in the breaking strain. The natural moisture content may be related to the proportions of hemicellulose and lignin. It is necessary to figure out what are the dominant factors to the UCS of natural fiber-reinforced soil. The chemical components and the physical properties of coir, jute and water hyacinth are shown in Table 9.1 and Table 9.2,

Table 9.1 Chemical composition (%) of plant fibers

Fiber	Cellulose	Hemicellulose	Lignin	Ash
Coir	43	15	38	2
Jute	60	22	16	1
Water hyacinth	46	21	11	11

Table 9.2 Physical properties of concerned fibers

Fiber	Natural moisture content (%)	Specific gravity (g/cc)	Breaking tensile strength (MPa)	Breaking strain (%)
Coir	7	1.24	150	26
Jute	14	1.12	500	10
Water hyacinth	12	0.70	313	14
Polypropylene	0	0.91	153	125

respectively. It should be noted that the state of the chemical composition in a natural fiber is assumed to be stable. The natural moisture contents of fibers in Table 9.2 were measured after three days of drying in the air. Under these conditions, the other physical properties were measured and the fiber-reinforced soil samples were made. Additionally, the polypropylene fiber was used as a comparison to plant fibers, and the physical properties are shown in Table 9.2.

9.2.3 Soil Properties

The experiments were performed at Indian Institute of Technology Guwahati, and the soil samples were collected from a site of the campus. The majority component of the sample is silt with the proportion of 50% and the others are clay and sand. Based on the unified soil classification system [61], the adopted soil sample is labeled as clayed silt. The grain size distribution is shown in Fig. 9.2. The liquid limit and the plastic limit are 42% and 24%, respectively, according to the ASTM D4318 [62]. The optimum water content is 17% and the maximum dray density is 1.69 g/cc as shown in Table 9.3 based on the ASTM D1557 [63]. The internal friction angle and

Fig. 9.2 Grain-size distribution of soil sample

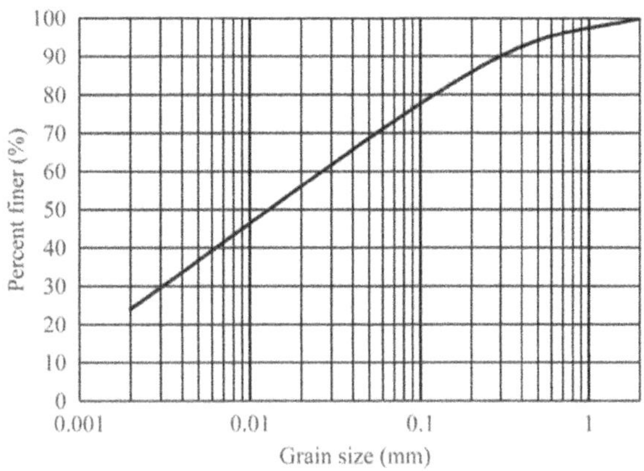

Table 9.3 Physical properties of soil

Property	Value
Specific gravity	2.63
Liquid limit	42
Plastic limit	24
Shrinkage index	21
Plasticity index	18
Optimum moisture content (%)	17
Maximum dry density (g/cc)	1.69
Cohesion (kPa)	16.4
Angle of internal friction (°)	14.4

the cohesion are 14.4° and 16.4 kPa, respectively, according to the ASTM D3080 [64]. The other soil properties are shown in Table 9.3.

To minimize the possible difference in individual fiber, the three plant fibers and the polypropylene fiber were derived from the same sources, respectively. The plant fiber reinforcement adopted was produced from the stem of water hyacinth, the coir and the jute. The same aspect ratio of 70 (defined as the ratio of the length over the diameter of fiber) was adopted for all the fibers. In this study, the mean diameter of jute was 0.2 mm and the length of the fiber was produced as 14 mm. The same geometry was applied to the coir fiber. As the water hyacinth is monofilament, the mean diameter was 0.4 mm with the length of 28 mm.

9.3 Bayesian Nonparametric General Regression Method

The novel Bayesian nonparametric general regression (BNGR) algorithm was proposed by Yuen and Ortiz [53]. This method takes advantages of the Bayesian inference of selecting proper input variables and the general regression neural network of powerful prediction. Two advantages of the BNGR method were concluded by Yuen and Ortiz [53]. One is that the prior distribution of the regression coefficients has no effect on model class selection. The other is the number of model competitors is diminished dramatically compared with the traditional regression approaches.

The general regression neural network is used to obtain the regression relationship between the inputs and outputs without prescribing an explicit parametric formula. Assuming that the input variables are written in vector, X, the output y can be expressed as the conditional mean given X:

$$E(y|X) = \frac{\int_{-\infty}^{\infty} yp(X, y)dy}{\int_{-\infty}^{\infty} p(X, y)dy} \tag{9.1}$$

where $p(X, y)$ is the joint probability density function (PDF) of X and y. Because it is unknown for most cases, the kernel density approximation $\hat{p}(X, y)$ was adopted to calculate it. The common choice is the Gaussian mixture distribution as shown below:

$$\hat{p}(X, y) = \frac{1}{N\left(2\pi\sigma_1^2\right)^{(d+1)/2}} \sum_{n=1}^{N} \exp\left[-\frac{(X - X_n)^T (X - X_n) + (y - y_n)^2}{2\sigma_1^2}\right] \tag{9.2}$$

Substituting Eq. (9.2) into Eq. (9.1), the expected value of y can be written as:

$$\hat{y}(X) = E_{\hat{p}}(y|X) = \frac{\sum_{n=1}^{N} y_n \exp\left[-(X - X_n)^T (X - X_n)/\left(2\sigma_1^2\right)\right]}{\sum_{n=1}^{N} \exp\left[-(X - X_n)^T (X - X_n)/\left(2\sigma_1^2\right)\right]} \tag{9.3}$$

where σ_1^2 is an unknown parameter, which is used to balance the smoothness of the regression model and the precision of the fitness.

To select the proper value of σ_1^2 and overcome the over-fitting, the point to be predicted is dismissed in the summations of Eq. (9.3):

$$\hat{y}(X_m) = \frac{\sum_{\substack{n=1\\ n\neq m}}^{N} y_n \exp\left[-(X_m - X_n)^T (X_m - X_n)/\left(2\sigma_1^2\right)\right]}{\sum_{\substack{n=1\\ n\neq m}}^{N} \exp\left[-(X_m - X_n)^T (X_m - X_n)/\left(2\sigma_1^2\right)\right]} \tag{9.4}$$

Then, fitting the predictions from Eq. (9.4) with the measured data, the optimal value of σ_1^2 can be obtained.

According to GRNN, the parameter σ_1^2 is an unknown parameter that should be determined based on Eq. (9.4). Similarly, there should exist an unknown vector, $\boldsymbol{\theta}$, in the Bayesian framework of general regression. By using the Bayes' theorem, the posterior PDF of $\boldsymbol{\theta}$ is expressed as:

$$p(\boldsymbol{\theta}|y, X, C) = \frac{p(y|\boldsymbol{\theta}, X, C)p(\boldsymbol{\theta}|C)}{p(y|X, C)} \tag{9.5}$$

where $p(y|X, C)$ is the normalizing constant. $p(\boldsymbol{\theta}|C)$ is the prior PDF of the uncertain parameters to represent the prior knowledge of the researchers' awareness. $p(y|\boldsymbol{\theta}, X, C)$ is the likelihood function to reflect the fitting of the measured data given the parameter vector, $\boldsymbol{\theta}$. C is the general regression model.

It is important to determine the likelihood function in Eq. (9.5), and it can be written in the form of conditional PDF as:

$$p(y|\boldsymbol{\theta}, X, C) = \prod_{m=1}^{N} p(y_m|y_1, \ldots, y_{m-2}, y_{m-1}, \boldsymbol{\theta}, X, C) \tag{9.6}$$

where

$$p(y_m|y_1, \ldots, y_{m-2}, y_{m-1}, \boldsymbol{\theta}, X, C) = \left(2\pi\sigma_{2,m}^2\right)^{-1/2} \exp\left[-\frac{\left(y_m - \hat{y}_{m|m-1}(X_m)\right)^2}{2\sigma_{2,m}^2}\right] \tag{9.7}$$

where $\hat{y}_{m|m-1}(X_m)$ is the regression of y on X according to the first $m-1$ measured data. It can be calculated as:

$$\hat{y}_{m|m-1}(X_m) = \frac{\sum_{n=1}^{m-1} y_n \exp\left[-\left((X_m - X_n)^T(X_m - X_n)\right)/2\sigma_{1,m}^2\right]}{\sum_{n=1}^{m-1} \exp\left[-\left((X_m - X_n)^T(X_m - X_n)\right)/2\sigma_{1,m}^2\right]} \tag{9.8}$$

The parameter, $\sigma_{1,m}^2$, in Eq. (9.8) can be written in the following form.

$$\sigma_{1,m}^2 = \frac{v_1}{m-1} \sum_{n=1}^{m-1} (X_m - X_n)^T(X_m - X_n) \tag{9.9}$$

where v_1 is defined as smoothing scale parameter to be calculated [53]:

$$\sigma_{2,m}^2 = \frac{v_2}{\sum_{n=1}^{m-1} \exp\left[-2(X_m - X_n)^T(X_m - X_n)\right]} \tag{9.10}$$

where v_2 is defined as prediction error scale parameter to be calculated [53].

Up to now, the unknown vector, $\boldsymbol{\theta} = \left[v_1\ v_2\right]^T$, is formulated based on the framework of BNGR method.

Then, the posterior PDF of the unknown vector, $\boldsymbol{\theta}$, can be expressed as:

$$\begin{aligned} p(v_1, v_2|y, X, C) &\propto p(v_1, v_2)p(y|v_1, v_2, X, C) \\ &\propto (v_2)^{-(N/2)} \exp\left[-\frac{1}{2v_2}\sum_{m=1}^{N} \Omega_m\left(y_m - \hat{y}_{m|m-1,v_1}(X_m)\right)^2\right] \end{aligned} \tag{9.11}$$

where Ω_m is written as:

$$\Omega_m = \sum_{n=1}^{m-1} \exp\left[-2(X_m - X_n)^T(X_m - X_n)\right] \tag{9.12}$$

By letting the partial differentiation of the posterior PDF with respect to the parameter ν_2 be zero, one has:

$$\frac{\partial p(\nu_1, \nu_2 | y, X, C)}{\partial \nu_2} = 0 \tag{9.13}$$

Then,

$$\nu_2^*(\nu_1) = \frac{1}{N}\sum_{m=1}^{N} \Omega_m \left(y_m - \hat{y}_{m|m-1,\nu_1}(X_m)\right)^2 \tag{9.14}$$

The parameter, ν_1^*, can be obtained by maximizing the following function:

$$g(\nu_1) = p\left(\nu_1, \nu_2^*(\nu_1) | y, X, C\right) \tag{9.15}$$

With the calculated ν_1 and ν_2, the regression model is the relationship between a subset of candidate input variables and the interested output. The proper set of design variables can be determined from the optional models using the principle of Bayesian inference stated in the following. Different combinations of these candidate input variables produce different regression models, such as C_1, C_2, ..., C_k.

Based on the Bayes' principle, the plausibility of a model can be given as:

$$P(C_k | y, X) = \frac{p(y|X, C_k)P(C_k)}{\sum_{k=1}^{N_c} p(y|X, C_k)P(C_k)} \tag{9.16}$$

Finally, $p(y|X, C_k)$ can be readily obtained [53]:

$$p(y|X, C_k) \approx \frac{2\Gamma(N/2+1)\sqrt{2\pi \prod_{m=1}^{N} \left(\Omega_m / |\hbar_k(\nu_1^*)|\right)}}{(B_{U1} - B_{L1})(B_{U2} - B_{L2})\pi^{N/2}} \times \left[\sum_{m=1}^{N} \Omega_m \left(y_m - \hat{y}_{m|m-1,\nu_1}(X_m)\right)^2\right]^{-(N/2+1)} \tag{9.17}$$

This is the Bayesian formulated general regression neural network. It should be noted that, with the same set of input variables, the function structure is automatically generated according to the principle of general regression neural network. Thus, the Bayesian model class selection process becomes very efficient because it does not require generating different function forms using the same set of input variables in contrast to the traditional regression methods. The readers may refer to Appendix G for a MATLAB code of the implementation of the method.

9.4 Results and Discussion

9.4.1 Data Preparation and Model Class Selection

The unconfined compression strength (UCS) test is used to estimate the bearing strength and it is popular in engineering due to its simple and fast test procedure. In the present study, the UCS tests were carried out to analyze the strength of fiber reinforced soil. The volume of fiber (Shukla, 2017) can be a significant variable which would influence the UCS of fiber reinforced soil. However, in this study, the weight of fiber is adopted because this parameter can be easily used by the engineers. The manufactured fibers were mixed into the soil sample with 0.50%, 0.75% and 1% of dry soil weight to reveal the influence of fiber content. The proportion of fiber was limited to 1% to eliminate the possible non-uniform density of soil-fiber mixture as the fibers tend to bond together. The moisture content and soil density of the fiber reinforced soil were designed and controlled based on the optimal moisture content (OMC) and the maximum dry density (MDD) of soil. Thus, the moisture contents were OMC-5%, OMC, OMC+5% and the soil densities were set as 0.95MDD, MDD, and 1.05MDD. Based on these design parameters, the weighted fiber was carefully mixed with the oven dried soil sample so that the fibers are randomly distributed. Then, the required amount of water was added to create the fiber reinforced soil sample. After that, the sample was sealed and kept for 24 hours so that the water distributes in soil sample uniformly. The specimen was molded to be 50 mm in diameter and 100 mm in height. A constant strain rate of 1.25 mm/min was applied to the soil specimen for all the tests. Three samples were made and tested for each compaction state. Because the uniform mix of the fiber and soil samples is not easy to be controlled. The three repeated samples could have relatively large discrepancy. Average of the three samples may lead the reasonable data have larger error. Hence, all the measured data were summarized to form the database. The detailed experimental program is illustrated in Fig. 9.3. A series of 324 laboratory tests were carried out to measure the UCS. Additionally, the bare soil samples without fiber were made and 3 (OMC-5%, OMC, OMC+5%) $\times$3 (0.95MDD, MDD, 1.05MDD) = 9 scenarios were designed. The measured UCSs of these bare soil samples were used to normalize the measured UCSs of fiber reinforced soil samples and to minimize the impact of soil type on the UCS, so that a general regression model can be obtained to predict the cases with other soil types and different fibers.

Based on the previous description of the experimental procedure and the properties of natural fibers, the database was prepared to obtain a regression model using the BNGR method. A total of 11 parameters were summarized to evaluate their contributions to the UCS. They are related to the physical properties of fibers, such as the natural moisture content, x_1, the specific gravity, x_2, the breaking tensile strength, x_3, and breaking strain, x_4. Besides, they are related to the chemical component of plant fiber, such as the cellulose, x_5, hemicellulose, x_6, lignin, x_7, and ash, x_8, as well as the characteristics of fiber-reinforced soil, such as soil density, x_9, moisture content, x_{10}, and fiber percentage, x_{11}. The output is the UCS measured in the experiments. Thus, the size of the database is 324 according to the discussion in the previous section. The statistics of the input variables and the UCS is presented in Table 4.

On the basis of the BNGR algorithm described in Sect. 3, $2^{11} - 1 = 2047$ models were generated as the model candidates. It should be noted that the number of potential models would increase dramatically if the traditional regression methods were adopted as there will be huge combinations of possible function forms from the same set of input variables. The BNGR method overcomes this weakness of the traditional approaches and renders the model class selection feasibly. According to the size of the database, 70% of these records (226 points) were randomly selected as the training database which is used for obtaining the optimal model and the smoothing scale parameter in Eq. (9.9) as well as the prediction error scale parameter in Eq. (9.10).

Table 9.4 Statistics of the 11 inputs for fiber-reinforced soil

Parameter	Mean	Standard deviation	Skewness	Kurtosis	Minimum	Median	Maximum
x_1	8.25	5.4109	−0.5083	1.7390	0	9.5	14
x_2	0.9925	0.2064	−0.2386	1.5778	0.7	1.015	1.24
x_3	279	143.8484	0.5642	1.7130	150	233	500
x_4	43.75	47.3509	1.1025	2.2898	10	20	125
x_5	37.25	22.4777	−0.8637	2.1681	0	44.5	60
x_6	14.5	8.8028	−0.8660	2.0592	0	18	22
x_7	16.25	13.8484	0.5536	2.0127	0	13.5	38
x_8	3.5	4.3943	1.0656	2.2656	0	1.5	11
x_9	1.7	0.0695	-1.44^{-13}	1.5	1.615	1.7	1.785
x_{10}	17	4.0888	0.0	1.5	12	17	22
x_{11}	0.75	0.2044	0.0	1.5	0.5	0.75	1
y	1.0414^3	838.29	1.6121	6.0143	70.0383	752.8709	42384

Note x_1: natural moisture content (%); x_2: specific gravity (g/cc); x_3: breaking tensile strength (MPa); x_4: elongation at break; x_5: cellulose; x_6: hemicellulose (%); x_7: lignin (%); x_8: ash (%); x_9: soil density (g/cc); x_{10}: soil moisture content (%); x_{11}: percentage fiber (%); y: UCS

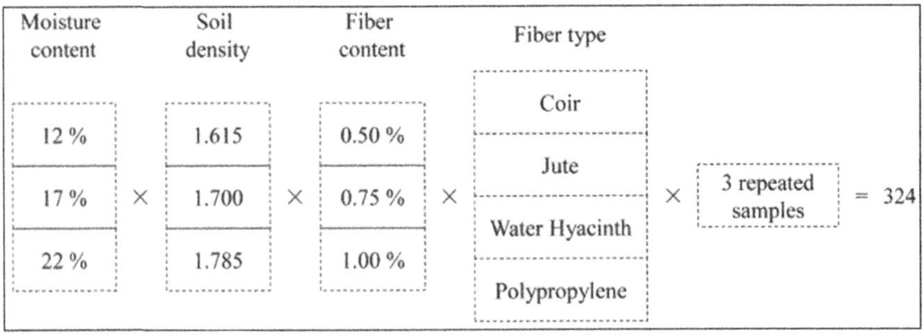

Fig. 9.3 Experimental program for measuring UCS of fiber-reinforced soil

Figure 9.4 shows that the learning ability of the BNGR algorithm is quite efficient and the obtained regression model gave a high correlation coefficient on the training phase. Results of the BNGR method for some selected models are shown in Table 9.5. As the choice of prior distribution for the vector $\boldsymbol{\theta}$ has no influence on the model class selection [53], the uniform prior distributions of [0, 100] were used for both the smoothing scale parameter and the prediction error scale parameter. Six models with the selected parameters in the BNGR method are tabulated. The first column denotes the selected input variables. For instance, (9 10 11) represents the selected model containing the input variables of x_9, x_{10} and x_{11}. The second and third columns show the smoothing scale parameter, v_1, and the prediction error scale parameter, v_2. The fourth column is the maximum likelihood. The fifth and sixth columns show the evidence and the plausibility of the corresponding model. To verify the fitness of the selected model, the mean squared errors (MSEs) of these models are shown

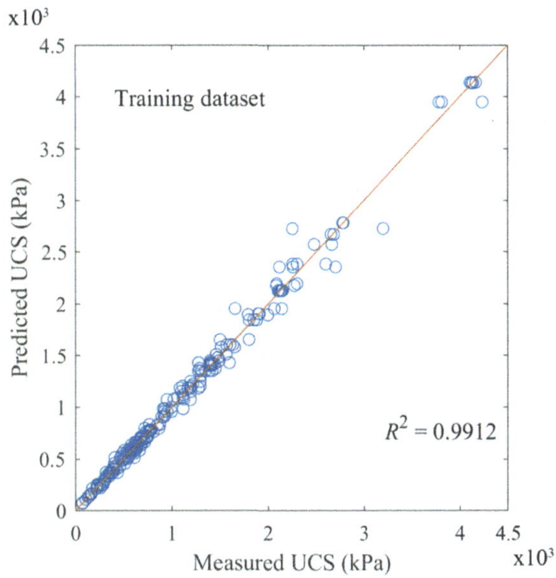

Fig. 9.4 Predicted UCS based on training data set

Table 9.5 Model class selection results

Model	v_1	v_2	Likelihood	$p(y \mid X, C_k)$	$P(C_k \mid y, X)$
(4 5 8 9 10 11)	0.0012	0.0435	2.2123×10^{-16}	2.0202×10^{-09}	0.9997
(1 5 8 9 10 11)	0.0014	0.0384	4.0388×10^{-20}	4.1658×10^{-13}	2.0614×10^{-4}
(4 5 6 9 10)	0.0119	0.3365	8.5895×10^{-38}	3.0366×10^{-42}	1.5026×10^{-33}
(3 9 10 11)	1.25×10^{-6}	0.4899	1.8495×10^{-111}	4.1485×10^{-117}	2.0528×10^{-108}
(9 10 11)	0.0020	2.1227	2.2578×10^{-124}	4.2221×10^{-119}	2.0892×10^{-110}
(9)	0.0078	89.6807	2.7578×10^{-177}	8.7536×10^{-184}	4.3315×10^{-175}

Note v_1: Smoothing scale parameter; v_2: Prediction error scale parameter

Table 9.6 Mean squared error of testing data for each model

Model	MSE
(4 5 8 9 10 11)	1.3826×10^4
(1 5 8 9 10 11)	1.5261×10^4
(4 5 6 9 10)	3.4008×10^4
(3 9 10 11)	1.3339×10^5
(9 10 11)	1.6573×10^5
(9)	5.6045×10^5

in Table 9.6. The model ranking is in the same order as in Table 9.5. The smallest MSE is obtained for the selected model and a smaller model plausibility produces a larger MSE.

It can be found in Table 9.5 that the most plausible model, (4 5 8 9 10 11), contains the parameters of breaking strain (x_4), cellulose (x_5), ash (x_8), soil density (x_9), soil moisture content (x_{10}) and fiber percentage (x_{11}). These selected input variables can be generally separated into two groups: *group A* contains the last three influencing factors, x_9, x_{10} and x_{11}, which have been evaluated in the literature; *group B* includes the three indices (i.e., breaking strain, cellulose and ash) with respect to the chemical and physical properties of fibers. This means that the inherent properties of natural fibers have significant effects on the variation of the UCS. From the scanning electron microscope (SEM) images of the four different types of fibers (Fig. 9.5), there is no doubt that the breaking strain contributes to the UCS, because the fibers at the possible

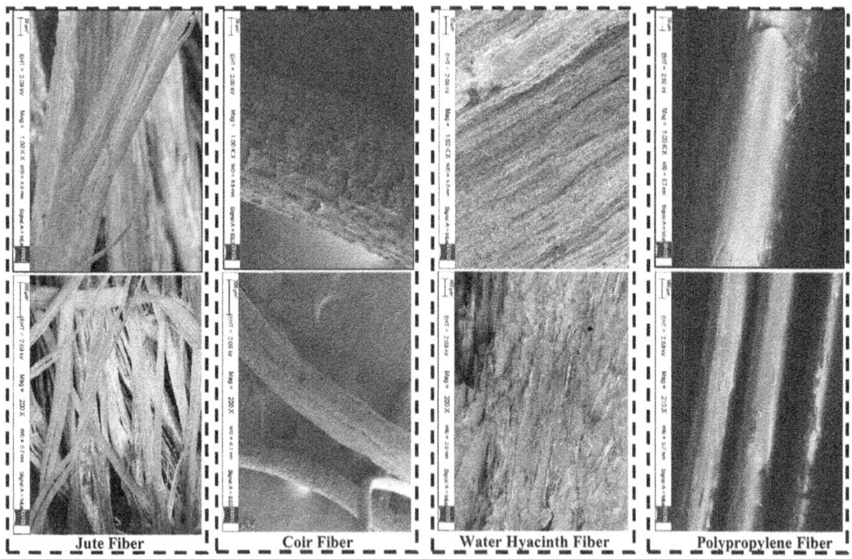

Fig. 9.5 FE-SEM images showing fiber morphology for selected fiber type at two magnifications (1,000× and 200×)

yield surface resist the relative movement of soil particles and postpone the shear failure at that surface. The cellulose as a chosen input variable contributes to the structural strength of natural fibers [65] and consequently affects the UCS. This is consistent with the results in Bordoloi et al.'s analysis [38, 66] for the case of jute fiber. The ash is another influencing factor to the UCS and this factor can be used as an index to select the natural fiber. The specific gravity of fiber is not included into the model. The reason may be that the percentage of the mixed fiber is less than 1%. The natural moisture content is an inherent property of fibers. The interchange of water between fibers and the surrounding soil may be tiny and this has less contribution to the UCS.

9.4.2 *Validation of the Optimal Model*

9.4.2.1 Using the Remaining 30% of the Testing Database

In this part, the optimal model was verified using the rest of 30% testing database (98 points). Figure 9.6 illustrates the measured UCS from the testing database versus the predicted UCS as well as the perfect match line, that is, 45° line. Both the results using BNGR method and the traditional GRNN method were plotted in this figure. It should be noted that the same set of input variables was adopted in the GRNN method for comparison. The results predicted by the BNGR algorithm distribute around the 45° line in a small range. While the ones predicted by the traditional GRNN method scatter in a relatively larger range surrounding the 45° line. The

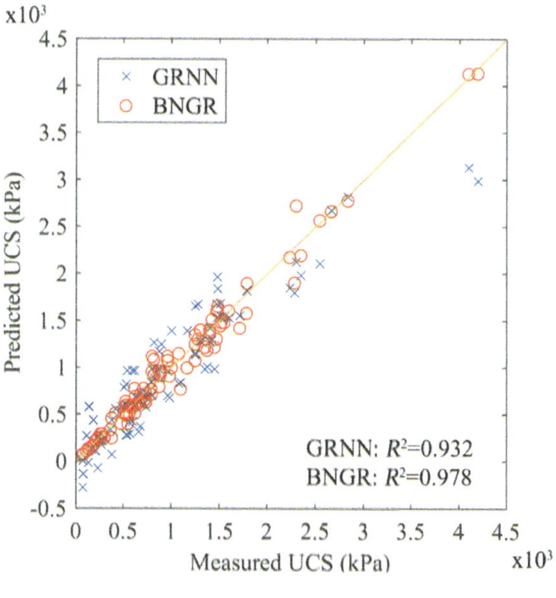

Fig. 9.6 Measured UCS versus predicted UCS based on testing data

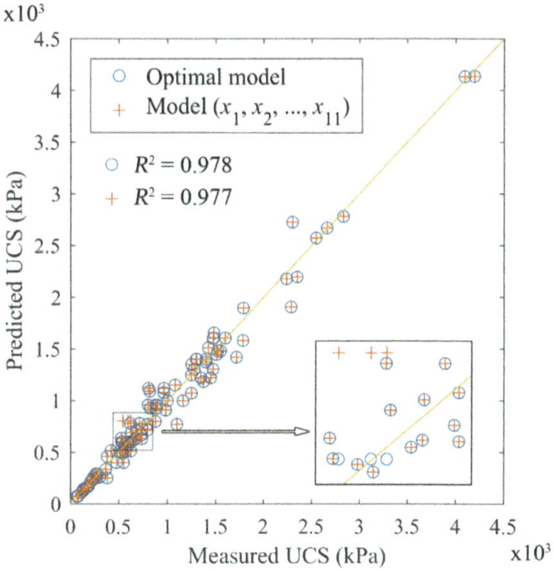

Fig. 9.7 Predicted UCS using model with eleven inputs ($x_1, x_2, \ldots, x_{11}$)

holdout method was used to obtain the smoothing parameter in the traditional GRNN based on Eq. (9.4). Moreover, the coefficient of determination R^2 is 0.978 with the BNGR method and that is 0.932 with the traditional GRNN. Given that the GRNN method cannot select a proper set of input variables, the merits of small error and model class selection pronounce the advantages of the BNGR method. Figure 9.7 shows that the model with all 11 input variables (i.e., $x_1, x_2, \ldots, x_{11}$) is not quite good in the local view window, even though the fitting capability is good. Furthermore, the model uncertainty increases with a greater numbers of input variables. Figure 9.8 presents the predicted results using the model with the input variables of soil density, moisture content, and fiber percentage in *group A*. Because of missing significant input variables related to the fiber type, the estimated points scatter in a larger area compared to that from the optimal model. This means that the UCS of fiber-reinforced soil is closely related to the fiber type. The understanding of influencing factors that cause the variation in the UCS of soil with different types of natural fibers may be useful for engineers to further optimize their designs when using natural fiber reinforcement.

9.4.2.2 Using the Data in the Literature

To further verify the obtained regression model, three cases related to the fiber reinforced soil were reviewed and the measured UCSs were collected to make a comparison. It should be noted that, when collecting data, the input variables of the testing data should have the some domains with that of the training data in the regression model based on the BNGR method. The adopted fiber reinforcements were polyester

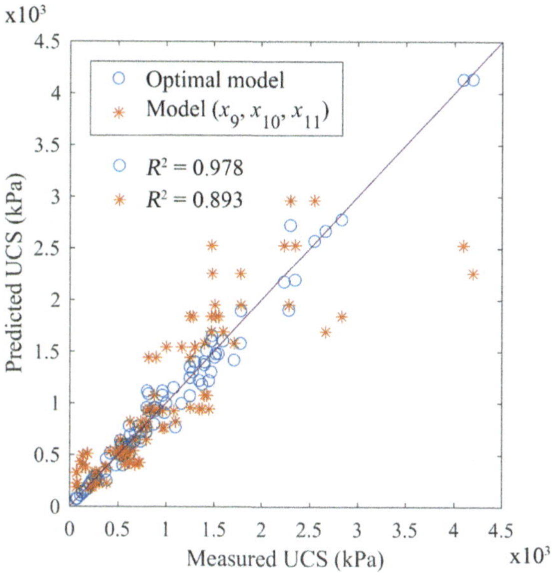

Fig. 9.8 Predicted UCS using model with three inputs (x_9, x_{10} and x_{11})

fiber [66], coir fiber [18] and synthetic fiber [67], respectively. Three types of soil were presented in these studies and the associated soil properties, such as soil classification, plastic limit, liquid limit, optimum moisture content (OMC) and maximum dry density, are summarized in Table 9.7 and the measured UCSs are presented in Table 9.8. Because the intrinsic properties (such as grain-size distribution) have great difference in different soils which show various shear strengths, the measured UCSs of fiber-reinforced soil were normalized using that of the corresponding bare soil to lighten the influence of soil type. The predicted results are shown in Fig. 9.9. It can be found that the predicted results are acceptable, which means that the proposed regression model is reasonable and can be applied to make a prediction.

Table 9.7 Soil properties of the soil samples in the three cases

Soil classification	Plastic limit	Liquid limit	Optimum moisture content (%)	Maximum dry density (g/mm^3)	References
CH (USCS)	28	52	16.2	1.766	Kumar et al. [68]
CL (USCS)	18	32	20	1.68	Dasaka and Sumesh [18]
Fly ash (30%) + Silty sand (70%) (Indian Standard)	–	–	14.25	1.71	Chauhan et al. [67]

Table 9.8 Measured UCSs of fiber-reinforced soil in the three cases

UCS (kPa)	Kumar et al. [68]		Dasaka and Sumesh [18]		Chauhan et al. [67]		
	Fiber content (%)						
	0.5	1.0	0.5	1.0	0.50	0.75	1.00
Fiber-reinforced soil	283	323	340	443	217	239	256
Bare soil	192	192	206	206	101.8	101.8	101.8

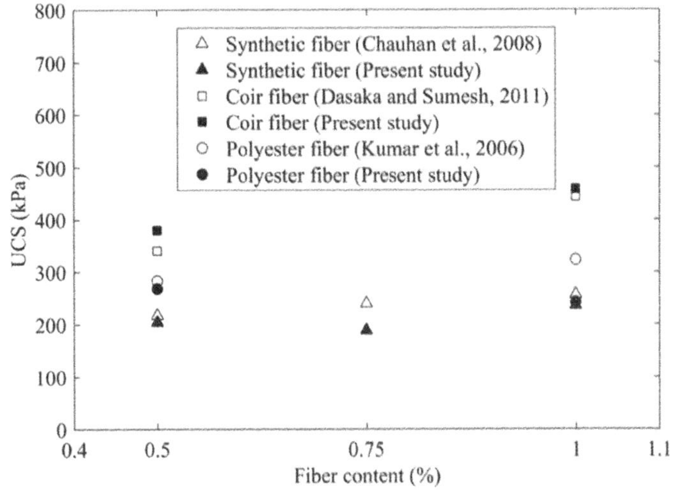

Fig. 9.9 Comparison between measured and predicted UCS

9.4.3 *Robustness of the BNGR Algorithm Using the K-Fold Cross-Validation Method*

This part examined the robustness of the BNGR algorithm through evaluating the root-mean-square-error (RMSE) using the *K-fold* cross-validation method. The database was split into five data sets with equal (or approximately equal) proportions. $k - 1$ folds were adopted as training database and the last fold was used for evaluation. The RMSE was calculated based on the predicted results on the testing phase. Figure 9.10 presents the RMSE of the regression model on each fold that varies in a small range. This means that the optimal model is independent of the selection of the training database, and the BNGR algorithm is robust.

9.4.4 Sensitivity Analysis of Different Influencing Factors

Based on the robust performance of the BNGR algorithm, it is attempted to investigate the contribution of each input variable to the UCS of fiber-reinforced soil. The concept of sensitivity analysis (SA) was introduced to evaluate the significant input variables [37, 68]. The contribution percentage of each input variable to the UCS is calculated using the following equations:

$$N_i = f_{\max}(x_i) - f_{\min}(x_i) \tag{9.18}$$

$$S_i = \frac{N_i}{\sum_{j=1}^{n} N_j} \times 100 \tag{9.19}$$

where x_i means the ith input variable. Through keeping other input variables as their mean value, $f_{\min}(x_i)$ and $f_{\max}(x_i)$ are the calculated minimum and maximum, respectively, of the predicted UCS over the domain of x_i. Figure 9.11 presents the sensitivity analysis for the input variables in the optimal model. The reader can observe that the soil moisture content stands out from the three input variables (x_9: soil moisture content, x_{10}: soil density and x_{11}: fiber percentage) in *group A*. This finding is in close agreement with the previous research work by Kurugodu et al. [37]. The sensitivity analysis of the other three input variables (x_4: breaking strain,

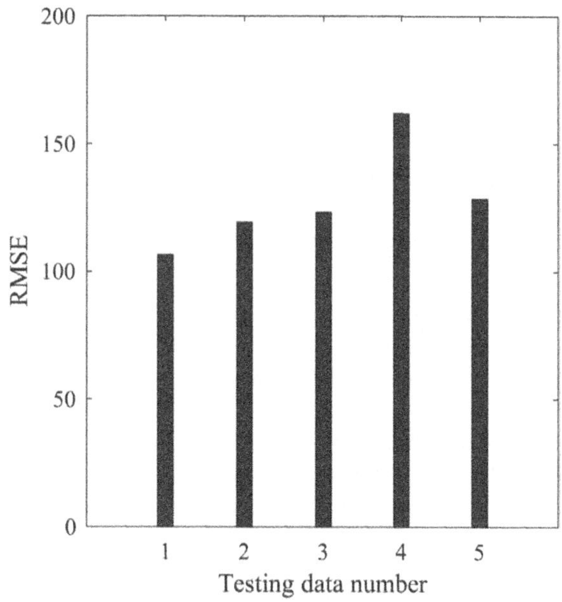

Fig. 9.10 Performance of BNGR algorithm using five different testing data sets based on k-fold cross-validation

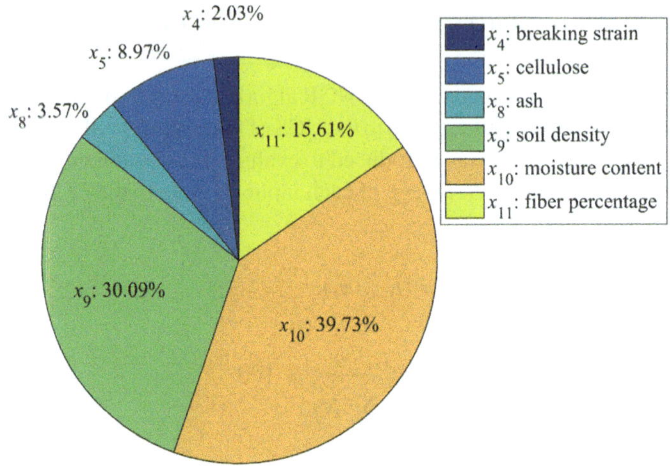

Fig. 9.11 Sensitivity analysis of input variables in the optimal model

x_5: cellulose and x_8: ash) in *group B* indicates that the cellulose has the highest effect on the UCS of fiber-reinforced soil followed by ash and breaking strain.

This solution is reasonable because the cellulose is a type of polymer with long chains of sugar units. This can provide the strength of fiber and consequently improve the UCS of fiber-reinforced soil. To further examine the three input variables in *group B*, each of them was associated with the three input variables in *group A* to form three sets of input variables. Based on the BNGR algorithm, Fig. 9.12 presents the predicted UCS on the testing phase. The largest R^2 was obtained in Fig. 9.12b, which means that the factor of cellulose is the dominant factor among the three input variables in *group B* to select the type of natural fibers. This is consistent with the conclusion from the above sensitivity analysis.

9.4.5 Relationships Between the UCS and the Input Variables

According to the discussion in the previous sections, the BNGR algorithm is robust and the significant input variables are selected properly. In the following, a parametric study is carried out to detect the variation of the UCS regarding the six input variables. Figure 9.13 illustrates the calculated UCS using the optimal regression model. In the calculation, except for the target input variable, the other five remaining input variables were fixed at their mean values. One can observe that with the increase in the soil density, the UCS increases accordingly (Fig. 9.13a). An increase in the soil moisture content results in a decrease in the UCS (Fig. 9.13b). These two input variables are related to the soil properties and the observed phenomena are coincident with that in the literature [46, 69, 70]. With a larger percentage of natural fiber in soil, the UCS of fiber-reinforced soil will be improved to a higher level. This is because

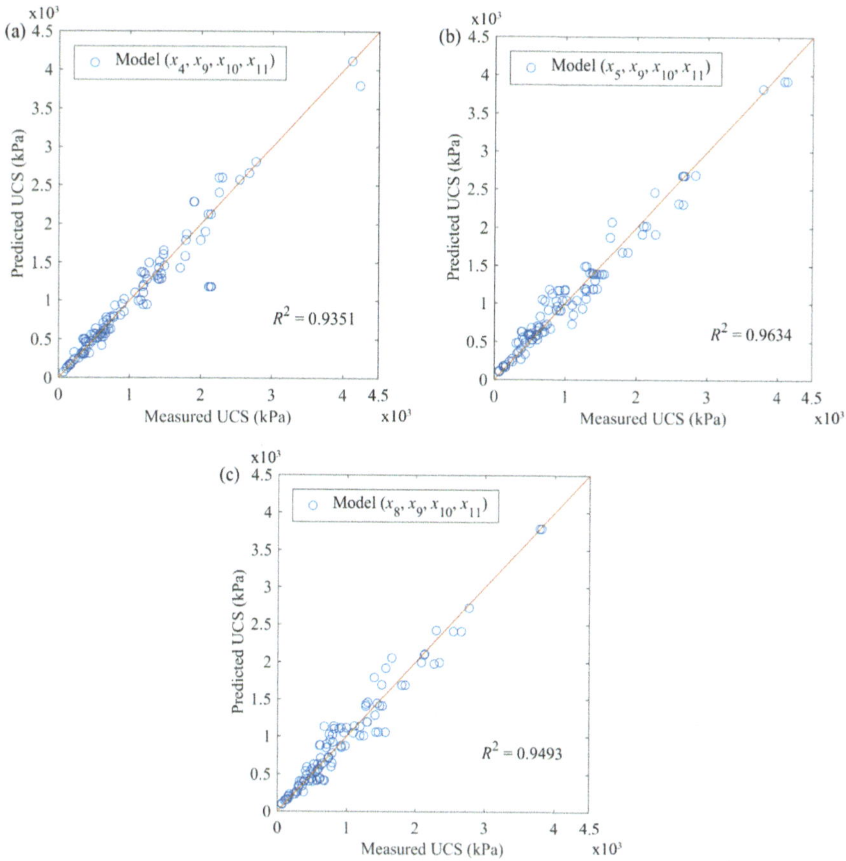

Fig. 9.12 Influence of chemical and physical properties on UCS: **a** elongation at break; **b** cellulose; and **c** ash

the strength of natural fibers is relatively stronger than the shear resistance among soil particles. A higher fiber content means more load is sustained by the natural fibers and produces a higher UCS. The nonlinear behavior of the UCS with respect to the breaking strain, the cellulose and the ash can be observed in Figs. 9.12d-f. The breaking strain can enhance the interaction among soil particles, but a quite large value in this factor cannot improve the UCS because a much large elongation has little contribution to the shear strength at possible slip surfaces (Fig. 9.13d). A larger proportion of cellulose can enhance the strength of natural fiber-reinforced soil as shown in Fig. 9.13e. This can be related to the inherent feature of this organic polymer. The cellulose can be used as an index to select the type of plant fiber in real practice. Figure 9.13f indicates that a relatively large percentage of ash can decrease the UCS because the ash itself cannot affect the strength of a natural fiber. It always exists with other chemical components, and a larger proportion in the ash would

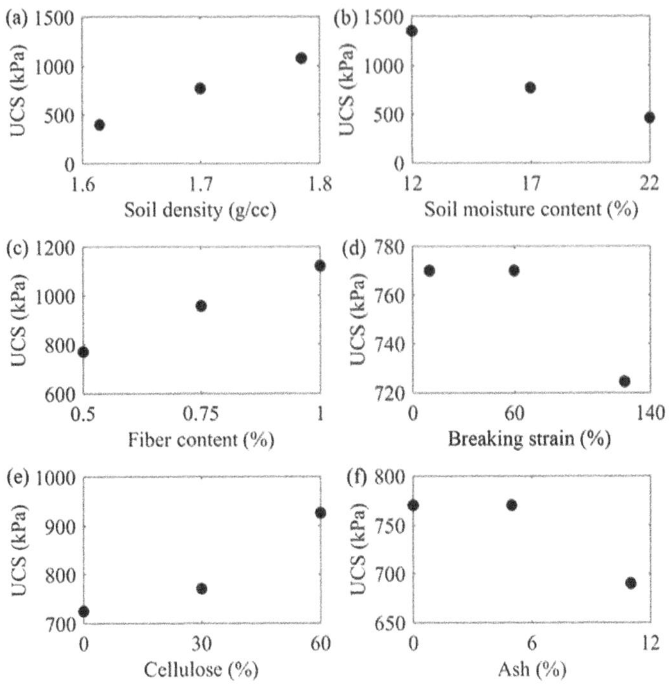

Fig. 9.13 Variation of UCS with respect to six input variables: **a** soil density; **b** soil moisture content; **c** fiber content; **d** breaking strain; **e** cellulose; and **f** ash

produce less proportion in other components; for instance, cellulose, consequently, resulting in a smaller UCS.

9.5 Conclusions

This study presented an extensive understanding of possible influencing factors (related to the chemical components and physical properties of natural fibers and the properties of soil) to the unconfined compressive strength (UCS) of natural fiber-reinforced soil. Six significant input variables (the breaking strain of fibers, the proportions of cellulose and ash in fibers, the soil density, the soil moisture content and the fiber content) were effectively selected from 11 potential candidates by using the Bayesian nonparametric general regression (BNGR) algorithm. The generated optimal model in the BNGR method was verified on the testing phase as well as was validated using the measured data in the literature. The predictions are reasonable. The robustness of the BNGR algorithm was examined using the *k-fold* cross-validation. The sensitivity analysis and the parametric study show that the soil moisture content plays a dominant role in the UCS followed by the soil density and the fiber percentage. The chemical component of cellulose is far beyond the factors of the breaking strain and the ash. Hence, the cellulose can be used to choose the type of fibers. A relatively larger value in the soil density, the fiber percentage or the proportion of cellulose can enhance the strength of fiber-reinforced soil. However, the behavior of the soil moisture content is opposite. The breaking strain can increase the UCS but a much larger value in this makes little effect. The proportion of ash in fiber reinforcement should be in a small value.

References

1. Ranjan G, Vasan RM, Charan HD (1996) Probabilistic analysis of randomly distributed fiber-reinforced soil. J Geotech Eng 122(6):419–426
2. Khattak MJ, Alrashidi M (2006) Durability and mechanistic characteristics of fiber reinforced soil-cement mixtures. Int J Pavement Eng 7(1):53–62
3. Tang CS, Shi B, Zhao LZ (2010) Interfacial shear strength of fiber reinforced soil. Geotext Geomembr 28(1):54–62
4. Hejazi SM, Sheikhzadeh M, Abtahi SM, Zadhoush A (2012) A simple review of soil reinforcement by using natural and synthetic fibers. Constr Build Mater 30:100–116
5. Hamidi A, Hooresfand M (2013) Effect of fiber reinforcement on triaxial shear behavior of cement treated sand. Geotext Geomembr 36:1–9
6. Fatahi B, Fatahi B, Le TM, Khabbaz H (2013) Small-strain properties of soft clay treated with fibre and cement. Geosynth Int 20(4):286–300
7. Zhao LS, Zhou WH, Fatahi B, Li XB, Yuen KV (2016) A dual beam model for geosynthetic-reinforced granular fill on an elastic foundation. Appl Math Model 40(21–22):9254–9268
8. Mirzababaei M, Arulrajah A, Horpibulsuk S, Aldava M (2017) Shear strength of a fibre-reinforced clay at large shear displacement when subjected to different stress histories. Geotext Geomembr 45(5):422–429
9. Mirzababaei M, Arulrajah A, Horpibulsuk S, Soltani A, Khayat N (2018) Stabilization of soft clay using short fibers and poly vinyl alcohol. Geotext Geomembr 46(5):646–655
10. Qi Y, Indraratna B, Heitor A, Vinod JS (2018) Effect of rubber crumbs on the cyclic behavior of steel furnace slag and coal wash mixtures. J Geotech Geoenviron Eng 144(2):1–11
11. Soltani A, Deng A, Taheri A, Mirzababaei M (2018) Rubber powder–polymer combined stabilization of South Australian expansive soils. Geosynth Int 25(3):304–321
12. Wang HL, Chen RP, Cheng W, Qi S, Cui YJ (2019) Full-scale model study on variations of soil stress in geosynthetic-reinforced pile-supported track bed with water level change and cyclic loading. Can Geotech J 56(1):60–68
13. Kumar A, Walia BS, Bajaj A (2007) Influence of fly ash, lime, and polyester fibers on compaction and strength properties of expansive soil. J Mater Civ Eng 19(3):242–248
14. Tang CS, Wang DY, Cui YJ, Shi B, Li J (2016) Tensile strength of fiber-reinforced soil. J Mater Civ Eng 28(7):04016031
15. Sharma V, Kumar A (2017) Influence of relative density of soil on performance of fiber-reinforced soil foundations. Geotext Geomembr 45(5):499–507
16. Han J, Guo J (2017) Geosynthetics used to stabilize vegetated surfaces for environmental sustainability in civil engineering. Front Struct Civil Eng 11(1):56–65
17. Fatahi B, Khabbaz H, Fatahi B (2012) Mechanical characteristics of soft clay treated with fibre and cement. Geosynth Int 19(3):252–262
18. Dasaka SM, Sumesh KS (2011) Effect of coir fiber on the stress-strain behavior of a reconstituted fine-grained soil. J Nat Fibers 8(3):189–204
19. Anggraini V, Asadi A, Farzadnia N, Jahangirian H, Huat BBK (2016) Reinforcement benefits of nanomodified coir fiber in lime-treated marine clay. J Mater Civ Eng 28(6):06016005

20. Wambua P, Ivens J, Verpoest I (2003) Natural fibres: can they replace glass in fibre reinforced plastics? Compos Sci Technol 63(9):1259–1264
21. Mesbah A, Morel JC, Walker P, Ghavami K (2004) Development of a direct tensile test for compacted earth blocks reinforced with natural fibers. J Mater Civ Eng 16(1):95–98
22. John MJ, Anandjiwala RD (2008) Recent developments in chemical modification and characterization of natural fiber-reinforced composites. Polym Compos 29(2):187–207
23. Singh HP, Bagra M (2013) Improvement in CBR value of soil reinforced with jute fiber. Int J Innov Res Sci Eng Technol 2(8):3447–3452
24. Bordoloi S, Hussain R, Garg A, Sreedeep S, Zhou WH (2017) Infiltration characteristics of natural fiber reinforced soil. Trans Geotech 12:37–44
25. Garg A, Hazra B, Zhu H, Wen Y (2019) A simplified probabilistic analysis of water content and wilting in soil vegetated with non-crop species. CATENA 175:123–131
26. Sivakumar Babu GL, Vasudevan AK (2008) Strength and stiffness response of coir fiber-reinforced tropical soil. J Mater Civ Eng 20(9):571–577
27. Subaida EA, Chandrakaran S, Sankar N (2008) Experimental investigations on tensile and pullout behaviour of woven coir geotextiles. Geotext Geomembr 26(5):384–392
28. Wang Y, Guo P, Ren W, Yuan B, Yuan H, Zhao Y (2017) Laboratory investigation on strength characteristics of expansive soil treated with jute fiber reinforcement. Int J Geomech 17(11):04017101
29. Jing XY, Zhou WH, Zhu HX, Yin ZY, Li Y (2018) Analysis of soil-structural interface behavior using three-dimensional DEM simulations. Int J Numer Anal Meth Geomech 42(2):339–357
30. Zhu H, Zhou WH, Yin ZY (2018) Deformation mechanism of strain localization in 2D numerical interface tests. Acta Geotech 13(3):557–573
31. Chen WB, Zhou WH, Jing XY (2019) Modeling geogrid pullout behavior in sand using discrete-element method and effect of Tensile Stiffness. Int J Geomech 19(5):04019039
32. Wang HL, Chen RP, Liu QW, Kang X, Wang YW (2019) Soil–geogrid interaction at various influencing factors by pullout tests with applications of FBG sensors. J Mater Civ Eng 31(1):04018342
33. Zhao LS, Zhou WH, Geng X, Yuen KV, Fatahi B (2019) A closed-form solution for column-supported embankments with geosynthetic reinforcement. Geotextiles and Geomembranes, in press. https://doi.org/10.1016/j.geotexmem.2019.01.006
34. Zhou WH, Jing XY, Yin ZY, Geng X (2019) Effects of particle sphericity and initial fabric on the shearing behavior of soil–rough structural interface. Acta Geotech. https://doi.org/10.1007/s11440-019-00781-2
35. Zhu HH, Zhang CC, Tang CS, Shi B, Wang BJ (2014) Modeling the pullout behavior of short fiber in reinforced soil. Geotext Geomembr 42(4):329–338
36. Mirzababaei M, Mohamed M, Miraftab M (2017) Analysis of strip footings on fiber-reinforced slopes with the aid of particle image velocimetry. J Mater Civ Eng 29(4):04016243
37. Kurugodu HV, Bordoloi S, Hong Y, Garg A, Garg A, Sreedeep S, Gandomi AH (2018) Genetic programming for soil-fiber composite assessment. Adv Eng Softw 122(February):50–61
38. Bordoloi S, Kashyap V, Garg A, Sreedeep S, Wei L, Andriyas S (2018) Measurement of mechanical characteristics of fiber from a novel invasive weed: a comprehensive comparison with fibers from agricultural crops. Measurement 113:62–70
39. Kiefa MAA (1998) General regression neural networks for driven piles in cohesionless soils. J Geotech Geoenviron Eng 124(12):1177–1185
40. Yuen KV (2010) Bayesian methods for structural dynamics and civil engineering. Wiley, New York
41. Yuen KV, Kuok SC (2015) Efficient Bayesian sensor placement algorithm for structural identification: a general approach for multi-type sensory systems. Earthquake Eng Struct Dynam 44(5):757–774
42. Mu HQ, Yuen KV (2015) Novel outlier-resistant extended Kalman filter for robust online structural identification. J Eng Mech 141(1):04014100
43. Gamse S, Zhou WH, Tan F, Yuen KV, Oberguggenberger M (2018) Hydrostatic-season-time model updating using Bayesian model class selection. Reliab Eng Syst Saf 169:40–50

44. Jin YF, Yin ZY, Wu ZX, Zhou WH (2018) Identifying parameters of easily crushable sand and application to offshore pile driving. Ocean Eng 154(March):416–429
45. Tan F, Zhou WH, Yuen KV (2018) Effect of loading duration on uncertainty in creep analysis of clay. Int J Numer Anal Meth Geomech 42(11):1235–1254
46. Zhou WH, Liu F, Yi S, Chen YZ, Geng X, Zheng C (2018) Simultaneous stabilization of Pb and improvement of soil strength using nZVI. Sci Total Environ 651:877–884
47. Zhou WH, Qi XH (2019) Root cohesion estimation of riparian trees based on model uncertainty characterization. J Mater Civ Eng 31(2):04018389
48. Zhou WH, Yuen KV, Tan F (2013) Estimation of maximum pullout shear stress of grouted soil nails using Bayesian probabilistic approach. Int J Geomech 13(5):659–664
49. Tan F, Zhou WH, Yuen KV (2016) Modeling the soil water retention properties of same-textured soils with different initial void ratios. J Hydrol 542:731–743
50. Meulenkamp F, Grima MA (1999) Application of neural networks for the prediction of the unconfined compressive strength (UCS) from Equotip hardness. Int J Rock Mech Min Sci 36(1):29–39
51. Specht DF (1991) A general regression neural network. IEEE Trans Neural Networks 2(6):568–576
52. Ceryan N, Okkan U, Kesimal A (2012) Application of generalized regression neural networks in predicting the unconfined compressive strength of carbonate rocks. Rock Mech Rock Eng 45(6):1055–1072
53. Yuen K, Ortiz GA (2016) Bayesian nonparametric general regression. Int J Uncertain Quantif 6(3):195–213
54. Yuen KV, Ortiz GA, Huang K (2016) Novel nonparametric modeling of seismic attenuation and directivity relationship. Comput Methods Appl Mech Eng 311:537–555
55. Jenkins SH (1930) The determination of cellulose in straws. Biochem J 24(5):1428
56. Goering HK, Van Soest PJ (1970) Forage fiber analyses (apparatus, reagents, procedures, and some applications. USDA agricultural handbook
57. TAPPI Test Methods (1996). TAPPI T222. om-88 klason lignin
58. ASTM E1755-01 (2017) Standard method for the determination of ash in biomass. Annual book of ASTM standards, vol 11, no 5. ASTM, Philadelphia, PA
59. IS-2720-Part 3-1980 (1980) Methods of test for soils. Determination of Specific Gravity, Bureau of Indian Standards Publications, New Delhi
60. IS-1670 (1991) Determination of breaking load and elongation at break of single strand textiles and yarns. Bureau of Indian Standards Publications, New Delhi
61. ASTM D2487-11 (2011) Standard practice for classification of soils for engineering purposes (Unified Soil Classification System). ASTM, West Conshohocken, PA
62. ASTM D4318-10 (2010) Standard test methods for liquid limit, plastic limit, and plasticity index of soils. ASTM, West Conshohocken, PA
63. ASTM D1557-07 (2015) Standard test methods for laboratory compaction characteristics of soil using modified effort (56,000 ft-lbf/ft3 (2,700 kN-m/m3)). ASTM, West Conshohocken, PA, 2007, Annual Book of Standards 4, 2015
64. ASTM D3080/D3080M-11 (2011) Standard test method for direct shear test of soils under consolidated drained conditions. ASTM, West Conshohocken, PA
65. Methacanon P, Weerawatsophon U, Sumransin N, Prahsarn C, Bergado DT (2010) Properties and potential application of the selected natural fibers as limited life geotextiles. Carbohydr Polym 82(4):1090–1096
66. Kumar A, Walia BS, Mohan J (2006) Compressive strength of fiber reinforced highly compressible clay. Constr Build Mater 20(10):1063–1068
67. Chauhan MS, Mittal S, Mohanty B (2008) Performance evaluation of silty sand subgrade reinforced with fly ash and fibr. Geotext Geomembr 26:429–435

68. Gandomi AH, Yun GJ, Alavi AH (2013) An evolutionary approach for modeling of shear strength of RC deep beams. Mater Struct 46(12):2109–2119
69. Terzaghi K (1943) Theoretical soil mechanics. Wiley, New York
70. Su LJ, Zhou WH, Chen W Bin, Jie X (2018) Effects of relative roughness and mean particle size on the shear strength of sand-steel interface. Measurement 122:339–346

Chapter 10
An Efficient Probabilistic Back-Analysis Method for Braced Excavations

Abstract In this chapter, we present an efficient Bayesian back-analysis procedure for braced excavations using wall deflection data at multiple points. A response surface method is adopted to efficiently evaluate the wall response. Deflection data for 49 wall sections from 11 case histories are collected to characterize the model error of the finite element method for evaluating the deflections at various points. A braced excavation project in Hangzhou, China is chosen to illustrate the effectiveness of the proposed procedure. The results indicate that the soil parameters could be updated more significantly for the updating that uses the deflection data at multiple points than that only uses the maximum deflection data. The predicted deflections from the updated parameters agree fairly well with the field observations. The main significance of the proposed procedure is that it improves the updating efficiency of the soil parameters without adding monitoring effort compared with the traditional method that uses the maximum deflection data.

10.1 Introduction

Accurate estimation of soil parameters is vital to prediction of the wall and ground responses in a braced excavation project. Due to the inherent uncertainty and measurement error, soil parameters measured in laboratory testing normally suffer from errors [1–3]. Additionally, the soil samples tested in a laboratory might not be representative of the soil mass in the field [4]. Direct use of the soil parameters from laboratory tests in reliability analysis of a braced excavation problem might lead to an unrealistic evaluation of the safety of the excavation system. Hence, it is desirable to reduce the uncertainty in the soil parameter for more rational reliability-based design of the excavation system. Methods used to reduce the uncertainty in soil parameters can be generally classified into two categories. The first category adopts certain empirical relations or statistical correlations between the soil parameters (such as undrained shear strength, friction angle, etc.) and selected field or laboratory test results (such as over-consolidation ratio, cone tip resistance in the cone penetration test and standard penetration test value). The field or laboratory test results are used as conditional information to update the soil parameter distributions. Examples of

W. Zhou et al., *Practice of Bayesian Probability Theory in Geotechnical Engineering*,
https://doi.org/10.1007/978-981-15-9105-1_10

this category can be found in the works of Cao and Wang [5], Ching et al. [6], Ching et al. [7] and Wang et al. [8]. The second category adopts selected field observations to reduce the uncertainty in the soil parameters via back analysis. This category of methods has been widely applied [9–11] due to several attractive advantages. One remarkable merit is that back analysis using field observations partially reflects field conditions such as the effective normal stress and shear displacement rate of the soil mass [4]. These conditions cannot be reproduced in the laboratory. In addition, the back analysis incorporates certain important factors that might not be well represented by the laboratory samples, such as the structural fabric of the soil and pre-existing shear planes [4].

Many methods are available for back analysis of soil parameters, such as the least squares method [9], artificial neural networks [12], genetic algorithm [10], maximum likelihood method [13] and Bayesian method [14, 15]. Most of the methods (e.g., least squares method, artificial neural networks) treat the soil parameters as constants rather than random variables. The results for these back-analysis methods are a set of fixed values, but these fixed values are not necessarily the true values of the soil parameters because uncertainties usually exist in both the field observations and the calculation models [11, 16]. Recently, the Bayesian method has gained wide application in a variety of geotechnical problems, such as slope stability [16], pile capacity [17] and braced excavations [11, 15]. The Bayesian method is superior to the aforementioned back-analysis methods in several aspects. First, uncertainty in the soil parameters can be adequately considered in this method. In addition, the Bayesian method can logically incorporate other sources of information, such as prior knowledge or expert judgment on the soil parameters [18].

For application of Bayesian updating in the problem of braced excavation, Wang et al. [15] updated the model factor of a semi-empirical KJHH model for prediction of the maximum ground settlement using centrifuge data. The updated model was further used to predict the maximum ground settlement and calculate the reliability of the braced excavation problem. The KJHH model consists of three polynomial functions which are used to predict the maximum wall deflection (MWD), maximum ground settlement (MGS) and surface settlement profile (SSP). The functions are obtained by fitting the inputs (i.e., soil and structure parameters) and outputs (i.e., MWD, MGS and SSP) of some artificial finite element analyses of braced excavations. The finite element analyses used in the KJHH model merely consider one typical case of the braced excavation, that is, the case with a flat ground surface underlain by certain soft to medium clays. In addition, only the parameters of the softest soil and supporting structures are selected as the inputs of the KJHH model. Details of the model could be found in Kung et al. [19]. Juang et al. [11] proposed a Bayesian framework for updating of soil parameters using field observations in multistage braced excavations. The framework provides a reasonable method for adjustment of a possibly unsafe or uneconomical design scheme as determined prior to excavation. However, certain drawbacks exist in the framework that limits its applicability and updating efficiency. First, similar to Wang et al. [15], the semi-empirical KJHH model was adopted to evaluate the maximum wall deflection and maximum ground settlement. The model is only applicable to specific conditions, that is, excavations with a flat ground surface underlain by certain soft to medium clays. Excavation problems with complex boundary conditions (e.g., rotation fixity of the wall) or irregular geometries (e.g., excavation in an inclined slope) cannot be handled by this model. This limitation restricts the applicability of the framework. Secondly, the framework only uses the maximum wall deflection and maximum ground settlement data in the Bayesian updating because the outputs of the KJHH model are limited to the MWD and MGS. This amount of data might not be sufficient to update the soil parameter efficiently, as demonstrated later in the chapter. Thirdly, the framework could only update the parameters of the softest clay layer because the inputs of the KJHH model only contain these soil parameters. In reality, the MWD may be sensitive to the soil parameters of other soil layers. The importance of the soil parameters for layers other than the softest clay is neglected in the framework. Finally, measurement errors of the observation data are not considered and model errors are viewed as random variables with fixed probability distributions. These treatments are not realistic because any observation inevitably suffers from certain measurement errors. In addition, the distribution of the model error may not be accurate due to limited database and the model uncertainty needs to be updated when more observation information is available.

In practice, the deflection data for multiple points on the wall are always available. Acquisition of these data generally requires the same monitoring efforts as the maximum wall deflection. It is of interest to examine whether the efficiency of Bayesian updating improves if additional wall deflection data are used. Hence, this chapter presents a more versatile and efficient Bayesian back-analysis framework using the wall deflection data at multiple points. In the framework, model errors are also considered as random variables and are updated together with the soil parameters for various soil layers. To overcome the applicability problem mentioned previously, a response surface method is adopted to rapidly calculate the wall response. The response surface acts as a surrogate for any numerical method (such as the finite element method or finite difference method). It appears in the form of polynomial functions and can be readily constructed using the results of several trial analyses with the numerical method. Similar to Juang et al. [11], Bayesian updating is conducted in a stage-by-stage manner. The updated soil parameter and model errors in the current excavation stage are used to predict the maximum wall deflection of the subsequent excavation stages. The efficiency of the Bayesian updating is evaluated by comparing the predicted maximum wall deflection with the observed maximum wall deflection. The model uncertainty of finite element method is characterized using wall deflection data for 49 wall sections from 11 case histories to obtain the prior distribution and correlation structure of the model errors. A multilayer excavation problem in Hangzhou, China is chosen to illustrate the effectiveness of the proposed framework. The spatial variability of the soil parameters is not considered in this chapter. The effects of the prior coefficient of variance (COV) of the soil parameters on Bayesian updating are also discussed.

10.2 Method for Updating the Soil Parameters

This section describes the method used in updating the soil parameters and model errors. First, the framework for Bayesian updating is briefly reviewed. Note that numerical analyses of geotechnical problems customarily require inputs of multiple soil parameters, such as cohesion (c), friction angle (ϕ) and elastic modulus (E). The number of input parameters might increase if multiple soil layers exist in a soil profile (each soil layer possesses one set of soil parameters). Hence, the information to be updated in the Bayesian framework is a joint probability density function (PDF) of multiple soil parameters. For each parameter, calculation of the marginal distribution or the statistics (such as the mean and standard deviation) always involves a high-dimensional integral. This task is quite difficult or even impossible to accomplish. Although analytical solutions or approximate solutions have been proposed in the past [18], those solutions apply only to limited cases. Specifically, the solutions are effective only if the prior distribution of the soil parameter is a normal distribution and the geotechnical responses [such as the factor of safety (FOS) or displacement of geotechnical structures] vary linearly with the input parameters. Hence, a highly efficient sampling technique is adopted in this chapter, namely the Markov chain Monte Carlo (MCMC) simulation method. Another difficulty that was frequently encountered in the Bayesian updating framework is repeated evaluation of the geotechnical response. It is costly and time-consuming to calculate the geotechnical response hundreds of thousands of times using traditional numerical methods such as the finite element method (FEM) or finite difference method (FDM). To cope with this problem, a response surface method is adopted to ease the computational burden.

10.2.1 Framework for Bayesian Updating

The framework begins with the uncertainty of the geotechnical models. It is well recognized that model error exists in any geotechnical model due to oversimplification of the real conditions of geotechnical structures, such as boundary conditions and constitutive models of the soils [20]. Due to the model error, the responses of one geotechnical structure evaluated from any model generally do not agree with the observed performances of the structure. The model error is usually represented by a model bias factor, which is the ratio of the actual geotechnical response to the prediction from geotechnical models. In this chapter, the wall deflections are evaluated using a finite element method (FEM). The model error of the FEM analysis is expressed as follows:

$$y_{\max} = \varepsilon_{\max} \cdot \delta_{\max}(\boldsymbol{\theta}) \tag{10.1}$$

$$y = \varepsilon \cdot \delta(\boldsymbol{\theta}) \tag{10.2}$$

where y_{max} is the actual maximum wall deflection; ε_{max} means the model bias factor for the maximum wall deflection; $\boldsymbol{\theta}$ is input soil parameters for finite element analysis; $\delta_{max}(\boldsymbol{\theta})$ is the maximum wall deflection evaluated from the FEM; $y = [y_{z1}, y_{z2}, \ldots, y_{z,np}]^T$ is the actual wall deflections of the points at depth $z = [z_1, z_2, \ldots, z_{np}]^T$, where z_i = the depth from the ground surface to the ith point of the wall and np = number of the observed points used for Bayesian updating; $\varepsilon = [\varepsilon_1, \varepsilon_2, \ldots, \varepsilon_{np}]^T$, where ε_i is the model bias factor for the deflection of the point at depth z_i and $\delta(\boldsymbol{\theta}) = [\delta(\boldsymbol{\theta}, z_1), \delta(\boldsymbol{\theta}, z_2), \ldots, \delta(\boldsymbol{\theta}, z_{np})]^T$, where $\delta(\boldsymbol{\theta}, z_i)$ denotes the wall deflection of the point at depth z_i evaluated from FEM analysis. The model bias factor of FEM for evaluation of the wall deflection is not well documented in the literature. Although Zhang et al. [21] characterized the model uncertainty of FEM using the deflection data from 45 real excavation case histories, the characterization was mainly for cantilever deflections and the maximum deflection. The cantilever deflection refers to the deflection of the wall without any supporting structures such as struts. Hence the model bias factors for strut-braced excavations (whose deformation patterns are different from that of the cantilever deflections) and for the deflections at various points are characterized in Sect. 10.3.1. It will be shown that ε is lognormally distributed. It is worth noting that the model error could be further updated if more deflection data are available. The updating of model error could be well incorporated in the Bayesian framework. In this work, the mean and standard deviation of the model errors are also viewed as random variables or vectors and are updated along with the soil parameters.

Besides the model error, the measurement error is another reason that causes the disparity between numerical results and the observed performances of geotechnical structures. Due to the presence of measurement errors, the observed deflections may not be the real response of the wall in a braced excavation project. The measurement error is considered by multiplying the deflections evaluated from FEM analysis by a measurement uncertainty factor:

$$d_{max} = \chi_{max} \cdot y_{max} \tag{10.3}$$

$$d = \chi \cdot y \tag{10.4}$$

where d_{max} is the observed maximum deflection, d is the observed deflections at various depths, χ_{max} and χ are the measurement uncertainty factor for the maximum deflection and the deflections at various depths, respectively. To avoid negative values, the measurement uncertainty factor is always assumed to be a lognormally distributed random variable [22]. The measurement errors are also independent of the model bias factors. Different instruments have different correlation structure of χ. The standard deviation and correlation structure of the inclinometer instrument are discussed in Sect. 10.3.2.

Based on Eqs. (10.1) and (10.4), it can be easily shown that the natural logarithm of the observed deflections, that is, $\ln(d_{max})$ and $\ln(d)$ are normally distributed, because they are the sum of some independent normally distributed random variables or vectors, $\ln(\chi_{\max})$, and $\ln(\varepsilon)$:

$$\ln(d_{\max}) = \ln(\chi_{\max}) + \ln(\varepsilon_{\max}) + \ln(\delta_{\max}(\boldsymbol{\theta})) \tag{10.5}$$

$$\ln(d) = \ln(\chi) + \ln(\varepsilon) + \ln(\delta(\boldsymbol{\theta})) \tag{10.6}$$

The mean of ln(d) is $\mu_{\ln(d)} = \mu_{\ln(\chi)} + \mu_{\ln(\varepsilon)} + \ln(\delta(\boldsymbol{\theta}))$, where $\mu_{\ln(\chi)}$ and $\mu_{\ln(\varepsilon)}$ are the mean values of $\ln(\chi)$ and $\ln(\varepsilon)$, respectively. The covariance of ln(d), $C_{\ln(\mathrm{d})} = C_{\ln(\chi)} + C_{\ln(\varepsilon)}$, where $C_{\ln(\chi)}$ and $C_{\ln(\varepsilon)}$ are the covariance of ln (χ) and $\ln(\varepsilon)$, respectively. The value of $\mu_{\ln(\chi)}$, $\mu_{\ln(\varepsilon)}$, $C_{\ln(\chi)}$ and $C_{\ln(\varepsilon)}$ can be calculated from the statistics of the corresponding lognormally distributed vectors, that is, χ and ε [23]. Note that the covariance of one random vector includes the information of both standard deviation and correlation coefficient of the random vector. The covariance and the (standard deviation, correlation coefficient) information could be mutually converted based on the following relations, that is, $\rho_{\mathrm{XY}} = \mathrm{COV(X, Y)}/\sigma_{\mathrm{X}}\sigma_{\mathrm{Y}}$ and $\sigma_{\mathrm{X}}=\sqrt{\mathrm{COV(X, X)}}$, where ρ_{XY} is the correlation coefficient of two random variables X and Y, COV(X, Y) is the covariance of the X and Y and σ_{X} is the standard deviation of the X. Likewise, the mean and variance of $\ln(d_{\max})$ are, respectively, given by $\mu_{\ln(d_{\max})} = \mu_{\ln(\chi_{\max})} + \mu_{\ln(\varepsilon_{\max})} + \ln(\delta_{\max}(\boldsymbol{\theta}))$ and $\sigma^2_{\ln(d_{\max})} = \sigma^2_{\ln(\chi_{\max})} + \sigma^2_{\ln(\varepsilon_{\max})}$, where $\mu_{\ln(\chi_{\max})}$ is the mean of $\ln(\chi_{\max})$, $\mu_{\ln(\varepsilon_{\max})}$ is the mean of $\ln(\varepsilon_{\max})$, $\sigma^2_{\ln(\chi_{\max})}$ is the variance of $\ln(\chi_{\max})$ and $\sigma^2_{\ln(\varepsilon_{\max})}$ is the variance of $\ln(\varepsilon_{\max})$. The value of $\mu_{\ln(\chi_{\max})}$ and $\sigma^2_{\ln(\chi_{\max})}$ could be calculated from the mean and variance of $\chi_{\max}$ while the value of $\mu_{\ln(\varepsilon_{\max})}$ and $\sigma^2_{\ln(\varepsilon_{\max})}$ could be calculated from the mean and variance of $\varepsilon_{\max}$, that is, $\mu_{\varepsilon_{\max}}$ and $\sigma^2_{\varepsilon_{\max}}$.

Using the results in Eq. (10.5), the likelihood that the observed maximum deflection is equal to $D_{\max}$ given $\mu_{\varepsilon_{\max}}$, $\sigma_{\varepsilon_{\max}}$ and $\boldsymbol{\theta}$ is expressed by:

$$\mathrm{L}(\mu_{\varepsilon_{\max}}, \sigma_{\varepsilon_{\max}}, \boldsymbol{\theta}|\mathrm{D}_{\max}) = \frac{1}{\sqrt{2\pi}\sigma_{\ln(d_{\max})}} \exp[-\frac{1}{2\sigma^2_{\ln(d_{\max})}}\left(\ln(\mathrm{D}_{\max}) - \mu_{\ln(d_{\max})}\right)^2] \tag{10.7}$$

Let $f(\boldsymbol{\theta})$ and $f(\mu_{\varepsilon_{\max}}, \sigma_{\varepsilon_{\max}})$ denote the prior probability density function of $\boldsymbol{\theta}$ and $(\mu_{\varepsilon_{\max}}, \sigma_{\varepsilon_{\max}})$, which respectively represent our prior knowledge of the soil properties (mean and standard deviation of $\varepsilon_{\max}$). In a Bayesian framework, the posterior probability density function of $(\boldsymbol{\theta}, \mu_{\varepsilon_{\max}}, \sigma_{\varepsilon_{\max}})$ is given by Zhang et al. [16]:

$$f\left(\mu_{\varepsilon_{\max}}, \sigma_{\varepsilon_{\max}}, \boldsymbol{\theta}|\mathrm{D}_{\max}\right) = k_1 \cdot f(\boldsymbol{\theta}) \cdot f\left(\mu_{\varepsilon_{\max}}, \sigma_{\varepsilon_{\max}}\right) \cdot \mathrm{L}(\mu_{\varepsilon_{\max}}, \sigma_{\varepsilon_{\max}}, \boldsymbol{\theta}|\mathrm{D}_{\max}) \tag{10.8}$$

where k_1 is a normalization constant that makes the probability density function valid.

Similarly, if observations of multiple points are available, the likelihood that the observed deflection at depth $z = [z_1, z_2, \ldots, z_{np}]^T$ are equal to $D = [D_{z1}, D_{z2}, \ldots,$

$D_{z,np}]^T$ given μ_ε, σ_ε and $\boldsymbol{\theta}$ is expressed by:

$$\mathrm{L}(\mu_\varepsilon, \sigma_\varepsilon, \boldsymbol{\theta}|\mathrm{D}) = \frac{1}{2\pi^{\frac{np}{2}}\left|C_{\ln(d)}\right|^{\frac{1}{2}}} \exp[-\frac{1}{2}\left(\ln(\mathrm{D}) - \mu_{\ln(d)}\right)^T C_{\ln(d)}^{-1}(\ln(\mathrm{D}) - \mu_{\ln(d)})] \tag{10.9}$$

where $\left|C_{\ln(d)}\right|$ is the determinant of $\left|C_{\ln(d)}\right|$. Let $f(\mu_\varepsilon, \sigma_\varepsilon)$ denotes the prior distribution $(\mu_\varepsilon, \sigma_\varepsilon)$, which represent the prior knowledge of $(\mu_\varepsilon, \sigma_\varepsilon)$. The posterior probability density function of $(\boldsymbol{\theta}, \mu_\varepsilon, \sigma_\varepsilon)$ is expressed by:

$$f(\mu_\varepsilon, \sigma_\varepsilon, \boldsymbol{\theta}|\mathrm{D}) = k_2 \cdot f(\boldsymbol{\theta}) \cdot f(\mu_\varepsilon, \sigma_\varepsilon) \cdot \mathrm{L}(\mu_\varepsilon, \sigma_\varepsilon, \boldsymbol{\theta}|\mathrm{D}) \tag{10.10}$$

where k_2 is a normalization constant.

10.2.2 Procedure for the Markov Chain Monte Carlo Simulation Using the Metropolis–Hastings Algorithm

The procedure for MCMC sampling of $(\mu_\varepsilon, \sigma_\varepsilon, \boldsymbol{\theta})$ with a PDF of Eq. (10.8) or (10.10) using the Metropolis–Hasting algorithm is summarized as follows. Additional details can be found in Juang et al. [11] and Zhang et al. [16].

(1) Determine the first point $(\mu_\varepsilon, \sigma_\varepsilon, \boldsymbol{\theta})^{(1)}$ of the Markov chain, which can be simply sampled from the prior distribution of $(\mu_\varepsilon, \sigma_\varepsilon, \boldsymbol{\theta})$ or set to the mean value.

(2) Generate the next point of the Markov chain $(\mu_\varepsilon, \sigma_\varepsilon, \boldsymbol{\theta})^{(k)}$, based on the current point, $(\mu_\varepsilon, \sigma_\varepsilon, \boldsymbol{\theta})^{(k-1)}$. Note that the superscript k in $(\mu_\varepsilon, \sigma_\varepsilon, \boldsymbol{\theta})^{(k)}$ denote the kth sample of $(\mu_\varepsilon, \sigma_\varepsilon, \boldsymbol{\theta})$ rather than $(\mu_\varepsilon, \sigma_\varepsilon, \boldsymbol{\theta})$ to the kth power.

   a. Sample a candidate $(\mu_\varepsilon, \sigma_\varepsilon, \boldsymbol{\theta})^*$ from a jumping or kernel distribution $T((\mu_\varepsilon, \sigma_\varepsilon, \boldsymbol{\theta})^*|\ (\mu_\varepsilon, \sigma_\varepsilon, \boldsymbol{\theta})^{(k-1)})$. The jumping distribution is assumed to be a multivariate normal distribution with mean $= (\mu_\varepsilon, \sigma_\varepsilon, \boldsymbol{\theta})^{(k-1)}$ and covariance matrix $= s * C_{(\mu_\varepsilon, \sigma_\varepsilon, \boldsymbol{\theta})}$, where s is a scaling factor, and $C_{(\mu_\varepsilon, \sigma_\varepsilon, \boldsymbol{\theta})}$ is the covariance matrix of the prior distribution of $(\mu_\varepsilon, \sigma_\varepsilon, \boldsymbol{\theta})$.
   b. Generate a random number u from a uniform distribution $U(0, 1)$.
   c. Calculate the ratio of the densities r

$$r = \frac{q((\mu_\varepsilon, \sigma_\varepsilon, \boldsymbol{\theta})^*|\mathrm{D})}{q((\mu_\varepsilon, \sigma_\varepsilon, \boldsymbol{\theta})^{(k-1)}|\mathrm{D})} \tag{10.11}$$

   where $q(\mu_\varepsilon, \sigma_\varepsilon, \boldsymbol{\theta}|\mathrm{D})$ is the unnormalized posterior PDF, that is, the right-hand side of Eq. (10.8) without k_1 or the right-hand side of Eq. (10.10) without k_2.

d. Decide whether $(\mu_\varepsilon, \sigma_\varepsilon, \boldsymbol{\theta})^*$ is acceptable based on the following rule: If $u \leq \min(r, 1)$, then $(\mu_\varepsilon, \sigma_\varepsilon, \boldsymbol{\theta})^*$ is acceptable and set $(\mu_\varepsilon, \sigma_\varepsilon, \boldsymbol{\theta})^{(k)} = (\mu_\varepsilon, \sigma_\varepsilon, \boldsymbol{\theta})^*$; otherwise, set $(\mu_\varepsilon, \sigma_\varepsilon, \boldsymbol{\theta})^{(k)} = (\mu_\varepsilon, \sigma_\varepsilon, \boldsymbol{\theta})^{(k-1)}$.
e. Stop the MCMC simulation if $k = n$; otherwise, set $k = k+1$ and repeat steps (a–d).

The Metropolis–Hastings algorithm randomly generates samples from the posterior PDF of $(\mu_\varepsilon, \sigma_\varepsilon, \boldsymbol{\theta})$. It should be noted that initial samples do not strictly obey the posterior PDF because the Markov chain has not yet reached the steady state. It is reasonable to discard certain initial samples as burn-in samples [11]. In addition, the efficiency of the MCMC depends on many factors such as the jumping function and the MCMC length. These factors should be carefully selected. A MATLAB code for MCMC sampling is presented in Appendix H.

10.2.3 Response Surface Method

As mentioned previously, the repeated evaluation of the wall deflection using the finite element method is time-consuming. Hence, a response surface method is adopted to bypass the tedious FEM analysis. The basic concept of the response surface method is to relate the geotechnical response and input parameters using a polynomial function. The function is used to simply calculate the geotechnical response with less computational effort. The method has been applied in many geotechnical problems such as slope stability analysis and underground tunneling [24–26]. It is also used to evaluate the wall deflection in this work. The maximum wall deflection is given by

$$\delta_{\max}(\boldsymbol{\theta}) = b_0 + \sum_{i=1}^{n} b_i\theta_i + \sum_{i=1}^{n} b_{n+i}\theta_i^2 \tag{10.12}$$

where n is the number of the input parameters; $\theta_1, \theta_2, \ldots, \theta_n$ denote input parameters for the FEM analysis, which are components of the random vector $\boldsymbol{\theta}$; and b_0, b_1, $b_2, \ldots, b_{2n}$ are regression coefficients of the polynomial function. Note that a second-order polynomial function without cross terms [i.e., the terms $\theta_i\theta_j$ $(i \neq j)$] is adopted to calculate the maximum deflection. This order is sufficient to obtain a satisfactory accuracy, as illustrated in Sect. 4.3. Although a higher order polynomial function could also be applied for a more precise evaluation of the maximum deflection, the effort required to obtain the unknown regression coefficients (i.e., $b_0, \ldots, b_{2n}$) can increase significantly. This treatment agrees well with the model proposed by Kung et al. [19]. Similarly, the wall deflection at any depth z_j, that is, $\delta(\boldsymbol{\theta}, z_j)$, could also be evaluated using the same form of function

$$\delta(\boldsymbol{\theta}, z_j) = b_{j0} + \sum_{i=1}^{n} b_{ji}\theta_i + \sum_{i=1}^{n} b_{j,n+i}\theta_i^2 \tag{10.13}$$

where b_{j0}, b_{j1}, b_{j2},...,$b_{j,2n}$ are regression coefficients of the polynomial function.

The regression coefficients for the response surfaces are obtained using a nonlinear least squares curve-fitting procedure. The basic steps are described as follows. First, three representative values, that is, $[\mu-2\sigma, \mu, \mu+2\sigma]$ (μ = mean, σ = standard deviation), for each input parameter are selected as the input for the curve-fitting procedure. One realization of FEM analysis (also referred to as one experiment) requires one combination of all the parameters as inputs. Theoretically, 3^n combinations could be designed, which is quite a large number. Hence, an orthogonal technique [27] is adopted to produce a more effective experimental scheme. A total of 81 combinations of the n input parameters are designed according to the orthogonal design table, which is generated in the statistical analysis software SPSS. Second, 81 experiments of the FEM analysis for the braced excavation problem are conducted based on the 81 combinations of input parameters. Each experiment generates Ns maximum deflections and Ns deflection profiles for the diaphragm wall, where Ns is the number of excavation stages considered in the finite element analysis. Finally, based on the inputs and outputs for the 81 FEM experiments, the regression coefficients are estimated using a nonlinear regression procedure in MATLAB, that is, *nlitfit*. It should be noted that the response surfaces for the deflection at various depths and various excavation stages are different, which means that multiple response surfaces should be constructed. For example, if the deflection data at three points and three excavation stages are used in the Bayesian updating, the number of required response surfaces is $3 \times 3 = 9$.

10.3 Measurement Errors and Model Errors

As illustrated in Sect. 10.2.1, statistics and correlation coefficient of the measurement error, χ and the model error, ε, are indispensable to evaluate the likelihood function of the observed deflections. Meanwhile, the prior input of Bayesian updating also requires the information of the statistics of the model error. As clarified by Ledesma et al. [28], a good correlation structure of the measurement error is crucial for an effective back analysis. Hence, those χ and ε information could not be simply assumed in the Bayesian updating. This section characterizes the model errors of the wall deflections at various points and the maximum deflection by collecting deflection data for 49 wall sections from 11 case histories. In addition, inclinometer is usually used to measure the deflection of a wall in practice. The standard deviation and correlation structure of the measurement error for this instrument are evaluated.

10.3.1 Model Errors of the Wall Deflections in Braced Excavations

To evaluate the model errors of wall deflections, deflection profiles for 49 wall sections from 11 case histories are collected. These cases are from tunnel or metro line construction projects in Thailand, China, America, Australia and Singapore. All these case histories involve multiple soil layers and the ranges for the installation depth (ID_i), final excavation ($H_{e,f}$) and half of the excavation width ($B/2$) are, respectively, [22 m, 45 m], [12 m, 26 m] and [8.8 m, 16.45 m]. Details of the excavation cases are summarized in Table 10.1. Finite element analyses are conducted to numerically evaluate the wall deflections for each case. With the measured data and the FEM analyses results, the model bias factors for the wall deflections at different points are calculated in two steps: (1) The location of the wall points is normalized by the excavation depth, that is, DP/H_e, where DP is the depth from the ground surface and H_e is the excavation depth. For each wall section, one set of model bias factors are calculated by dividing the observed deflections at depths $[DP_1/H_e, \ldots, DP_{np}/H_e]^T$ by the corresponding deflections evaluated from FEM analysis. 49 model bias factor vectors will be obtained for the 49 wall sections. As some of the case histories have been numerically analyzed in the corresponding references, the given numerical results from the literatures are directly used for the evaluation of ε. (2) One vector is viewed as one sample of the model bias factors. The statistics (i.e., mean and standard deviation) and correlation coefficient matrix for the model bias factors at depths $[DP_1/H_e, \ldots, DP_{np}/H_e]^T$ are directly calculated from the 49 vectors. The mean and standard deviation for the maximum deflection are $\mu_{\varepsilon_{\max}} = 0.905$ and $\sigma_{\varepsilon_{\max}} = 0.303$, respectively.

Table 10.2 summarizes the statistics and correlation structures for the model bias factors at certain locations. The range of DP/H_e is set to [0.5, 1.2] in Table 10.1b. The reason is that the DP/H_e for the points associated with the maximum deflection falls within this range in the collected case histories. Moreover, some hypothesis tests are also conducted to evaluate the distribution of the ε for the maximum deflections and the deflections at various points. The results show that all the ε could be described by lognormal distributions. Hence, lognormally distributed ε with the mean and standard deviation summarized in Table 10.2a is used as prior information and be further updated in the Bayesian updating process. The correlation structure in Table 10.1a is used to evaluate the likelihood function in Sect. 4.4. The results in Table 10.2b could be used in back analysis of other braced excavation problems. Note that the model bias factors in Table 10.2 are obtained from finite element analysis with the Mohr–Coulomb failure criterion. The model bias factors for different models should be different because different models induce different numerical responses of geotechnical structures. If another constitutive model is used in the back analysis of braced excavations, the model bias factors need to be re-evaluated.

Table 10.1 Database of braced excavation case histories

Case number	Case name	ID_i (m)	$H_{e,f}$ (m)	$B/2$ (m)	References
1	Bangkok Sukhumvit Station	27.9	20.9	11.5	Likitlersuang et al. [29]
2	Boston Courthouse Station (I05N1)	26.2	17.3	16.45	Whittle et al. [30]
3	Boston Courthouse Station (I06N1)	26.2	17.3	16.45	Whittle et al. [30]
4	Boston Courthouse Station (I06S1)	26.2	17.3	16.45	Whittle et al. [30]
5	Boston Courthouse Station (I07N1)	26.2	17.3	16.45	Whittle et al. [30]
6	Boston Courthouse Station (I07S1)	26.2	17.3	16.45	Whittle et al. [30]
7	Kaohsiung O6 station	36.0	19.6	10.35	Hsiung [31]
8	Perth Esplanade Station (East wall)	22.0	12.0	11.75	Lehane and Zaremba [32]
9	Perth Esplanade Station (West wall)	22.0	12.0	11.75	Lehane and Zaremba [32]
10	Shanghai Changqing Road Station (North wall)	31.2	16.4	13.25	Lu et al. [33]
11	Shanghai Changqing Road Station (South wall)	31.2	16.4	13.25	Lu et al. [33]
12	Shanghai East Nanjing Road Station (maximum)	45.0	26.0	11.5	Lu et al. [33]
13	Shanghai East Nanjing Road Station (minimum)	45.0	26.0	11.5	Lu et al. [33]
14	Shanghai Shangcheng Road Station (North Wall)	34.0	16.0	10.0	Lu et al. [33]
15	Shanghai Shangcheng Road Station (South wall)	34.0	16.0	10.0	Lu et al. [33]
16	Shanghai Metro Line 10 (Monitored value 1)	28.0	14.6	8.8	Liu et al. [34]
17	Shanghai Metro Line 10 (Monitored value 2)	28.0	14.6	8.8	Liu et al. [34]
18	Shanghai Metro Line 10 (Monitored value 3)	28.0	14.6	8.8	Liu et al. [34]
19	Shanghai Metro line 9 (Section 1, B14)	33.0	18.0	11.5	Tan and Wei [35]

(continued)

Table 10.1 (continued)

Case number	Case name	ID_i (m)	$H_{e,f}$ (m)	$B/2$ (m)	References
20	Shanghai Metro line 9 (Section 2, B10)	33.6	16.0	10.0	Tan and Wei [35]
21	Shanghai Metro line 9 (Section 2, B11)	33.6	16.0	10.0	Tan and Wei [35]
22	Shanghai Metro line 9 (Section 2, B12)	33.6	16.0	10.0	Tan and Wei [35]
23	Shanghai Metro line 9 (Section 2, B13)	33.6	16.0	10.0	Tan and Wei [35]
24	Shanghai Metro line 9 (Section 2, B16)	33.6	16.0	10.0	Tan and Wei [35]
25	Shanghai Metro line 9 (Section 2, B17)	33.6	16.0	10.0	Tan and Wei [35]
26	Shanghai Metro line 9 (Section 2, B18)	33.6	16.0	10.0	Tan and Wei [35]
27	Shanghai Metro line 9 (Section 2, B19)	33.6	16.0	10.0	Tan and Wei [35]
28	Shanghai Metro line 9 (Section 3, B20)	33.6	16.0	10.0	Tan and Wei [35]
29	Shanghai Metro line 9 (Section 3, B21)	33.6	16.0	10.0	Tan and Wei [35]
30	Shanghai Metro line 9 (Section 3, B7)	33.6	16.0	10.0	Tan and Wei [35]
31	Shanghai Metro line 9 (Section 3, B8)	33.6	16.0	10.0	Tan and Wei [35]
32	Shanghai Metro line 9 (Section 4, B25)	31.2	18.0	5.0	Tan and Wei [35]
33	Shanghai Metro line 9 (Section 4, B3)	31.2	18.0	5.0	Tan and Wei [35]
34	Shanghai Metro line 9 (Section 4, B2)	31.2	18.0	5.0	Tan and Wei [35]
35	Shanghai Pudian Road	26.5	16.5	10.2	Wang et al. [36]
36	Shanghai Pudongdadao Road	27.0	16.0	9.25	Wang et al. [36]
37	Shanghai South Pudong Road	27.0	17.3	10.4	Wang et al. [36]
38	Shanghai South Xizan Road	38.0	20.6	11.4	Wang et al. [36]
39	Shanghai Yanchang Road	27.0	15.2	9.05	Wang et al. [36]
40	Shanghai Yishan Road	28.0	15.5	8.65	Wang et al. [36]

(continued)

Table 10.1 (continued)

Case number	Case name	ID_i (m)	$H_{e,f}$ (m)	$B/2$ (m)	References
41	Singapore Hougang–Buangkok tunnel	18.0	13.3	8.0	Teo and Wong [37]
42	Singapore Lavender Station (Measurement 1)	28.0	15.66	11.5	Teo and Wong [37]
43	Singapore Lavender Station (Measurement 2)	28.0	15.66	11.5	Teo and Wong [37]
44	Singapore Lavender Station (Measurement 3)	28.0	15.66	11.5	Teo and Wong [37]
45	Singapore new Circle Line tunnel (I-104 April 17)	42.0	33.3	10.0	Corral and Whittle [38]
46	Singapore new Circle Line tunnel (I-104 April 20)	42.0	33.3	10.0	Corral and Whittle [38]
47	Singapore new Circle Line tunnel (I-65 April 17)	42.0	33.3	10.0	Corral and Whittle [38]
48	Singapore Rochor Complex	24.0	6.3	47.5	Teo and Wong [37]
49	Taipei TNEC	35.0	19.7	21.5	Lim et al. [39]

Note ID_i = the installation depth of wall; $H_{e,f}$ = the final excavation depth; $B/2$ = half of the excavation width

10.3.2 Measurement Errors of Inclinometer Instruments

As discussed in Sect. 10.2.1, the measurement error is expressed by a measurement uncertainty factor, χ, which is the ratio of the measured deflection and the real deflection. χ is usually modeled as a lognormally distributed random variable or vector in back analysis [22, 28]. The mean of the χ is assumed to be 1 which means that the measurement of the inclinometer is unbiased [40]. Hence the only unknown information of χ is its covariance. Ledesma et al. [28] reported the covariance structure of the measurement error for various instruments. In their study, the measurement error is defined as the different value between the measured and the real response of a system, that is, $e = d\text{-}y$. The covariance matrix, C, of the measurement error for inclinometers is given by Eq. (10.14) [28].

$$C_{ij} = \sigma^2 \sum_{m=1}^{\min(i,j)} l_m^2 \tag{10.14}$$

where C_{ij} is the covariance of the measurement error at point i and j, σ^2 is the variance of the slope measurement performed with the inclinometer device and l_m is the

Table 10.2 Statistics and correlation structures for the model bias factors at various locations

DP/H_e	μ_ε	σ_ε	Correlation coefficient matrix										
a. Modell bias factors at DP/H_e = [0.55 0.75 0.96 1.16 1.23 1.30 1.44 1.58 1.71]													
0.55	0.923	0.290	1.00	0.92	0.72		0.55	0.51	0.46	0.37	0.31		0.24
0.75	0.924	0.289	0.92	1.00	0.91		0.77	0.74	0.69	0.60	0.54		0.46
0.96	0.868	0.331	0.72	0.91	1.00		0.94	0.91	0.87	0.78	0.71		0.58
1.16	0.812	0.325	0.55	0.77	0.94		1.00	0.99	0.98	0.92	0.84		0.66
1.23	0.825	0.327	0.51	0.74	0.91		0.99	1.00	0.99	0.95	0.87		0.67
1.30	0.809	0.335	0.46	0.69	0.87		0.98	0.99	1.00	0.97	0.88		0.68
1.44	0.743	0.330	0.37	0.60	0.78		0.92	0.95	0.97	1.00	0.93		0.74
1.58	0.711	0.339	0.31	0.54	0.71		0.84	0.87	0.88	0.93	1.00		0.92
1.71	0.717	0.374	0.24	0.46	0.58		0.66	0.67	0.68	0.74	0.92		1.00
b. Model bias factors at DP/H_e = [0.5 0.6 0.7 0.8 0.9 1.0 1.1 1.2 1.3 1.4 1.5]													
0.50	0.931	0.353	1.00	0.98	0.94	0.85	0.74	0.63	0.49	0.38	0.27	0.20	0.10
0.60	0.928	0.321	0.98	1.00	0.98	0.92	0.81	0.71	0.58	0.47	0.37	0.29	0.19
0.70	0.930	0.302	0.94	0.98	1.00	0.97	0.90	0.82	0.70	0.59	0.48	0.39	0.28
0.80	0.916	0.303	0.85	0.92	0.97	1.00	0.97	0.91	0.81	0.72	0.62	0.53	0.42
0.90	0.888	0.308	0.74	0.81	0.90	0.97	1.00	0.98	0.92	0.83	0.73	0.63	0.52
1.00	0.856	0.303	0.63	0.71	0.82	0.91	0.98	1.00	0.97	0.91	0.83	0.74	0.64
1.10	0.819	0.293	0.49	0.58	0.70	0.81	0.92	0.97	1.00	0.97	0.92	0.84	0.75
1.20	0.803	0.288	0.38	0.47	0.59	0.72	0.83	0.91	0.97	1.00	0.98	0.94	0.86
1.30	0.811	0.303	0.27	0.37	0.48	0.62	0.73	0.83	0.92	0.98	1.00	0.98	0.92
1.40	0.758	0.306	0.20	0.29	0.39	0.53	0.63	0.74	0.84	0.94	0.98	1.00	0.97
1.50	0.751	0.317	0.10	0.19	0.28	0.42	0.52	0.64	0.75	0.86	0.92	0.97	1.00

Note DP/H_e = depth from the ground surface/excavation depth

Table 10.3 Standard deviation and correlation coefficient of the measurement error at various points

z-z_f (m)	σ_e (mm)	σ_χ	Correlation coefficient matrix								
7	1.35	0.050	1.00	0.90	0.81	0.75	0.72	0.70	0.64	0.59	0.55
9	1.50	0.045	0.90	1.00	0.90	0.83	0.80	0.78	0.71	0.65	0.61
11	1.66	0.042	0.81	0.90	1.00	0.92	0.89	0.86	0.78	0.72	0.68
13	1.80	0.041	0.75	0.83	0.92	1.00	0.96	0.93	0.85	0.79	0.74
14	1.87	0.042	0.72	0.80	0.89	0.96	1.00	0.97	0.88	0.81	0.76
15	1.93	0.044	0.70	0.78	0.86	0.93	0.97	1.00	0.91	0.84	0.79
18	2.11	0.054	0.64	0.71	0.78	0.85	0.88	0.91	1.00	0.92	0.87
21	2.29	0.073	0.59	0.65	0.72	0.79	0.81	0.84	0.92	1.00	0.94
24	2.45	0.103	0.55	0.61	0.68	0.74	0.76	0.79	0.87	0.94	1.00

Note $z-z_f$ = the distance from the measuring point to the fixed point of an inclinometer

distance between two consecutive measurement points. With the covariance matrix, *C*, the standard deviation and correlation structure of the measurement error, *e*, at different points could be readily derived using the relations: $\rho_{XY} = \mathrm{COV(X, Y)}/\sigma_X\sigma_Y$ and $\sigma_X=\sqrt{\mathrm{COV(X, X)}}$. The standard deviation of χ, σ_χ is then represented by σ_e/d, which is approximately equal to $\sigma_\chi = \sigma_e/y$. The correlation structure of χ is assumed to be equal to that of *e* for simplicity. With the standard deviation and correlation structure of χ, the covariance of χ could be derived using the same relations, that is, $\rho_{XY} = \mathrm{COV(X, Y)}/\sigma_X\sigma_Y$ and $\sigma_X=\sqrt{\mathrm{COV(X, X)}}$.

Table 10.3 summarizes the standard deviation and correlation coefficient of the measurement errors for $\sigma = 0.0005$ and $l_m = 1$ m. As shown in Table 10.3, the standard deviation of *e* is directly proportional to the distance from the fixed point (normally at the bottom of the inclinometer) to the measurement point. This trend is consistent with the measurement error results given by Calvello [40]. In addition, the σ_e values in Table 10.2 are in the same level of σ_e as those given by Calvello [40], which justifies the selection of the $\sigma = 0.0005$. For example, at a distance of 18 m from the fixed point, the resulting standard deviation of *e* is 2.11 mm. This value is close to the value given by Calvello [40], that is, 0.00727 feet (= 2.21 mm) at a distance of 57 feet (= 17.4 m). The results in Table 2 are used to evaluate the likelihood of the observed deflections.

10.4 Illustrative Example

A braced excavation in a metro project in Hangzhou, China is chosen to illustrate the effectiveness of Bayesian updating using wall deflection data at multiple points. In the following, the prior distribution of the input soil parameters is discussed. Then, the response surfaces are constructed for evaluation of the maximum deflection and deflections for the points at different depths. Next, the parameters for the MCMC simulation method are investigated to ensure convergence of the Bayesian updating. The effectiveness of Bayesian updating using different amounts of data is evaluated by comparing the predicted maximum deflection with the observed maximum deflection. Finally, the effect of the prior COV on the Bayesian updating is analyzed.

10.4.1 Excavation of a Metro Project in Hangzhou, China and FEM Simulation

An excavation problem [41, 42] for a metro project at the intersection of Pengbu road and Qianchao road in Hangzhou, China is adopted as an example to illustrate the Bayesian updating. The excavation length (L) and width (B) for this component of the project are $L = 192$ m and $B = 44.5$ m, respectively. The soil profile and excavation stages for the metro project are plotted in Fig. 10.1. As shown in Fig. 10.1, six soil

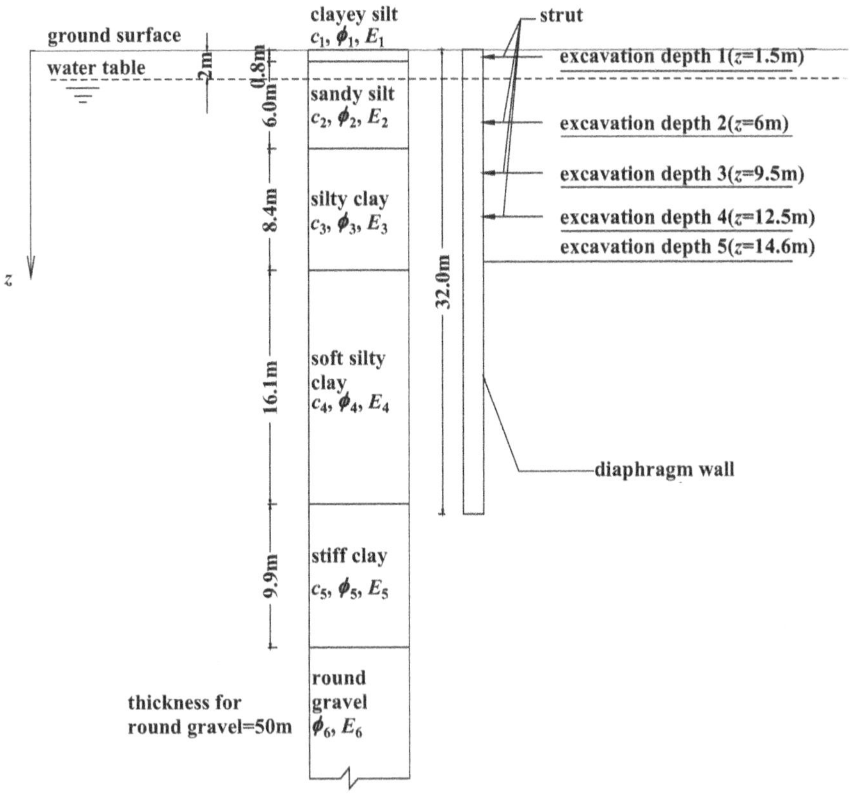

Fig. 10.1 Soil profile and excavation stages for the metro project in Hangzhou

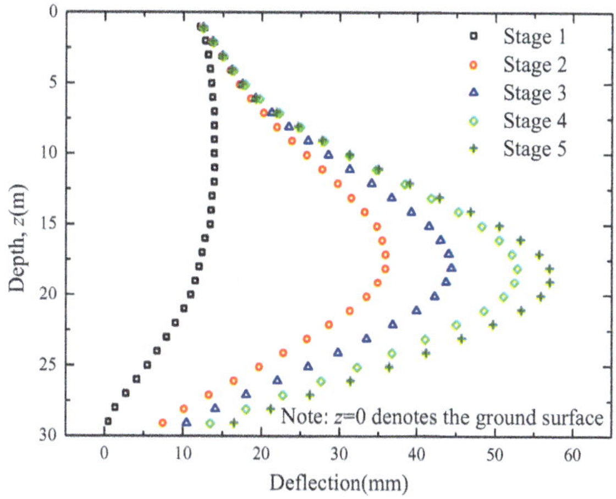

Fig. 10.2 Observed deflection data for the diaphragm wall at different excavation stages

layers exist in the area, that is, clayey silt, sandy silt, silty clay, soft silty clay, stiff clay and round gravel. The excavation supporting system consists of a 32-m-high diaphragm wall and four steel struts installed at different elevations (see Fig. 10.1). The lateral movements of the diaphragm wall were recorded by inclinometers at each excavation stage [42]. The observed deflection data for the diaphragm wall are plotted in Fig. 10.2.

The widely used finite element commercial software PLAXIS [43] is adopted to evaluate the wall deflection. Because the excavation length in the railway run direction is much longer than the excavation width, a two-dimensional plane strain model is adopted to simulate the excavation problem. In the model, each soil layer is assumed to be homogeneous. The finite element mesh for FEM analysis is plotted in Fig. 10.3. As shown in Fig. 10.3, approximately 1700 15-noded triangular elements are generated. The meshes surrounding the diaphragm wall are more refined than those in other sections to reduce possible computation errors due to the large displacement gradients in this area. For the boundary conditions, horizontal restraints are set for the left and right boundaries, and total restraints are set for the bottom boundary. It is worth noting that rotational fixity is applied on the top of the wall to constrain its rotation around the z-axis (the direction perpendicular to the book), and a virtual soil layer is constructed above the ground surface. The rationale for these choices is that the diaphragm wall is connected by rigid steel on the component above the ground. This boundary condition is important for a more accurate simulation of the behavior of the diaphragm wall. Otherwise, excess rotation will be generated on the top section of the diaphragm wall, which is inconsistent with the observed deflection data in Fig. 10.2. Moreover, the structure line representing the diaphragm wall is not allowed to be located outside the model area in the PLAXIS software. Hence, a weightless virtual soil layer is constructed on the top of the FEM model to cover the entire diaphragm wall.

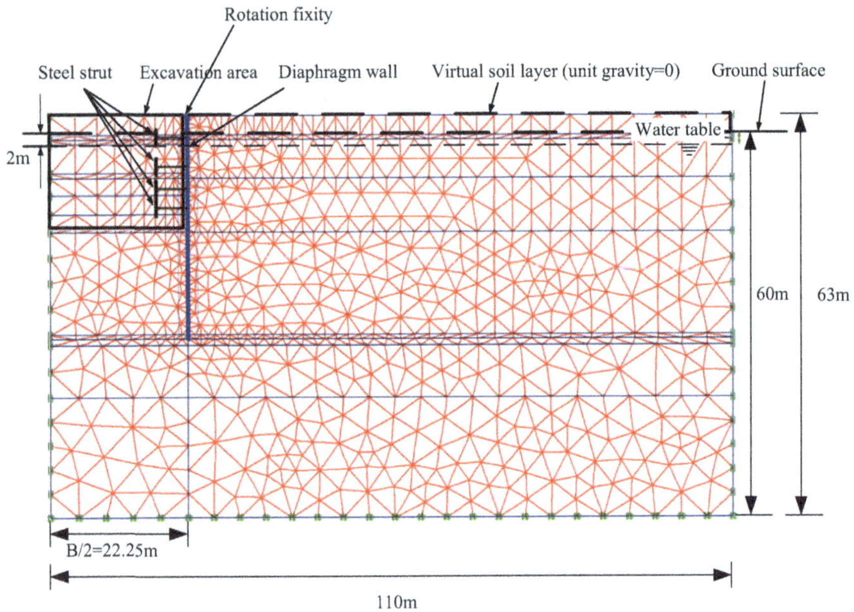

Fig. 10.3 Finite element mesh for the braced excavation problem

A Mohr–Coulomb constitutive model is adopted to simulate the behavior of the soil. This model is simple and easy to use. Although more advanced models [e.g., hardening-soil model [9] and models considering the small strain stiffness] have been proposed to realistically simulate the soil behavior under excavation conditions, more input parameters are required in those models. The added soil parameters make the back analysis more complicated and computational expensive. For example, if two more parameters are considered for each soil layer, $2 \times 6 = 12$ (6 = the number of soil layers) more parameters will be induced. In addition, more laboratory tests are required to estimate the parameters, which are extraordinarily time-consuming for multilayered excavation problems. These reasons may explain why the simple Mohr--Coulomb model is still in wide use in many excavation problems [44, 45]. In this chapter, only the parameters for the Mohr–Coulomb constitutive model are measured in the laboratory. Hence, the Mohr–Coulomb model is adopted in the FEM analysis. Details of the measured soil parameters can be found in Table 10.3. Note that the effective stress parameters, rather than total stress parameters, are used in the analysis. The reason is that the safety of an excavation problem is dominated by the long-term condition, that is, the drained condition [46].

Table 10.4 Soil parameters measured in the laboratory test

Layer number	Soil type	Unit weight, γ (kN·m^{-3})	Cohesion, c (kPa)	Friction angle, ϕ (°)	Elastic modulus, E (MPa)	Poisson's ratio, v
1	Clayey silt	18.9	11	32	25	0.49
2	Sandy silt	19.4	8.6	32.2	20	0.49
3	Silty clay	19.5	8.1	32.7	15	0.49
4	Soft silty clay	17.9	13	17.6	10	0.31
5	Stiff clay	18.1	25.25	16	150	0.2
6	Round gravel	20.0	0	35	300	0.15

10.4.2 Prior Knowledge of Soil Parameters

Generally, when the Mohr–Coulomb constitutive model is adopted, four parameters are applied (i.e., c, ϕ, E and v) for each soil layer. The shear strength (c and ϕ) and elastic modulus (E) for all soil layers are viewed as random variables. Because the cohesion for round gravel is 0, there are $3 \times 6-1 = 17$ ($6 =$ the number of soil layers) random variables in total. The mean values of these random variables are assumed to be the same as the values measured from laboratory tests (see Table 10.4). The prior COVs of the cohesion, friction angle and the elastic modulus are assumed to be 0.3, 0.2 and 0.3, respectively. These values fall within the range suggested by Phoon and Kulhawy [1]. Based on an extensive literature review by Phoon and Kulhawy [1], the COV for the friction angle is generally smaller than those of the cohesion and elastic modulus. Hence, this chapter adopts a smaller COV for the friction angle than for the other two parameters. The random variables are assumed to be independent and lognormally distributed to avoid negative values. Poisson's ratio and unit gravity are considered to be constant due to their low variabilities [1].

10.4.3 Construction of Response Surfaces

Second-order response surfaces with the forms of Eqs. (10.12) and (10.13) are constructed to establish the relationships among the 17 random variable parameters and the wall deflections. Following the preceding steps in Sect. 2.3, the response surface for the maximum deflection at stage 2 is obtained as follows

$$\begin{aligned}\delta_{\max} =& 102.16 - 0.026c_1 + 0.027c_2 + 0.12c_3 + 0.17c_4 - 0.0036c_5\\ &+ 0.011\phi_1 - 0.071\phi_2 - 0.028\phi_3 - 0.0027\phi_4 - 0.098\phi_5 + 0.0092\phi_6\\ &- 0.019E_1 - 0.17E_2 - 1.01E_3 - 3.03E_4 - 0.0025E_5 - 0.0029E_6\end{aligned}$$

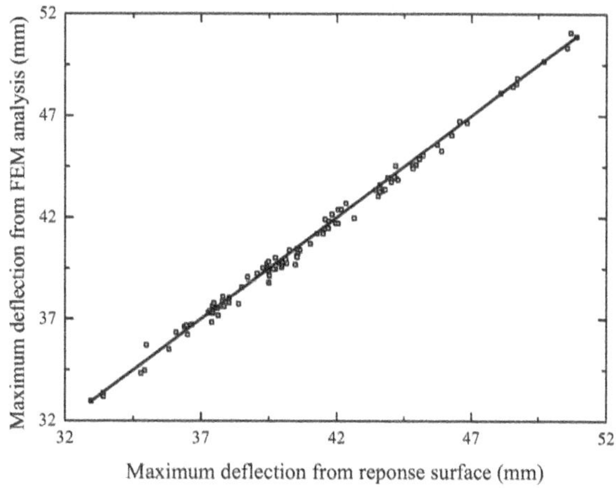

Fig. 10.4 Comparison of the maximum deflections calculated by the response surface and the FEM

$$
\begin{aligned}
&+0.00073c_1^2 - 0.0011c_2^2 - 0.0041c_3^2 - 0.0049c_4^2 + 0.000038c_5^2 \\
&- 0.00015\phi_1^2 + 0.00063\phi_2^2 - 0.000088\phi_3^2 - 0.000013\phi_4^2 \\
&+ 0.0023\phi_5^2 - 0.00011\phi_6^2 + 0.00017E_1^2 + 0.0011E_2^2 \\
&+ 0.010E_3^2 + 0.055E_4^2 + 0.0000013E_5^2 + 0.0000029E_6^2
\end{aligned} \quad (10.15)
$$

where c_1, c_2, c_3, c_4, c_5 are the cohesions (in kPa) for the upper five types of soils from up to down (see Fig. 10.1); ϕ_1, ϕ_2, ϕ_3, ϕ_4, ϕ_5, ϕ_6 are the friction angles (in degree) for the six types of soil from up to down (see Fig. 10.1); and E_1, E_2, E_3, E_4, E_5, E_6 are the elastic moduli (in MPa) for the six types of soils from up to down (see Fig. 10.1). To assess the effectiveness of the response surface in evaluating the maximum wall deflection, 100 additional combinations of the input parameters are randomly generated. The maximum deflections are evaluated using the FEM and response surface, and the results are plotted in Fig. 10.4. As shown in Fig. 10.4, the maximum deflections obtained from the two approaches are approximately the same. The response surfaces for the deflections at various depths and for other stages display similar accuracy. Hence, the second-order response surface is adopted in this chapter for evaluation of all deflections.

10.4.4 Bayesian Updating Using Prior Information from the Laboratory Test

In this section, selected parametric studies involving the scale factor and the length of the Markov chain are conducted first to obtain an efficient sampling of the MCMC simulation. Secondly, the Bayesian updating result is illustrated based on the proper values of scale factor and Markov chain length. The measurement error and model uncertainty of FEM discussed in Sect. 10.3 and prior distribution of soil parameters described in Sect. 10.4.2 are used in the parametric study. The prior means of the soil parameters are assumed according to the result of laboratory tests. The effects of the prior COV of the soil parameters on the Bayesian updating are discussed in Sect. 10.4.5. The observed deflections at nine points (i.e., the points at $z = [8, 11, 14, 17, 18, 19, 21, 23, 25$ m$]$) are used in the Bayesian updating. This data amount is one major difference from the previous studies, which usually adopt only the maximum deflection data.

The Bayesian updating is conducted in a stage-by-stage sequence. In other words, the updated parameters and model bias factors based on the field observations at the ith excavation stage are adopted as the prior distribution for the Bayesian updating at the $(i + 1)$th excavation stage. The field observation data at stage 1 are discarded because the deflection data in the early stages usually suffer from greater relative measurement errors. In addition, the deformation pattern of the diaphragm wall at the early stage is customarily different from those at the following stages, which makes the back analysis based on early stage data not meaningful [11].

To ensure a rapid convergence of the Monte Carlo Markov chain sampling, the scaling factor (a parameter determining the jumping function, see Sect. 10.2.2) and the length of Markov chain should be carefully selected. The values of these parameters are selected by conducting some parametric studies. The selection criteria for s and Markov chain length, respectively, are that the acceptance rate of the candidate samples should lie in [0.2, 0.4], and the variability of the sample statistics for a Markov chain is small (i.e., COV < 0.01). Based on these criteria, the values for the scaling factor and Markov chain length are, respectively, obtained as 0.05 and 100,000. The procedures for the parametric study are the same as Juang et al. [11] and are not repeated herein.

Based on the optimal values of the scale factor and Markov chain length, the Bayesian updating is conducted using the model uncertainty of FEM characterized in Sect. 10.3.1 and prior distribution discussed in Sect. 10.4.2. The relative frequency of the selected soil parameters (i.e., c_3, ϕ_3, and E_3) using the deflection data in stage 4 are plotted in Fig. 10.5. In this work, the relative frequency refers to the proportion of the sample in one interval to the total number of samples (i.e., 100,000). For comparison, the initial prior distributions assumed according to the laboratory test result are also plotted in Fig. 10.5. As shown in Fig. 10.5, the updating of the elastic modulus parameter is more significant than that for the cohesion and friction angle. The reason for this observation is that the wall deflection is more sensitive to the elastic modulus parameter than the cohesion and friction angle parameters, which is evident in Eq. (10.15).

To examine the effect of the data amount on the updating of soil parameters, Bayesian updating is also conducted using merely the maximum deflection. Figure 10.6 plots the relative frequency of the updated parameter E_3 for the Bayesian updating using the maximum deflection data at the 4th excavation stage. As shown in Figs. 10.5c and 10.6, the E_3 for the deflection data at nine points is updated more significantly than that for the maximum deflection data. For example, when deflection data at nine points are used, the mean of E_3 increases by 83% (i.e., from 15 to

Fig. 10.5 Prior and posterior distribution of c_3, ϕ_3, E_3 for the 4th excavation stage

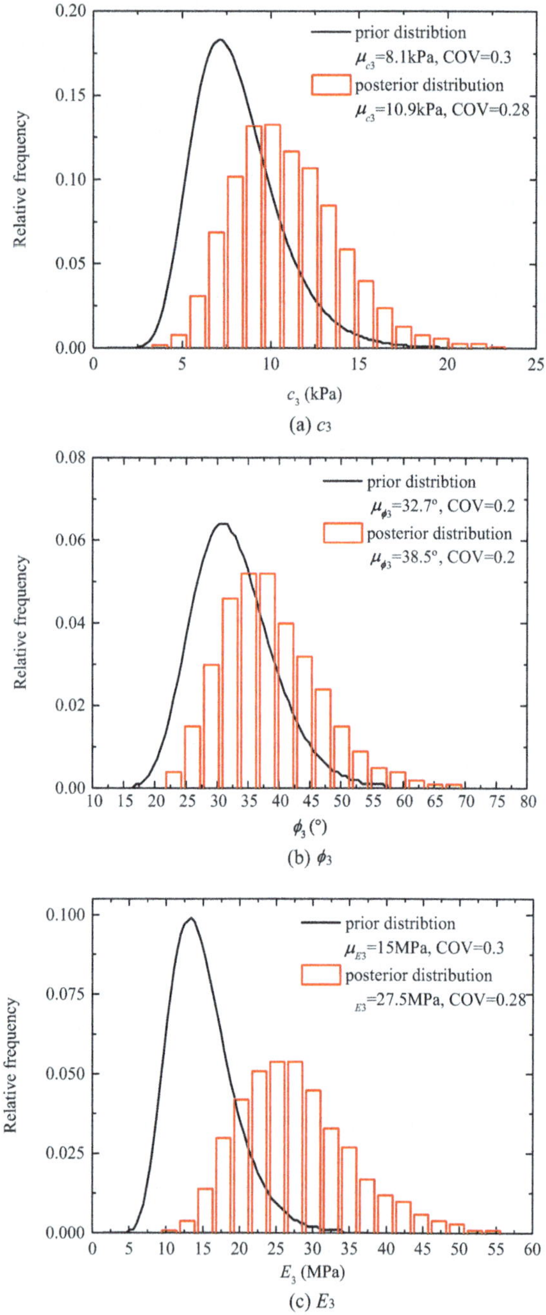

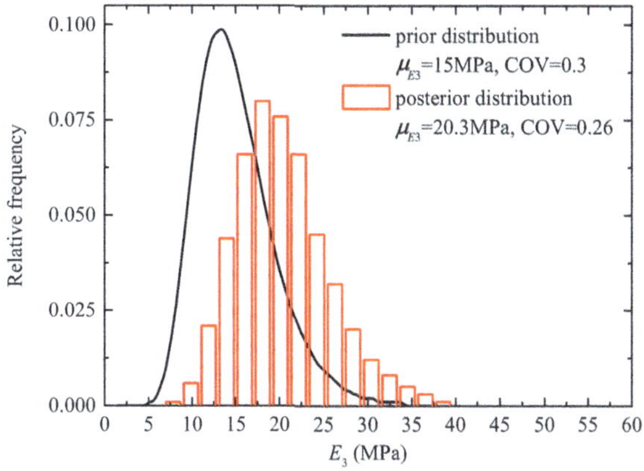

Fig. 10.6 Prior and posterior distribution of E_3 using the maximum deflection data for the 4th excavation stage

27.5 MPa) after updating. In contrast, the mean of E_3 increases by 35% (i.e., from 15 to 20.3 MPa) if only the maximum deflection data are used.

In addition to parameter calibration, another purpose of Bayesian updating is prediction of the wall response in future stages of braced excavation problems. Hence, the effect of the Bayesian updating on the prediction of the deflections in future excavation stages is illustrated. The predicted means of the maximum wall deflection using different amounts of data are plotted in Fig. 10.7. Similar to the Bayesian updating, the prediction of the maximum deflection is also conducted in a stage-by-stage sequence. In other words, the Bayesian updating results in the $(i-1)$th stage [i.e., the updated soil parameters and model bias factors using the field observations in the $(i-1)$th stage] are used to predict the deflections in the following excavation stages [i.e., the ith, $(i+1)$th, ... stage]. In addition, the mean of the maximum deflection is obtained by finding out the maximum element in the vector $[\mu_{p,y_{z1}} \cdot \mu_{p,\varepsilon_{z1}}, \mu_{p,y_{z1}} \cdot \mu_{p,\varepsilon_{z1}}, \ldots, \mu_{p,y_{z1}} \cdot \mu_{p,\varepsilon_{z1}}]^T$, where $\mu_{p,y_{zi}}$ is the posterior mean of the deflection at depth z_i and $\mu_{p,\varepsilon_{zi}}$ is the posterior mean of the model bias factor at depth z_i. The $\mu_{p,y_{zi}}$ is evaluated by computing the mean value of the deflections at depth z_i for 100,000 realizations rather than the deflection at depth z_i for the mean of the soil parameters. To clearly illustrate the effectiveness of the prediction, the predicted deflections are plotted versus the observed deflection data in Fig. 10.7. As shown in Fig. 10.7, the predicted maximum deflection approaches the observed maximum deflection in a stage-by-stage manner. For example, if the soil parameter from the laboratory test and model bias factors for different points are used, the predicted mean of the maximum wall deflection for the last excavation stage (i.e., excavation depth = 14.6 m) is 82.09 mm [see Fig. 10.7a]. This value is much different from the real performance of the diaphragm wall, that is, a maximum deflection of 56.98 mm. However, the prediction improves if the updated soil parameter is used in prediction. The predicted mean of the maximum deflection for the last stage decreases from

Fig. 10.7 Predicted mean of the maximum deflection using different amounts of data

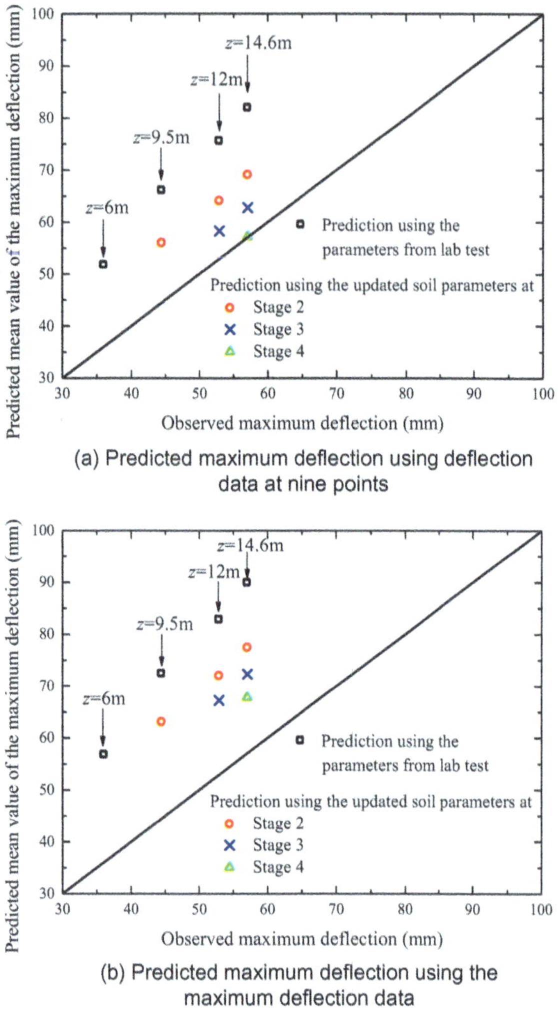

69.37 mm at stage 2 to 62.93 mm at stage 3 and 57.42 mm at stage 4 if multiple deflection data are used in the Bayesian updating [see Fig. 10.7a]. The predicted mean of the maximum deflection displays a similar trend if only the maximum deflection data are used in the Bayesian updating [see Fig. 10.7b]. Moreover, it is worth noting that the prediction accuracy varies with the field observation data amount used in Bayesian updating. As shown by Fig. 10.7b, a large disparity still exists between the predicted mean of the maximum deflection (i.e., 68.04 mm) and the observed data (i.e., 56.98 mm) for the last excavation stage even if the updated soil parameters are used at the fourth excavation stage. The relative error between the two is as large as $|68.04-56.98|/56.98 = 19.4\%$, indicating that Bayesian updating using only the maximum deflection data is not sufficient for a precise prediction. In contrast, if the deflection data at nine points are used, the prediction at stage 3 has achieved a relative high accuracy, that is, a relative error as small as $|62.93-56.98|/56.98 = 10.4\%$. Hence, if more deflection data are used in the Bayesian updating, the prediction of the maximum deflection will be much more effective. It is worth noting that the framework of Juang et al. [11] cannot handle this amount of data, because the outputs of the semi-empirical KJHH model limit only to the maximum wall deflection and the maximum ground settlement.

10.4.5 Effect of Prior Distribution of Soil Parameters

In the preceding discussions, the prior mean of the soil parameters is assumed according to the laboratory test and the prior COVs are assumed to be a set of relative low values. In practice, the COVs of the soil parameters might be site-specific. It is necessary to investigate the effect of the prior COV on the Bayesian updating. Hence, in this section, the Bayesian updating is conducted using different COVs of input parameters.

The COVs of the prior distribution of soil parameters (namely, all cohesions, friction angles and elastic moduli) are altered by multiplication with a coefficient of 1.5. In other words, the prior COV for the cohesion and elastic modulus is adjusted to $1.5 \times 0.3 = 0.45$, whereas the prior COV for the friction angle is adjusted to $1.5 \times 0.2 = 0.3$. These COV values still fall within the typical range given by Phoon and Kulhawy [1]. The updated means and COVs of the parameter E_3 for the two different prior COVs are plotted in Fig. 10.8. As shown in Fig. 10.8a, the updated mean of the E_3 increases with the excavation depth at which the updating is performed, but the mean for the larger prior COVs increases slightly more significantly. In addition, the updated COV of the E_3 decreases with the excavation depth at which the updating is performed. Similar to the mean, the updated COV for the larger prior COVs varies more significantly. These trends can be well explained based on the fact that the posterior PDF is directly proportional to the product of the prior PDF and PDF of the likelihood function. When the prior COV of an updating parameter is small, the prior PDF is highly centralized around the mean value and the posterior PDF is dominated by the prior PDF rather than the likelihood PDF. Consequently, the posterior (mean, COV) of the parameter is close to the prior (mean, COV). On the contrary, if the prior variability of the parameter is large, the posterior PDF is significantly influenced by the PDF of the likelihood function. The posterior (mean, COV) deviate far apart from the prior (mean, COV). In general, the updated mean and COV display a similar trend for different prior distributions. The results indicate that the soil parameters are effectively updated in the Bayesian updating regardless of the prior distribution.

Fig. 10.8 Bayesian updating using different prior COVs

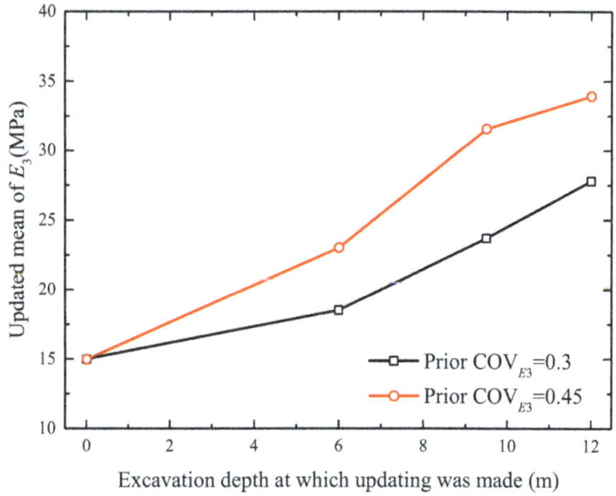

(a) Updated mean of E_3 for different prior COVs

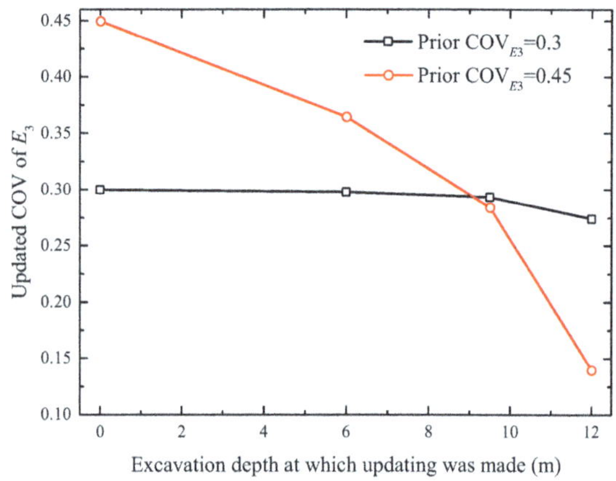

(b) Updated COV of E_3 for different prior COVs

10.5 Conclusions

In this chapter, an efficient Bayesian back-analysis method is proposed for braced excavation problems using the deflection data at multiple points. Response surfaces obtained from the finite element analysis results are adopted to rapidly evaluate the wall response in the process of braced excavation. A highly efficient sampling method, that is, the Markov chain Monte Carlo method, is applied to update the posterior probabilistic distributions of the soil parameters. Instead of the maximum deflection data, deflection data at multiple points on the wall are used in the updating process. Measurement errors of inclinometer are considered, and the model errors of finite element methods are updated along with the soil parameters. The updated soil parameters and model bias factors in the current excavation stage are used to predict the maximum wall deflection for the subsequent stages. A multilayered braced excavation problem in Hangzhou, China is used to illustrate the effectiveness of the proposed method. The effect of the prior distribution on the Bayesian updating is investigated. It is found that the accuracy of the predicted maximum deflection generally increases with the excavation stages. Use of only the maximum deflection data in the Bayesian updating might not be sufficient to obtain an accurate prediction. In contrast, the prediction accuracy improves significantly if the deflection data at multiple points of the wall are adopted.

References

1. Phoon KK, Kulhawy FH (1999) Characterization of geotechnical variability. Can Geotech J 36(4):612–24
2. Li DQ, Qi XH, Phoon KK, Zhang LM, Zhou CB (2014) Effect of spatially variable shear strength parameters with linearly increasing mean trend on reliability of infinite slopes. Struct Saf 49:45–55
3. Zhang J, Huang HW, Zhang LM, Zhu HH, Shi B (2014) Probabilistic prediction of rainfall-induced slope failure using a mechanics-based model. Eng Geol 168:129–40
4. Tang WH, Stark TD, Angulo M (1999) Reliability in back analysis of slope failures. Soils Found 39(5):73–80
5. Cao Z, Wang Y (2014) Bayesian model comparison and characterization of undrained shear strength. J Geotech Geoenviron Eng 140(6):04014018
6. Ching J, Phoon K-K, Chen Y-C (2010) Reducing shear strength uncertainties in clays by multivariate correlations. Can Geotech J 47(1):16–33
7. Ching J, Chen J, Yeh J, Phoon K (2012) Updating uncertainties in friction angles of clean sands. J Geotech Geoenviron Eng 138(2):217–29
8. Wang Y, Au S-K, Cao Z (2010) Bayesian approach for probabilistic characterization of sand friction angles. Eng Geol 114(3–4):354–63
9. Finno R, Calvello M (2005) Supported excavations: observational method and inverse modeling. J Geotech Geoenviron Eng 131(7):826–36
10. Hashash YMA, Levasseur S, Osouli A, Finno R, Malecot Y (2010) Comparison of two inverse analysis techniques for learning deep excavation response. Comput Geotech 37(3):323–33
11. Juang CH, Luo Z, Atamturktur S, Huang HW (2013) Bayesian updating of soil parameters for braced excavations using field observations. J Geotech Geoenviron Eng 139(3):395–406
12. Hashash Y, Marulanda C, Ghaboussi J, Jung S (2006) Novel approach to integration of numerical modeling and field observations for deep excavations. J Geotech Geoenviron Eng 132(8):1019–31
13. Wang L, Luo Z, Xiao J, Juang CH (2014) Probabilistic inverse analysis of excavation-induced wall and ground responses for assessing damage potential of adjacent buildings. Geotech Geol Eng 32(2):273–85
14. Huang J, Kelly R, Li D, Zhou C, Sloan S (2016) Updating reliability of single piles and pile groups by load tests. Comput Geotech 73:221–30
15. Wang L, Ravichandran N, Juang CH (2012) Bayesian updating of KJHH model for prediction of maximum ground settlement in braced excavations using centrifuge data. Comput Geotech 44:1–8
16. Zhang LL, Zhang J, Zhang LM, Tang WH (2010) Back analysis of slope failure with Markov chain Monte Carlo simulation. Comput Geotech 37(7–8):905–12
17. Park JH, Kim D, Chung CK (2012) Implementation of Bayesian theory on LRFD of axially loaded driven piles. Comput Geotech 42:73–80

18. Zhang J, Wilson T, Zhang LM (2010) Efficient probabilistic back-analysis of slope stability model parameters. J Geotech Geoenviron Eng 136(1):99–109
19. Kung GT, Juang CH, Hsiao EC, Hashash YM (2007) Simplified model for wall deflection and ground-surface settlement caused by braced excavation in clays. J Geotech Geoenviron Eng 133(6):731–47
20. Zhang J, Zhang LM, Wilson T (2009) Bayesian framework for characterizing geotechnical model uncertainty. J Geotech Geoenviron Eng 135(7):932–40
21. Zhang DM, Phoon KK, Huang HW, Hu QF (2015) Characterization of model uncertainty for cantilever deflections in undrained clay. J Geotech Geoenviron Eng 141(1):04014088
22. Feng X, Jimenez R (2014) Bayesian prediction of elastic modulus of intact rocks using their uniaxial compressive strength. Eng Geol 173:32–40
23. Phoon KK, Santoso A, Quek ST (2010) Probabilistic analysis of soil-water characteristic curves. J Geotech Geoenviron Eng 136(3):445–55
24. Li DQ, Jiang SH, Cao ZJ, Zhou W, Zhou CB, Zhang LM (2015) A multiple response-surface method for slope reliability analysis considering spatial variability of soil properties. Eng Geol 187:60–72
25. Li DQ, Zheng D, Cao ZJ, Tang XS, Phoon KK (2016) Response surface methods for slope reliability analysis: Review and comparison. Eng Geol 203:3–14
26. Mollon G, Dias D, Soubra A (2009) Probabilistic analysis of circular tunnels in homogeneous soil using response surface methodology. J Geotech Geoenviron Eng 135(9):1314–25
27. Tan O, Zaimoglu AS, Hinislioglu S, Altun S (2005) Taguchi approach for optimization of the bleeding on cement-based grouts. Tunn Undergr Space Technol 20(2):167–73
28. Ledesma A, Gens A, Alonso EE (1996) Estimation of parameters in geotechnical backanalysis—I maximum likelihood approach. Comput Geotech 18(1):1–27
29. Likitlersuang S, Surarak C, Wanatowski D, Oh E, Balasubramaniam A (2013) Finite element analysis of a deep excavation: a case study from the Bangkok MRT. Soils Found 53(5):756–73
30. Whittle AJ, Corral G, Jen LC, Rawnsley RP (2015) Prediction and performance of deep excavations for Courthouse Station, Boston. J Geotech Geoenviron Eng 141(4):04014123
31. Hsiung B-CB (2009) A case study on the behaviour of a deep excavation in sand. Comput Geotech 36(4):665–75
32. Lehane B, Zaremba N (2007) The performance of steel struts in a braced excavation in Perth CBD. Aust Geomech J 42(3):121–32
33. Lu Y, Tan Y, Peng F, Liao S (2012) FE simulation of deep excavations in sensitive soft clays. In: Hryciw RD, Athanasopoulos-Zekkos A, Yesiller N (eds) GeoCongress 2012: state of the art and practice in geotechnical engineering. Oakland, pp 750–9
34. Liu X-L, Ma Y, Guo G-Q, Tao T, Zhou H (2012) Applicability of PLAXIS2D used for numerical simulation in foundation pit excavations. Periodic Ocean Univ China 42(4):19–25 (in Chinese)
35. Tan Y, Wei B (2012) Observed behaviors of a long and deep excavation constructed by cut-and-cover technique in shanghai soft clay. J Geotech Geoenviron Eng 138(1):69–88
36. Wang ZW, Ng CWW, Liu GB (2005) Characteristics of wall deflections and ground surface settlements in Shanghai. Can Geotech J 42(5):1243–54
37. Teo PL, Wong KS (2012) Application of the Hardening Soil model in deep excavation analysis. IES J Part A Civil Struct Eng 5(3):152–65
38. Corral G, Whittle AJ (2010) Re-analysis of deep excavation collapse using a generalized effective stress soil model. In: Finno R, Hashash YMA, Arduino P (eds) 2010 Earth Retention Conference. American Society of Civil Engineers, Bellevue, Washington, DC
39. Lim A, Ou C-Y, Hsieh P-G (2010) Evaluation of clay constitutive models for analysis of deep excavation under undrained conditions. J GeoEng 5(1):9–20
40. Calvello M (2002) Inverse analysis of a supported excavation through Chicago glacial clays. Northwestern University, Evanston
41. Ran L, Ye XW, Zhu HH (2011) Long-term monitoring and safety evaluation of a metro station during deep excavation. Procedia Eng 14:785–92
42. Zhu HH, Ye XW, Ran L (2011) Reliability analysis of deep excavation based on field monitoring data. J Railw Eng Soc 28(9):7–11 (in Chinese)

43. Brinkgreve R (2002) PLAXIS (version 8) user's manual. Delft University of Technology and PLAXIS BV, Netherlands
44. Chowdhury SS, Deb K, Sengupta A (2013) Estimation of design parameters for braced excavation: numerical study. Int J Geomech 13(3):234–47
45. Orazalin ZY, Whittle AJ, Olsen MB (2015) Three-dimensional analyses of excavation support system for the stata center basement on the MIT Campus. J Geotech Geoenviron Eng 141(7):05015001
46. Atkinson J (2007) The mechanics of soils and foundations. Taylor & Francis

Appendix A
Matlab Code for Linear Model Class Selection

References:

Zhou WH, Yuen KV, Tan F (2014) Estimation of soil–water characteristic curve and relative permeability for granular soils with different initial dry densities. Eng Geol 179:1–9.

Zhou WH, Yuen KV, Tan F (2013) Estimation of maximum pullout shear stress of grouted soil nails using Bayesian probabilistic approach. Int J Geomech 13(5):659–664.

Yuen KV (2010) Recent developments of Bayesian model class selection and applications in civil engineering. Struct Saf 32(5):338–346.


```
function [B_matrix, posterior] = model_class_selection(X, Y, all_prior)
% X: Independent variables, N*NP
% Y: Dependent variable, N*2
% All_prior:  Give the prior pdf values for all unknown parameters
% B_matrix:   Regression coefficients obtained for all model classes
% posterior:  Posterior plausibility
    N  = size(X,1);    % Number of measured data
    NP = size(X,2);    % Number of regression coefficient
    a  = 2^(NP)-1;     % Number of model class candidates
    j  = 0;
    b_matrix      = zeros(NP,a); % all model class candidates
    B_matrix      = zeros(NP,a);
    sigma_matrix = zeros(a,1);
    y_matrix      = zeros(N,a);
    log_maximum_likelihood_matrix = zeros(a,1);
    prior_matrix     = zeros(a,1);
    evidence_matrix = zeros(a,1);
    log_O_matrix     = zeros(a,1);
    H_ni_matrix      = zeros(size(X,2)+1,size(X,2)+1,a);
    for nb = 1:(NP)
        b  = nchoosek(1:(NP),nb);
```


W. Zhou et al., *Practice of Bayesian Probability Theory in Geotechnical Engineering*,
https://doi.org/10.1007/978-981-15-9105-1

```
for i = 1:size(b,1)
    b_matrix(b(i,:),j+i)=b(i,:);
    x = X(:,b(i,:));
    prior = [all_prior(:,b(i,:)),all_prior(1,NP+1)];
    C     = ones(size(x,2),1,size(b,1));
    C(:,:,i) = (1/N)*x'*Y;
    A         = 1/N*(x')*x;
    B_sigma   = A\C(:,:,i);
    B_matrix(b(i,:),j+i) = B_sigma;
    sigma = (1/N)*(Y'-B_sigma'*x')*(Y'-B_sigma'*x')';
    sigma_matrix(j+i,1) = sigma;
    y = B_sigma'*x';
    y_matrix(1:N,j+i) = y;
    log_maximum_likelihood = log((2*pi*sigma)^(-N/2)*exp(-0.5*N));
    log_maximum_likelihood_matrix(j+i,1) = log_maximum_likelihood;
    H = N*[sigma^(-1)*A zeros(size(B_sigma));zeros(size(B_sigma'))
    (1/2)*(sigma^(-2))];
    H_ni = inv(H);
    H_ni_matrix(1:size(H_ni,1),1:size(H_ni,1),j+i) = H_ni;
    Nb = size(B_sigma,1);
    prior = prod(prior);
    prior_matrix(j+i,1) = prior;
    evidence =
    prior*(2*sigma)^((Nb+2-N)/2)*exp(-0.5*N)*(pi)^((Nb+1-N)/2)*N^(-(Nb+1)/2)*(
    det(A))^(-0.5);

            evidence_matrix(j+i,1) = evidence;
            log_O = log(prior*(det(A)^(-1/2))*(pi/N)^((Nb+1)/2)*(2*sigma)^((Nb+2)/2));
            log_O_matrix(j+i,1) = log_O;
        end
        j = i+j;
    end
    posterior        = evidence_matrix*(1/a)/sum(evidence_matrix*(1/a));
    [values,index]   = sort(posterior,'descend');
    posterior_top10 = posterior(index(1:10,1),1);
    b_top10          = b_matrix(:,index(1:10,1));
    B_top10          = B_matrix(:,index(1:10,1));
    log_O_top10      = log_O_matrix(index(1:10,1),1);
    log_maximum_likelihood_top10 = log_maximum_likelihood_matrix(index(1:10,1),1);
    dlmwrite('b_top10.txt', b_top10);
    dlmwrite('B_top10.txt', B_top10);
    dlmwrite('posterior_top10.txt', posterior_top10);
    dlmwrite('log_O_top10.txt', log_O_top10);
    dlmwrite('log_maximum_likelihood_top10.txt', log_maximum_likelihood_top10);
end
```

Appendix B
Matlab Code for Soft Soil Creep Models with Finite Difference Method

References:

Zhou WH, Tan F, Yuen KV (2018) Model updating and uncertainty analysis for creep behavior of soft soil. Comput Geotech 100:135–143.
Tan F, Zhou WH, Yuen KV (2018) Effect of loading duration on uncertainty in creep analysis of clay. Int J Numer Anal Methods Geomech 42:1235–1254.

```
% Close open windows, clear memory and screen
close all; clear; clc;
Model_n = input('Please enter a model number:');
% Model_Para_Vector_... is the model parameter vector in Table 2 on Page 1237
% from the reference: Tan F, Zhou WH, Yuen KV. (2018). Effect of
% loading duration on uncertainty in creep analysis of clay. Int. J. Numer.
% Anal. Methods Geomech., 42: 1235-1254.
switch Model_n
    case 1
        % Input the values for the vector Model_Para_Vector_YG_1989
        Pl_fun = M_YG_1989(Model_Para_Vector_YG_1989);
        [strain, pressure_b, elapsed_time, epsilon_vpi] =
        FiniteDiff(P,Pl_fun,u0,sigma0,epsilon0,epsilon_vp0, k0);
    case 2
        % Input the values for the vector Model_Para_Vector_Yin_1999a
        Pl_fun = M_Yin_1999a(Model_Para_Vector_Yin_1999a);
        [strain, pressure_b, elapsed_time, epsilon_vpi] =
        FiniteDiff(P,Pl_fun,u0,sigma0,epsilon0,epsilon_vp0, k0);
    case 3
        % Input the values for the vector Model_Para_Vector_Yin_1999b
         Pl_fun = M_Yin_1999b(Model_Para_Vector_Yin_1999b);
         [strain, pressure_b, elapsed_time, epsilon_vpi] =
         FiniteDiff(P,Pl_fun,u0,sigma0,epsilon0,epsilon_vp0, k0);
    case 4
        % Input the values for the vector Model_Para_Vector_Yin_1999c
         Pl_fun = M_Yin_1999c(Model_Para_Vector_Yin_1999c);
         [strain, pressure_b, elapsed_time, epsilon_vpi] =
         FiniteDiff(P,Pl_fun,u0,sigma0,epsilon0,epsilon_vp0, k0);
```


W. Zhou et al., *Practice of Bayesian Probability Theory in Geotechnical Engineering*,
https://doi.org/10.1007/978-981-15-9105-1

```
case 5
    % Input the values for the vector Model_Para_Vector_Yin_1999d
     Pl_fun = M_Yin_1999d(Model_Para_Vector_Yin_1999d);
     [strain, pressure_b, elapsed_time, epsilon_vpi] =
     FiniteDiff(P,Pl_fun,u0,sigma0,epsilon0,epsilon_vp0, k0);
case 6
    % Input the values for the vector Model_Para_Vector_Vermeer_et_al_1997
     Pl_fun = M_Vermeer_et_al_1997(Model_Para_Vector_Vermeer_et_al_1997);
     [strain, pressure_b, elapsed_time, epsilon_vpi] =
     FiniteDiff(P,Pl_fun,u0,sigma0,epsilon0,epsilon_vp0, k0);
case 7
    % Input the values for the vector Model_Para_Vector_KL_2001
     Pl_fun = M_KL_2001(Model_Para_Vector_KL_2001);
     [strain, pressure_b, elapsed_time, epsilon_vpi] =
     FiniteDiff(P,Pl_fun,u0,sigma0,epsilon0,epsilon_vp0, k0);
case 8
    % Input the values for the vector Model_Para_Vector_Yin_et_al_2011
     Pl_fun = M_Yin_et_al_2011(Model_Para_Vector_Yin_et_al_2011);
     [strain, pressure_b, elapsed_time, epsilon_vpi] =
     FiniteDiff(P,Pl_fun,u0,sigma0,epsilon0,epsilon_vp0, k0);
otherwise
        % Input the values for the vector Model_Para_Vector_Vector_Yang_et_al_2015
         Pl_fun = M_Yang_et_al_2015(Model_Para_Vector_Yang_et_al_2015);
         [strain, pressure_b, elapsed_time, epsilon_vpi] =
         FiniteDiff(P,Pl_fun,u0,sigma0,epsilon0,epsilon_vp0, k0);
end

% Model 1: YG_1989 model
function Pl_fun = M_YG_1989(Model_Para_Vector)
    P = Model_Para_Vector;
    P1=P(1); P2=P(2); P3=P(3); P4=P(4); P5=P(5)/60;
    %P1:Fai/V; P2:Lambda/V; P3:Kappa/V; P4:Sigma_zr; P5:t0
    Pl_fun = P1/P5*exp(-epsilon/P1).*((sigma0-u*sigma0)/P4).^(P2/P1);
end

% Model 2: Yin_1999a model
function Pl_fun = M_Yin_1999a(Model_Para_Vector)
    P = Model_Para_Vector;
    P1=P(1); P2=P(2); P3=P(3); P4=P(4); P5=P(5); P6=P(6)/60;
    %P1:LAMBDA/V; P2:KAPPA/V; P3:epsilon_lm; P4:FAI; P5:SIGMA_ZR; P6:t0
    Pl_fun =
    P4/P6.*(1+((epsilon_r0+P1*log((sigma0-u.*sigma0)/P5))-epsilon)/P3).^2.*...
             exp(P3*((epsilon_r0+P1*log((sigma0-u.*sigma0)/P5))-epsilon)./(P3+(...
             epsilon_r0+P1*log((sigma0-u.*sigma0)/P5))-epsilon)./P4);
end

% Model 3: Yin_1999b model
function Pl_fun = M_Yin_1999b(Model_Para_Vector)
    P = Model_Para_Vector;
    P1=P(1);P2=P(2);P3=P(3);P4=P(4);P5=P(5);P6=P(6);P7=P(7)/60;
    %P1:Lambda/V; P2:Kappa/V; P3:a; P4:b; P5:Fai; P6:Sigma_zr; P7:t0
    epsilon_r0   = 0;
    epsilon_r    = epsilon_r0+P1*log((sigma0-u.*sigma0)/P6);
    epsilon_lm   = P3-P4*log((sigma0-u.*sigma0)/sigma_u);
    Pl_fun       = P5/P7.*(1+(epsilon_r-epsilon)./epsilon_lm).^2.*...

                   exp(epsilon_lm.*(epsilon_r-epsilon)./(epsilon_lm+epsilon_r-epsilon)/
                   P5);
end
```

```
% Model 4: Yin_1999c model
function Pl_fun = M_Yin_1999c(Model_Para_Vector)
    P = Model_Para_Vector;
    P1=P(1);P2 = P(2);P3=P(3);P4=P(4);P5=P(5);P6=P(6);P7=P(7)/60;
    %P1:Lambda/V; P2:Kappa/V; P3:a; P4:b; P5:Fai; P6:Sigma_zr; P7:t0
    epsilon_p = epsilon0*ones(1,n);
    epsilon    = epsilon_p;
    epsilon_r  = epsilon_r0+P1*log((sigma0-u.*sigma0)/P6);
    fai0       = P4-P5*log((sigma0-u.*sigma0)/sigma_u);
    Pl_fun     = fai0/P7.*(1+(epsilon_r-epsilon)/P3).^2.*...
                 exp(P3*(epsilon_r-epsilon)./(P3+epsilon_r-epsilon)./fai0);
end

% Model 5: Yin_1999d model
function Pl_fun = M_Yin_1999d(Model_Para_Vector)
    P = Model_Para_Vector;
    P1=P(1);P2=P(2);P3=P(3);P4=P(4);P5=P(5);P6=P(6);P7=P(7)/60;P8=P(8);
    %P1:Lambda/V; P2:Kappa/V; P3:a; P4:b; P5:c; P6:d; P7:t0;P8:k0
    epsilon_p = epsilon0*ones(1,n);
    epsilon     = epsilon_p;
    epsilon_r   = epsilon_r0+P1*log((sigma0-u.*sigma0)/P7);
    fai0        = P5-P6*log((sigma0-u.*sigma0)/sigma_u);
    epsilon_lm  = P3-P4*log((sigma0-u.*sigma0)/sigma_u);
    Pl_fun      = fai0/P8.*(1+(epsilon_r-epsilon)./epsilon_lm).^2.*...

                  exp(epsilon_lm.*(epsilon_r-epsilon)./(epsilon_lm+epsilon_r-epsilon).
                  /fai0);
end

% Model 6: Vermeer et al_1997 model
function Pl_fun = M_Vermeer_et_al_1997(Model_Para_Vector)
    P  = Model_Para_Vector;
    P1 =P(1);P2=P(2);P3=P(3);P4=P(4);
    %P1:Lambda/V; P2:Kappa/V; P3:a; P4:b

    Pl_fun      =
    P3/t_d.*((sigma0-u.*sigma0)./exp(epsilon_vp/(P1-P2))/P4).^((P1-P2)/P3);
end

% Model 7: KL_2001 model
function Pl_fun = M_KL_2001(Model_Para_Vector)
    P  = Model_Para_Vector;
    P1 = P(1); P2 = P(2); P3 = P(3); P4 = P(4); P5=P(5);
    %P1:Lambda/V; P2:Kappa/V; P3:a; P4:b; P5:Fai
    Pl_fun      = 10.^((log10(sigma0-u.*sigma0)-P4-P5-P1.*epsilon_vp)/P3);
end

% Model 8: Yin et al_2011 model
function Pl_fun = M_Yin_et_al_2011(Model_Para_Vector)
    P = Model_Para_Vector;
    P1=P(1);P2=P(2);P3=P(3);P4=P(4);P5=P(5);P6=P(6);
    %P1:Lambda/V; P2:Kappa/V; P3:a; P4:b; P5:Fai; P6:Sigma_zr
    Pl_fun      = P5/t_d.*((sigma0-u.*sigma0)*(1+P3)./(1+P3*exp(-P4*epsilon_vp))...
                  ./exp(epsilon_vp/(P1-P2))/P6).^((P1-P2)/P5);
end

% Model 9: Yang et al_2015 model
function Pl_fun = M_Yang_et_al_2015(Model_Para_Vector)
    P = Model_Para_Vector;
    P1=P(1);P2=P(2);P3=P(3);P4=P(4);P5=P(5);
    %P1:Lambda/V; P2:Kappa/V; P3:a; P4:b; P5:Fai
    if (sigma0-u*sigma0)<P5
        Pl_fun = P1;
    else
        Pl_fun = P2*(1+P3*exp(-P4*epsilon_pl));
    end
end
```

```
% Finite difference simulation of Viscoplastic strain rate combined with 1D
% consolidation governing equation shown in Eq. (2), the hydraulic conductivity
% Eq.(3)on Page 1239 (Tan et al. 2018)
function [strain, pressure_b, elapsed_time, epsilon_vpi] = FiniteDiff(P,Pl_fun,
u0,sigma0,epsilon0,epsilon_vp0, k0)
    P2=P;
    H=0.0075;  %soil depth
    delt_z=0.00075/H;
    n=10;
    e0=3.429;
    rw=9.8;
    u_p=(u0/sigma0*ones(1,n));
    epsilon_p=epsilon0*ones(1,n);
    epsilon_0=epsilon0;
    epsilon_vp=epsilon_vp0;
    depsilon_vp=0;
    A=rw*P2*H^2/k0/sigma0;
    allowable_error=2e-4;
    N_t=1;
    T=[0.1 1 10 100 1530]/A/60;
    Delt=[10 10 20 100 300];
    NN=1;
    for j=1:5
        D_T=T(j)/Delt(j);
        t_m(N_t,1)=D_T;
        while (sum(t_m)-D_T)<T(j)

            delt_t=D_T;
            epsilon_ref=zeros(1,n);
            ui=u_p;
            epsilon_i=epsilon_p; %initial vertical strain with the depth
            relative_error=max(abs(epsilon_p-epsilon_ref)./epsilon_p);
            nd=1;
            while relative_error>allowable_error
                u=u_p;   %initial pore water pressure with the depth, row vector
                epsilon_ref=epsilon_p;
                epsilon=epsilon_p;
                kr=(1-epsilon*(1+e0)/e0).^3./(1-epsilon);
                r=kr.*(1-u_p)*delt_t/delt_z^2;
                h=(1-u_p)*delt_t.*Pl_fun*A/P2;

    A_j=zeros(n,n);
    A_u1=diag(diag(A_j,1)-0.5*r(1,1:(n-1))',1);
    A_l1=diag(diag(A_j,-1)-0.5*r(1,2:(n))',-1);
    A_d1=diag(diag(A_j,0)+1+r(1,1:(n))',0);
    A_J1=A_u1+A_l1+A_d1;
    A_J1(n,n-1)=-r(1,n);
    A_u2=diag(diag(A_j,1)+0.5*r(1,1:(n-1))',1);
    A_l2=diag(diag(A_j,-1)+0.5*r(1,2:(n))',-1);
    A_d2=diag(diag(A_j,0)+1-r(1,1:(n))',0);
    A_J2=A_u2+A_l2+A_d2;
    A_J2(n,n-1)=r(1,n);
    u_p=((A_J1)\(A_J2*ui'+h'))'; %obtained the pore water pressure at
    next time, row vector
    depsilon_vp=Pl_fun*delt_t*A;
    epsilon_p=epsilon_i-P2./(1-u_p).*(u_p-ui)+depsilon_vp;
    %delt_t*A*Pl_fun;
    relative_error=max(abs(epsilon_p-epsilon_ref)./epsilon_p);
    em(nd,N_t)=relative_error;
    nd=nd+1;
    if nd>20 || (isnan(relative_error))==1
        nd=21;
        break
    end
end
```

```
        if max(abs((epsilon_i-epsilon_p)./epsilon_i))<0.01 && nd<20 &&
        sum(depsilon_vp~=0)~=0

            S_m(N_t,1)=(0.5*(epsilon_0+epsilon_p(1,n))+sum(epsilon_p(1,1:n-1)))*de
            lt_z*100;
            epsilon_vp=epsilon_vp+depsilon_vp;
            gm(N_t,1)=P1_fun(1,n);
            DM(N_t,1)=delt_t*A;
            %all chaged delt_t for each time point
            TM(N_t,1)=delt_t;
            t(N_t)=sum(t_m); %dimensionless time
            tm(1,N_t)=t(N_t)*A*60;% time, min
            D_T=delt_t*1.1;  %larger increase, more time
            up_m(1:n,N_t)=u_p';
            ep_m(1:n,N_t)=epsilon_p';
            Nm(N_t,1)=NN;  %Number for decreasing delt_t
            N_t=N_t+1;
            t_m(N_t,1)=D_T;
        elseif max(abs((epsilon_i-epsilon_p)./epsilon_i))>0.01 || nd>20
            D_T=delt_t/5;
            t_m(N_t,1)=D_T;
            em(1:nd-1,N_t)=0;
            DM(N_t,1)=delt_t*A;
            u_p=ui;
            epsilon_p=epsilon_i;
        end
        if D_T<1e-6/60/A || N_t>3000 || max(u_p)>=1 || isreal(ui)==0 ||
        epsilon_0>0.774 || sum(depsilon_vp~=0)==0
            break
        end
    end
        if D_T<1e-6/60/A || N_t>3000 || max(u_p)>=1 || isreal(ui)==0 ||
        epsilon_0>0.774 || sum(depsilon_vp~=0)==0
            break
        end
end
            if N_t==1
                strain=0;
                pressure_b=0;
                elapsed_time=0;
                epsilon_vpi=0;
            elseif sum((tm(1,2:N_t-1)-tm(1,1:N_t-2))==0)>=1
                strain=0;
                pressure_b=0;
                elapsed_time=0;
                epsilon_vpi=0;
            elseif max(tm(1,1:size(up_m,2)))<1520
                strain=0;
                pressure_b=0;

                    elapsed_time=0;
                    epsilon_vpi=0;
                elseif max(tm(1,1:size(up_m,2)))>1520
                    %      index= up_m<1e-6;
                    %      up_m(index)=0;
                    strain=S_m';
                    pressure_b=up_m(n,:)*sigma0;
                    elapsed_time=(tm(1,1:size(up_m,2)));
                    epsilon_vpi=epsilon_vp;
                end
            end
```

Appendix C
Matlab Code for Transitional Markov Chain Monte Carlo Simulation

References:

Zhou WH, Tan F, Yuen KV (2018) Model updating and uncertainty analysis for creep behavior of soft soil. Comput Geotech 100:135–143.
Tan F, Zhou WH, Yuen KV (2018) Effect of loading duration on uncertainty in creep analysis of clay. Int J Numer Anal Methods GeomeFch 42:1235–1254.


```
function [xm, ym, um, pm, like_m, Posterior_m, am, Sm] = TMCMC(x_mean, x_std, NS, s,
UI, SIGMAI, MEASURED_S, TIME_S, N_data_s)
    % x_mean:     Mean of x;  x_std: Standerdiviation of x TMCMC samples
    % NS:         Number of TMCMC samples
    % N_data_s:   Number of measured data
    % s:          Stage number     % UI: Water pressure
    % SIGMAI:     Normal stress
    % MEASURED_S: Measured stress  % TIME_S: Time
    % N_data_s:   Number of measured data
    % xm:         Samples at all stages
    % ym:         Predicted strain
    % um:         Excess pore pressure
    % pm:         Ramdom variables of pdf
    % like_m:     Likelihood         % Posterior_m Posterior
    % am:         Accepted rate      % Sm:          Evidence
    b     = 0.5; % b is the scaling factor
    %% Data initialization
    ln_S = 0;    % ln_S is log evidence
    %% drawing prior samples
    nnn           = length(x_mean); % number of TMCMC samples
    ns            = 1;
    x             = zeros(nnn,NS);
    y             = zeros(NS,s);
    uuu           = zeros(NS,s);
    acm           = zeros(NS,1);
    posterior     = zeros(NS,1);
    am(1:NS,1,1)  = zeros(NS,1);
    log_like      = zeros(NS,1);
    nnx           = 1;
    xxxm(:,1,1)   =zeros(nnn,1);
```


W. Zhou et al., *Practice of Bayesian Probability Theory in Geotechnical Engineering*,
https://doi.org/10.1007/978-981-15-9105-1

```
while ns < (NS+1)
    xx    = (lognrnd(x_mean,x_std))';
    pc    = xx(2)*log(37.5/xx(4))/log(37.5);
    if xx(1)<xx(2) && xx(1)<pc && xx(1)>0.01*xx(2)
        for j  = 1:4
            u0 = UI(j);
            sigma0 = SIGMAI(j);
            C      = exp(xx(2)*log(xx(4))/(xx(2)-xx(1)));
            if (sigma0-u0)<C && j==1
                epsilon0=xx(1)*log(sigma0-u0);
            elseif (sigma0-u0)>C && j==1
                epsilon0=xx(2)*log((sigma0-u0)/xx(4));
            elseif j>1
                epsilon0=max(strain)/100;
            end
            [strain, pressure_b,
            elapsed_time]=solution_EP5(xx(1:nnn-1),u0,sigma0,epsilon0,C);
            if size(strain,2)>2
                time_s=TIME_S(1,N_data_s(j,1):N_data_s(j,2));

        time_u=time_s;
        [yy,uu]=interpolation(strain,
        pressure_b,elapsed_time,time_s,time_u);
    else
        xxxm(:,nnx,1)=xx;
        nnx=nnx+1;
        break
    end
    NAN1 = sum(isnan(yy));
    INF1 = sum(isinf(uu));
    if NAN1==0 && INF1==0
        yyy(1,N_data_s(j,1):N_data_s(j,2)) = yy;
        up(1,N_data_s(j,1):N_data_s(j,2))  = uu;
    else
        xxxm(:,nnx,1) = xx;
        nnx = nnx+1;
        break
    end
end
if j==4 && size(yyy,2)==s && size(strain,2)>2 && NAN1==0 && INF1==0
    log_like(ns,1)=-0.5*s*log(2*pi)-0.5*s*log(xx(nnn)) ...
        -0.5*(yyy-MEASURED_S)*(yyy-MEASURED_S)'/xx(nnn);
    NAN2=sum(isnan(log_like));
    INF2=sum(isinf(log_like));
    if NAN2==0 && INF2==0
        x(1:nnn,ns)=xx;
        y(ns,1:s)=yyy;
        uuu(ns,1:s)=up;
        prior_m(ns,1)=prod(lognpdf(xx',x_mean,x_std));
        ns=ns+1;
        clear yyy NAN1 INF1
    else
```

```
                    xxxm(:,nnx,1)=xx;
                    nnx=nnx+1;
                end
            end
        end
    end
    like_m(1:NS,1,1)=log_like;
    Posterior_m(1:NS,1,1)=prior_m;
    xm(1:nnn,1:NS,1)=x;
    ym(1:NS,1:s,1)=y;
    um(1:NS,1:s,1)=uuu;
    %% compute the weights
    np=1;
    p = 0;
    pm(np,1)=p;
    while p<1
        low_p = p;
        up_p = 2;
        old_p = p;
        while up_p-low_p > 1e-6
            current_p = (low_p + up_p)/2;
            temp = exp((current_p-p)*(log_like-max(log_like)));
            cov_temp = std(temp)/mean(temp);
            if cov_temp > 1
                up_p = current_p;
            else
                low_p = current_p;
            end

    end
    p = current_p;
    if p > 1
        break;
    end % breakout if approaching the final stage
    weight = temp/sum(temp); % weights are normalized
    ln_S = ln_S+log(mean(temp))+(p-old_p)*max(log_like);
%% compute the covariance matrix
    mu_x = x*weight;
    sigma = zeros(nnn,nnn);
    for i=1:NS
        sigma = sigma + b^2*weight(i)*(x(:,i)-mu_x)*(x(:,i)-mu_x)';
    end
```

```
    sqrt_s = sqrtm(sigma);
    sam_ind = deterministicR((1:NS),weight);
    current_x = x;
    current_log_like = log_like;
    current_y=y;
    current_u=uuu;
    an=0;
    nnx=1;
    for i=1:NS
        now_ind = sam_ind(i);
        Nn=1;
        while Nn<2
            x_c =(abs(mvnrnd((current_x(:,now_ind))',sigma,1)))';
            pc=x_c(2)*log(37.5/x_c(4))/log(37.5);
            if x_c(1)<x_c(2) && x_c(1)<pc && x_c(1)>0.01*x_c(2)
                for j=1:4
                    u0=UI(j);
                    sigma0=SIGMAI(j);
                    C=exp(x_c(2)*log(x_c(4))/(x_c(2)-x_c(1)));
                    if (sigma0-u0)<C && j==1
                        epsilon0=x_c(1)*log(sigma0-u0);
                    elseif (sigma0-u0)>C && j==1
                        epsilon0=x_c(2)*log((sigma0-u0)/x_c(4));
                    elseif j>1
                        epsilon0=max(strain)/100;

    end
    [strain, pressure_b,
    elapsed_time]=solution_EP5(x_c(1:nnn-1),u0,sigma0,epsilon0,C);
    if size(strain,2)>2
        time_s=TIME_S(1,N_data_s(j,1):N_data_s(j,2));
        time_u=time_s;
        [yy,uu]=interpolation(strain,
        pressure_b,elapsed_time,time_s,time_u);
    else
        xxxm(:,nnx,np+1)=x_c;
        nnx=nnx+1;
        break
    end
    NAN1=sum(isnan(yy));
    INF1=sum(isinf(uu));
    if NAN1==0 && INF1==0
        yyy(1,N_data_s(j,1):N_data_s(j,2))=yy;
        up(1,N_data_s(j,1):N_data_s(j,2))=uu;
    else
        xxxm(:,nnx,np+1)=x_c;
        nnx=nnx+1;
        break
    end
end
```

```
if j==4 && size(yyy,2)==s && size(strain,2)>2 && NAN1==0 &&
INF1==0
    log_like_c = -0.5*s*log(2*pi)-0.5*s*log(x_c(nnn)) ...
        -0.5*(yyy-MEASURED_S)*(yyy-MEASURED_S)'/x_c(nnn);
    NAN2=sum(isnan(log_like_c));
    INF2=sum(isinf(log_like_c));
    if NAN2==0 && INF2==0
        old_x=current_x(:,now_ind);
        current_prior=prod(lognpdf(old_x',x_mean,x_std));
        x_prior=prod(lognpdf(x_c',x_mean,x_std));
        r = exp(p*(log_like_c -
        current_log_like(now_ind))+log(x_prior)-log(current_prior)
        );
        if r >= rand
            x(:,i) = x_c;
            current_x(:,now_ind) = x_c;
            current_log_like(now_ind) = log_like_c;
            current_y(now_ind,:)=yyy;
            current_u(now_ind,:)=up;
            log_like(i) = log_like_c;
            posterior(i,1)=exp(p*log_like_c)*x_prior;
            prior_m(i,1)=x_prior;
            y(i,:)=yyy;
            uuu(i,:)=up;
                        an=an+1;
                        clear yyy NAN1 INF1
                    else
                        x(:,i) = current_x(:,now_ind);
                        log_like(i) = current_log_like(now_ind);

                        posterior(i,1)=exp(p*current_log_like(now_ind))*curren
                        t_prior;
                        prior_m(i,1)=current_prior;
                        y(i,:)=current_y(now_ind,:);
                        uuu(i,:)=current_u(now_ind,:);
                    end
                    Nn=Nn+1;
                else
                    xxxm(:,nnx,np+1)=x_c;
                    nnx=nnx+1;
                end
            end
        end
    end
    ac_ratio=an/i;
    acm(i,1)=ac_ratio;
  end
  np=np+1;
  pm(np,1)=p;
  xm(1:nnn,1:NS,np)=x;
  ym(1:NS,1:s,np)=y;
  um(1:NS,1:s,np)=uuu;
  am(1:NS,1,np)=acm;
  Sm(1,np)=ln_S;
  Posterior_m(1:NS,1,np)=posterior;
  like_m(1:NS,1,np)=log_like;
end
```

```
%% Final stage
temp = exp((1-old_p)*(log_like-max(log_like)));
weight = temp/sum(temp);
ln_S = ln_S+log(mean(temp))+(1-old_p)*max(log_like);
mu_x = x*weight;
sigma = zeros(nnn,nnn);
for i=1:NS
    sigma = sigma + b^2*weight(i)*(x(:,i)-mu_x)*(x(:,i)-mu_x)';
end
sqrt_s = sqrtm(sigma);
sam_ind = deterministicR((1:NS),weight);
current_x = x;
current_log_like = log_like;
current_y=y;
current_u=uuu;
an=0;
nnx=1;
for i=1:NS
    now_ind = sam_ind(i);
    Nn=1;
    while Nn<2
        x_c =(abs(mvnrnd((current_x(:,now_ind))',sigma,1)))';
        pc=x_c(2)*log(37.5/x_c(4))/log(37.5);
        if x_c(1)<x_c(2) && x_c(1)<pc && x_c(1)>0.01*x_c(2)
            for j=1:4
                u0=UI(j);
                sigma0=SIGMAI(j);
                C=exp(x_c(2)*log(x_c(4))/(x_c(2)-x_c(1)));

    if (sigma0-u0)<C && j==1
        epsilon0=x_c(1)*log(sigma0-u0);
    elseif (sigma0-u0)>C && j==1
        epsilon0=x_c(2)*log((sigma0-u0)/x_c(4));
    elseif j>1
        epsilon0=max(strain)/100;
    end
    [strain, pressure_b,
    elapsed_time]=solution_EP5(x_c(1:nnn-1),u0,sigma0,epsilon0,C);

    if size(strain,2)>2
```

```
                time_s=TIME_S(1,N_data_s(j,1):N_data_s(j,2));
                time_u=time_s;
                [yy,uu]=interpolation(strain,
                pressure_b,elapsed_time,time_s,time_u);
            else
                xxxm(:,nnx,np+1)=x_c;
                nnx=nnx+1;
                break
            end
            NAN1=sum(isnan(yy));
            INF1=sum(isinf(uu));
            if NAN1==0 && INF1==0
                yyy(1,N_data_s(j,1):N_data_s(j,2))=yy;
                up(1,N_data_s(j,1):N_data_s(j,2))=uu;
            else
                xxxm(:,nnx,np+1)=x_c;
                nnx=nnx+1;
                break
            end
        end
if j==4 && size(yyy,2)==s && size(strain,2)>2 && NAN1==0 && INF1==0
    log_like_c = -0.5*s*log(2*pi)-0.5*s*log(x_c(nnn)) ...
        -0.5*(yyy-MEASURED_S)*(yyy-MEASURED_S)'/x_c(nnn);
    NAN2=sum(isnan(log_like_c));
    INF2=sum(isinf(log_like_c));
    if NAN2==0 && INF2==0
        old_x=current_x(:,now_ind);
        current_prior=prod(lognpdf(old_x',x_mean,x_std));
        x_prior=prod(lognpdf(x_c',x_mean,x_std));
        r = exp(log_like_c -
        current_log_like(now_ind)+log(x_prior)-log(current_prior));
        if r >= rand
            x(:,i) = x_c;
            current_x(:,now_ind) = x_c;
            current_log_like(now_ind) = log_like_c;
            current_y(now_ind,:)=yyy;
            current_u(now_ind,:)=up;
            log_like(i) = log_like_c;
            posterior(i,1)=exp(log_like_c)*x_prior;
            y(i,:)=yyy;
            uuu(i,:)=up;
            an=an+1;
            clear yyy NAN1 INF1
        else
            x(:,i) = current_x(:,now_ind);
            log_like(i) = current_log_like(now_ind);

            posterior(i,1)=exp(current_log_like(now_ind))*current_prio
            r;
            y(i,:)=current_y(now_ind,:);
            uuu(i,:)=current_u(now_ind,:);
        end
        Nn=Nn+1;
    else
        xxxm(:,nnx,np+1)=x_c;
        nnx=nnx+1;
```

```
                    end
                end
            end
        end
        ac_ratio=an/i;
        acm(i,1)=ac_ratio;
    end
    xm(1:nnn,1:NS,np+1)=x;      ym(1:NS,1:s,np+1)=y;
    um(1:NS,1:s,np+1)=uuu;      pm(np+1,1)=1;
    am(1:NS,1,np+1)=acm;        Sm(1,np+1)=ln_S;
    Posterior_m(1:NS,1,np+1)=posterior;
    like_m(1:NS,1,np+1)=log_like;
end
```

Appendix D
Matlab Code for Identifying Parameters of a Given Model or Model Class Selection for Given Experimental Data

References:

Jin YF, Yin Z-Y, Zhou WH, Horpibulsuk S (2019) Identifying parameters of advanced soil models using an enhanced Transitional Markov chain Monte Carlo method. Acta Geotech 14(6):1925–1947.

Jin YF, Yin Z-Y, Zhou WH, Shao JF (2019) Bayesian model selection for sand with generalization ability evaluation. Int J Numer Anal Methods Geomech 43(14):2305–2327.

```
%%
dos('del Results*.dat'); % clean previous data
%%
soilmodel=2; % select one constitutive model you are interested in
             % for model class selection, all models can be gone through to compare
             each other
parameters=zeros(50,1); % maximum number of parameters is 50
switch soilmodel
    case 1    % MC
        parameters(1:8)=[0.8,10000,0.2,30,1.0,0,3,1]; % Trial values of parameters,
        some of them will
                                                       % be reserved if not selected for
                                                       identification
        x_low=[1000,20,0,0]; % low bound of selected parameters to be identified
        x_up=[50000,50,10,1]; % up bound of selected parameters to be identified

    case 2    % NLMC
        parameters(1:13)=[0.8,100,0.2,0.6,0.5,0.881,30,0.003,1,0,0,0,0];
        x_low=[20, -3, 0, 0];
        x_up=[50, -1, 10, 1];
```


W. Zhou et al., *Practice of Bayesian Probability Theory in Geotechnical Engineering*,
https://doi.org/10.1007/978-981-15-9105-1

```
    case 3    % SIMSAND
        parameters(1:16)=[0.8,60,40,0.5,0.785,0.5,0.003,0.45,30,0.003,1.0,1.0,2.0,2.0,
        250000,0.3];
        x_low=[50 0.1 0.5,-3,0.1,20,-3,0.1,0.1,0.1,0.0];
        x_up=[300 0.9 1.5,-1,1.0,50,-1,2.0,10,10,1];

    case 4    % for more constitutive models
       % ........

end
%%
N=2000;  % number of samples (e.g. 2000 sets of parameters)
ff=@loglikelihoodfun_SPMD4; % setting for parallel computing, just an option depending
on your computer
%%
        [x,ln_S]=TMCMC_DE_SPMD(ff,soilmodel,parameters,x_low,x_up,N,num_a,num_b);
        % x: joint distributions of selected parameters
        % ln_S: evidence of the model, bigger value represents the model more reliable
%%

function ObjVals=loglikelihoodfun_SPMD4(soilmodel,x,parameter)
% to computer log(likelihood)
switch soilmodel
    case 1    % MC
        % position of selected parameters to be identified in the array of parameters
        [parameter]
        posi=[2 4 5 31 32];
        % user defined for test type, e.g. here is conventional triaxial test by us
        testtype=2;
        % user defined for loading phase, e.g. here o means no IC or K0 compression phase
        preISO=0;
    case 2    % NLMC
        posi=[7 8 9 31 32];
        testtype=2;
        preISO=0;
    case 3    % SIMSAND
        posi=[5 7 8 9 10 11 12 13 31 32];
        testtype=2;
        preISO=0;
    case 4  % for more constitutive models
       % ........
end
```

```
%%
CoreNum=4;  % number of CPU cores called for computing
Num_p=length(x(1,:)); % number of samples
Num_p_average=Num_p/CoreNum; % number of samples calculated per core, must be integer
spmd
    switch labindex
        case 1
            for i=1:Num_p_average
                ObjVals(i,1)=Objective_error1(parameter,x(:,i+(labindex-1)*Num_p_average
                ),posi,testtype,preISO,soilmodel);
            end              % to computer log(likelihood) for samples of the 1st core
        case 2
            for i=1:Num_p_average
                ObjVals(i,1)=Objective_error2(parameter,x(:,i+(labindex-1)*Num_p_average
                ),posi,testtype,preISO,soilmodel);
            end
        case 3
            for i=1:Num_p_average
                ObjVals(i,1)=Objective_error3(parameter,x(:,i+(labindex-1)*Num_p_average
                ),posi,testtype,preISO,soilmodel);
            end
        case 4
            for i=1:Num_p_average
                ObjVals(i,1)=Objective_error4(parameter,x(:,i+(labindex-1)*Num_p_average
                ),posi,testtype,preISO,soilmodel);
            end
        % Attention: if CoreNum is bigger than 4, we need to put more up to this
        number
    end
end
%%
```

Appendix E
Matlab Code of DE-TMCMC Method

References:

DE-TMCMC: Deifferencial Evolutionary Transitional Markov Chain Monte Carlo:

Jin YF, Yin Z-Y, Zhou WH, Horpibulsuk S (2019) Identifying parameters of advanced soil models using an enhanced Transitional Markov chain Monte Carlo method. Acta Geotech 14(6):1925–1947.

```
%%
function [x,ln_S]=TMCMC_DE_SPMD(log_like_fun,soilmodel,y,x_low,x_up,N,num_a,num_b)
%% initialing
n=length(x_low);
b=0.2; %2.4/sqrt(n);
p=0; ln_S=0; AR_ratio=1;jj=1;
[delta,c,c_star,Pm,p_g] = deal(3,0.1,1e-12,0.9,0.2);
% CR = [1:n_CR]/n_CR;
% pCR = ones(1,n_CR)/n_CR;
%%  select the MCMC algorithm
algo=4;
%% drawing prior samples using the latin sampling
for i=1:n
    x(i,:)=LHS_uniform(x_low(i),x_up(i),N);% x_low(i)+(x_up(i)-x_low(i))*rand(1,N);
end
%% plot
Plot_figure(jj,x,x_low,x_up,num_a,num_b);
%%
% for i=1:N
log_like=feval(log_like_fun,soilmodel,x,y);
% end
%% write to file
Writedata(jj,p,ln_S,x,log_like);
%% compute the weigths
```


W. Zhou et al., *Practice of Bayesian Probability Theory in Geotechnical Engineering*,
https://doi.org/10.1007/978-981-15-9105-1

```
while p<1  % adaptively choose p
    low_p=p; up_p=2; old_p=p;
    while up_p-low_p>10e-6
        current_p=(low_p+up_p)/2;
        temp=exp((current_p-p)*(log_like-max(log_like)));
        cov_temp=std(temp)/mean(temp);
        if cov_temp>1
            up_p=current_p;
        else
            low_p=current_p;
        end
    end
    p=current_p;
    jj=jj+1;
    % break out if the final stage reaches.
    if p>1, break; end
    %
    weight=temp/sum(temp);  % weights are normalized
    ln_S=ln_S+log(mean(temp))+(p-old_p)*max(log_like);
    %% compute the covariance matrix
    mu_x=x*weight;
    sigma=zeros(n,n);
    for i=1:N
        sigma=sigma+b^2*weight(i)*(x(:,i)-mu_x)*(x(:,i)-mu_x)';  %(1/AR_ratio)* %
        modification AR according to Se-Hyeok Lee and Junho Song (2017)
    end
    sqrt_s=sqrtm(sigma);
    %% find the maximum weight
    [~,posi_max]=max(weight);
    %% do MCMC

 sam_ind=deterministicR((1:N),weight);  % tourament
 gamma_d = 2.38/sqrt(2*n);
 % Randomly generate coefficients
 lambda = unifrnd(-c,c,N,1);
 nn= unifrnd(0.5,1.0);
 %
 current_x=x; current_log_like=log_like;
 AR_num=0;                              % the acceptance number of samples
 for i=1:N
     now_ind=sam_ind(i);
     % Now randomly select 3 Points from SCEMPar.s
     [kk] = randperm(N);
     ii = kk(1:2); r1 = ii(1); r2 = ii(2);
     switch algo
         case 1     % Original TMCMC
             now_ind=sam_ind(i);
             x_c(:,i)=current_x(:,now_ind)+sqrt_s*rand(n,1);
         case 2     % DE-TMCMC
             g=randsample([gamma_d 1],1,'true',[0.9 0.1]);

             x_c(:,i)=current_x(:,now_ind)+g*(current_x(:,r1)-current_x(:,r2))+1e-6
             *randn(n,1);
```

```
        case 3    % DREAM-TMCMC
            z=rand(1,n);
            A =
            find(z<Pm);
                 % Derive subset A with dimensions to sample

            d_star=numel(A);
                     % How many dimensions are sampled?
            if d_star==0, [~,A]=min(z);  d_star=1; end
            gamma_d =
            2.38/sqrt(2*delta*d_star);
            % Calculate jump rate
            g=randsample([gamma_d 1],1,'true',[1-p_g
            p_g]);                              % Select gamma: 80/20 ratio
            (default)
            dX(1:n,i)=0;

            dX(A,i)=(1+lambda(i))*g*(current_x(A,r1)-current_x(A,r2))+c_star*randn
            (d_star,1);   % Compute ith jump with differential evolution
            x_c(:,i)=current_x(:,now_ind)+dX(:,i);

    case 4     % EDE-TMCMC from the NAG paper
        z=rand(1,n);
        A =
        find(z<Pm);
             % Derive subset A with dimensions to sample

        d_star=numel(A);
                 % How many dimensions are sampled?
        if d_star==0, [~,A]=min(z);  d_star=1; end
        gamma_d =
        2.38/sqrt(2*delta*d_star);
        % Calculate jump rate
        g=randsample([gamma_d 1],1,'true',[1-p_g
        p_g]);                              % Select gamma: 80/20 ratio
        (default)
        dX(1:n,i)=0;

        dX(A,i)=(1+lambda(i))*(g*(current_x(A,posi_max)-current_x(A,now_ind))+
        g*(current_x(A,r1)-current_x(A,r2)))+c_star*randn(d_star,1);   %
        Compute ith jump with differential evolution
        x_c(:,i)=current_x(:,now_ind)+dX(:,i);
end
% compute the likelihood
for j=1:n
    if x_c(j,i)>x_up(j)
        x_c(j,i)=x_low(j)+(x_up(j)-x_low(j))*rand(1);
    end

                if x_c(j,i)<x_low(j)
                    x_c(j,i)=x_low(j)+(x_up(j)-x_low(j))*rand(1);
                end
            end
        end
        %
        log_like_c=feval(log_like_fun,soilmodel,x_c,y);
        %
        for i=1:N
            now_ind=sam_ind(i);
            r=exp(log_like_c(i)-current_log_like(now_ind));
            if r>=rand
                x(:,i)=x_c(:,i);
                current_x(:,now_ind)=x_c(:,i);
                current_log_like(now_ind)=log_like_c(i);
                log_like(i)=log_like_c(i);
                AR_num=AR_num+1;
            else
```

```
                x(:,i)=current_x(:,now_ind);
                log_like(i)=current_log_like(now_ind);
            end
        end
        AR_ratio=AR_num/N;
        if (AR_ratio==0)
            AR_ratio=0.001;
        end
        %% plot
        Plot_figure(jj,x,x_low,x_up,num_a,num_b);
        %% write to file
        Writedata(jj,p,ln_S,x,log_like);
    end
    %% final stage
    temp=exp((1-old_p)*(log_like-max(log_like)));
    weight=temp/sum(temp);                        % weights are normalized
    ln_S=ln_S+log(mean(temp))+(1-old_p)*max(log_like);
    mu_x=x*weight; sigma=zeros(n,n);
    for i=1:N
        sigma=sigma+b^2*weight(i)*(x(:,i)-mu_x)*(x(:,i)-mu_x)'; %(1/AR_ratio)*
    end
    sqrt_s=sqrtm(sigma);

%% find the maximum weight
[~,posi_max]=max(weight);
sam_ind=deterministicR((1:N),weight);
gamma_d = 2.38/sqrt(2*n);
% Randomly generate coefficients
lambda = unifrnd(-c,c,N,1);
nn= unifrnd(0.5,1.0);
current_x=x; current_log_like=log_like;
AR_num=0;                                      % the acceptance number of samples
for i=1:N
    now_ind=sam_ind(i);
    % Now randomly select 3 Points from SCEMPar.s
    [kk] = randperm(N);
    ii = kk(1:2); r1 = ii(1); r2 = ii(2);
    switch algo
        case 1     % Original TMCMC
            now_ind=sam_ind(i);
            x_c(:,i)=current_x(:,now_ind)+sqrt_s*rand(n,1);
        case 2     % DE-TMCMC
            g=randsample([gamma_d 1],1,'true',[0.9 0.1]);

            x_c(:,i)=current_x(:,now_ind)+g*(current_x(:,r1)-current_x(:,r2))+1e-6*ran
            dn(n,1);
        case 3     % DREAM-TMCMC
            z=rand(1,n);
            A =
            find(z<Pm);
             % Derive subset A with dimensions to sample
```

```
    d_star=numel(A);
        % How many dimensions are sampled?
    if d_star==0, [~,A]=min(z);  d_star=1; end
    gamma_d =
    2.38/sqrt(2*delta*d_star);                                          %
    Calculate jump rate
    g=randsample([gamma_d 1],1,'true',[1-p_g
    p_g]);                                  % Select gamma: 80/20 ratio
    (default)
    dX(1:n,i)=0;

    dX(A,i)=(1+lambda(i))*g*(current_x(A,r1)-current_x(A,r2))+c_star*randn(d_s
    tar,1);   % Compute ith jump with differential evolution
    x_c(:,i)=current_x(:,now_ind)+dX(:,i);
case 4     % EDE-TMCMC from the NAG paper
    z=rand(1,n);
    A =
    find(z<Pm);
     % Derive subset A with dimensions to sample

        d_star=numel(A);
            % How many dimensions are sampled?
        if d_star==0, [~,A]=min(z);  d_star=1; end
        gamma_d =
        2.38/sqrt(2*delta*d_star);                                      %
        Calculate jump rate
        g=randsample([gamma_d 1],1,'true',[1-p_g
        p_g]);                                  % Select gamma: 80/20 ratio
        (default)
        dX(1:n,i)=0;

        dX(A,i)=(1+lambda(i))*(g*(current_x(A,posi_max)-current_x(A,now_ind))+g*(c
        urrent_x(A,r1)-current_x(A,r2)))+c_star*randn(d_star,1);   % Compute ith
        jump with differential evolution
        x_c(:,i)=current_x(:,now_ind)+dX(:,i);
end
% check the bounds
```

```
                for j=1:n
                    if x_c(j,i)>x_up(j)
                        x_c(j,i)=x_low(j)+(x_up(j)-x_low(j))*rand(1);
                    end
                    if x_c(j,i)<x_low(j)
                        x_c(j,i)=x_low(j)+(x_up(j)-x_low(j))*rand(1);
                    end
                end
            end
            %
            log_like_c=feval(log_like_fun,soilmodel,x_c,y);
            for i=1:N
                now_ind=sam_ind(i);
                r=exp(log_like_c(i)-current_log_like(now_ind));
                if r>=rand
                    x(:,i)=x_c(:,i);
                    current_x(:,now_ind)=x_c(:,i);
                    current_log_like(now_ind)=log_like_c(i);
                    log_like(i)=log_like_c(i);
                    AR_num=AR_num+1;
                else
                    x(:,i)=current_x(:,now_ind);
                    log_like(i)=current_log_like(now_ind);
                end
            end
            AR_ratio=AR_num/N;
            %% plot
            Plot_figure(jj,x,x_low,x_up,num_a,num_b);
            %% write to file
            Writedata(jj,1,ln_S,x,log_like);
            %%  SIT resampling
            function outIndex=deterministicR(inIndex, q)

%%
% purpose: to perform the resampling stage of the SIR in order (number of samples)
steps.
% It uses Kitagama's deterministic resampling algorithm;
%%
if nargin<2, error('Not enough input arguments.'); end;
[S,~]=size(q);  % S= number of particles;
%% RESIDUAL RESAMPLING
N_babies=zeros(1,S); u=zeros(1,S);
% generate the cumulative distribution
cumDist=cumsum(q'); aux=rand(1); u=aux:1:(S-1+aux); u=u./S;
j=1;
for i=1:S
    while (u(1,i)>cumDist(1,j))
        j=j+1;
    end
    N_babies(1,j)=N_babies(1,j)+1;
end
% COPY RESAMPLED TRAJECTORIES:
index=1;
for i=1:S
    if (N_babies(1,i)>0)
        for j=index:index+N_babies(1,i)-1
            outIndex(j)=inIndex(i);
        end
```

```
    end
    index=index+N_babies(1,i);
end
%%
function s=LHS_uniform(xmin,xmax,nsample)
% s=lhsu(xmin,xmax,nsample)
% LHS from uniform distribution
% Input:
%   xmin    : min of data (1,nvar)
%   xmax    : max of data (1,nvar)
%   nsample : no. of samples
% Output:
%   s       : random sample (nsample,nvar)
%   Budiman (2003)
nvar=length(xmin);
ran=rand(nsample,nvar);
s=zeros(nsample,nvar);
for j=1: nvar
    idx=randperm(nsample);
    P =(idx'-ran(:,j))/nsample;
    s(:,j) = xmin(j) + P.* (xmax(j)-xmin(j));
end
```

Appendix F
Fortran Code of SIMSAND Model in Format of UMAT of ABAQUS

References:

Jin Y-F, Yin Z-Y, Shen S-L, Hicher P-Y (2016) Selection of sand models and identification of parameters using an enhanced genetic algorithm. Int J Numer Anal Methods Geomech 40(8):1219–1240.

```
! ------------------------------------------------------------
C***************************************************************
      SUBROUTINE UMAT(STRESS,STATEV,DDSDDE,SSE,SPD,SCD,
     1 RPL,DDSDDT,DRPLDE,DRPLDT,
     2 STRAN,DSTRAN,TIME,DTIME,TEMP,DTEMP,PREDEF,DPRED,CMNAME,
     3 NDI,NSHR,NTENS,NSTATV,PROPS,NPROPS,COORDS,DROT,PNEWDT,
     4 CELENT,DFGRD0,DFGRD1,NOEL,NPT,LAYER,KSPT,KSTEP,KINC)
C
      INCLUDE 'ABA_PARAM.INC'
C

      CHARACTER*80 CMNAME
      DIMENSION STRESS(NTENS),STATEV(NSTATV),
     1 DDSDDE(NTENS,NTENS),
     2 DDSDDT(NTENS),DRPLDE(NTENS),
     3 STRAN(NTENS),DSTRAN(NTENS),TIME(2),PREDEF(1),DPRED(1),
     4 PROPS(NPROPS),COORDS(3),DROT(3,3),DFGRD0(3,3),DFGRD1(3,3)
C
      Double Precision  cm(NPROPS),CC(6,6),depsp(6)
      Double Precision  Sig(6),eps(6),dSig(6),dEps(6),Epsp(6)
      Double Precision  hsv(NSTATV)
      Double Precision  StVar(NSTATV),StVar_DDS(NSTATV)
      Double Precision  Sig0(6),Sig_DDS(6),Sig_new(6),Sig_temp(6)
      Double Precision  Stress_star(6),sig_star(6),stress_old(6)
      Double Precision  Eps0(6),dEps_dds(6),deps_temp(6)
      Double Precision  dstran_star_dt(6),deps_ini(6)
      Double Precision  D(6,6),DDS(6,6),DD(6,6),DDE(6,6)
      Double Precision  DDSDD(NTENS,NTENS)
      Double Precision  theta
      integer           ntens,ii,jj,kk,i,j
      integer           CSL,psi_selection,sub_nstep
      Parameter( zero = 0.D0, one = 1.D0, two = 2.D0, three = 3.D0,
     1  third = one/three, half = 0.5D0, twoThirds = two/three,
     2  threeHalfs = 1.5D0 )
```


W. Zhou et al., *Practice of Bayesian Probability Theory in Geotechnical Engineering*,
https://doi.org/10.1007/978-981-15-9105-1

```
c
c*******************************************************************************c
c       parameters definition
c*******************************************************************************c
        cm      = props
c------- Relating to Elasticity
       e0      = cm(1)    ! Initial void ratio
       dK0     = cm(2)    ! dK=dK0*pat*(2.97-ee)**2/(1.+ee)*((p_module+bond)/pat)**dnn
       dG0     = cm(3)    ! dG=bG0*pat*(2.97-ee)**2/(1.+ee)*((p_module+bond)/pat)**dnn

      dnn     = cm(4)    ! defaut value: 0.6667

c------- Relating to Critical State Lines: eec=ecu+(ec0-ecu)*dexp(-1.*(pp0/pat/dlamda)**dksi)
      ec0     = cm(5)    ! Initial critical void ratio for initial grading when p=0
      ecu0    = cm(6)    ! =emin, Initial critical void ratio for initial grading when p=inifinity
      dlamda  = cm(7)
      dksi    = cm(8)
      fi      = cm(9)    ! critical friction angle to computer M
      fi      = fi*3.14159265358979323239/180.   ! change the unity
      pat     = 101.325

c------- Relating to shear mechanism f1=q/(p+bond)-Mp*epspd/(akp+epspd), d=ad*(Mpt-q/(p+bond))
      akp     = cm(10)   ! f=q/(p+bond)-Mp*epspd/(akp+epspd)
      ad      = cm(11)   ! d = ad*(Mpt-q/(p+bond))
      dnp     = cm(12)   ! fi_b=atan(tan(fi)*(ec/e)**dnd) defaut value=1
      dnd     = cm(13)   ! fi_d=atan(tan(fi)*(e/ec)**dnd) defaut value=2

c------- Relating to normal mechanism f2= 1/dn*(eta/dM)^(dn+1)*p+p-pm, associate flow rule
      dn      = cm(14)   ! canbe calculated by K0=1-sin(fi)
      pm0     = cm(15)   !
      dr0     = cm(16)   ! compression coefficient in loge-logp
c------- Relating to grain breakage
      dBr     = cm(17)   ! initial grading index =0.
      dParBr  = cm(18)   ! Br=wp/(dParBr+wp)
      ecuf    = cm(19)   ! eminf, fractal critical void ratio for fractal grading when p=inifinity
      Parecu  = cm(20)   ! obtained from emin-Br, ecu = ecuf+(ecu0-ecuf)*dexp(-1.*Parecu*dBr)

c------- Relating to bondong and debonding of cemented sand
      bond0   = cm(21)   ! for cemented sand
      if (bond0 .lt. 0.01) bond0=0.01
      dw      = cm(22)   ! degradation: dbond=-dw*bond*dEpsDP
c-------------------------------------------------------------------------------
      psi_selection=cm(23)  !psi_selection=1, e/ec; psi_selection=2, e-ec
      eps70     = cm(24)
c-------------------------------------------------------------------------------
c ---------------- select the critical state line -------------------------------
      if (dlamda .lt. 1. .AND. dksi .eq. 1) then
          CSL=1       ! eec=ec0-dlamda*(pp0/pat)**dksi
      else if (dlamda .lt. 1. .AND. dksi .lt. 1) then
          CSL=2       ! eec=ec0-dlamda*(pp0/pat)**dksi
      else
          CSL=3       ! eec=ecu0+(ec0-ecu0)*exp(-1.*(pp0/pat/dlamda)**dksi)
      endif
c      write(*,*) 'CSL=',CSL
c*********************************************************************c

21    format(6(1x,1p,e12.5))
c          stop
! ------------------------------------------------------------------------------
! Read STRESS STRAIN
! ------------------------------------------------------------------------------
      Sig0(1:4)   = STRESS(1:4)
      Eps0(1:3)   = STRAN(1:3)
      Eps0(4)     = STRAN(4)*0.5  ! change engineering strain to normal strain
      dEps(1:3)   = DSTRAN(1:3)
      dEps(4)     = DSTRAN(4)*0.5  ! change engineering strain to normal strain
      Sig0(5:6)   = zero
      Eps0(5:6)   = zero
      dEps(5:6)   = zero
      if( NSHR .GT. 1 ) then
          Sig0(5:6)   = STRESS(5:6)
          Eps0(5:6)   = STRAN(5:6)*0.5  ! change engineering strain to normal strain
          dEps(5:6)   = DSTRAN(5:6)*0.5  ! change engineering strain to normal strain
      end if
```

```
c change sign due to EVP-MCC formulation (compression positive)
c       write(*,*)  NTENS
c       write(*,21) dEps
! ------------------------------------------------------------------------------------
! Read state variables
! ------------------------------------------------------------------------------------
      hsv=STATEV
!     write(*,21) STATEV
! ------------------------------------------------------------------------------------
! Initialize state variables
! ------------------------------------------------------------------------------------
      IF (KSTEP == 1 .AND. TIME(2) ==zero) THEN
! ------------------------------------------------------------------------------------
! Initialize state variables
! ------------------------------------------------------------------------------------
      ! just for print for triaxial condition
       Sig_temp=( -1.)*Sig0
       call stress_invariant(Sig_temp, dI1, dI2, dI3, pp0, qq0)
c        write(*,*) pp0,qq0
      ! initialization
       evol = 0.
       Eps  = 0.
       hsv  = 0.

c*************************************************************************
         select case(CSL)

         case(1)
          if (pp0 .le. 1.d-2) then
             eec= ec0-dlamda*log(0.01/pat)
          else
             eec= ec0-dlamda*log(pp0/pat)
          end if

         case(2)

          if (pp0 .le. 1.d-2) then
             eec=ec0-dlamda*(0.01/pat)**dksi
          else
             eec=ec0-dlamda*(pp0/pat)**dksi
          end if

         case(3)
          if (pp0 .le. 1.d-2) then
             eec=ecu0+(ec0-ecu0)*exp(-1.*(0.01/pat/dlamda)**dksi)
          else
             eec=ecu0+(ec0-ecu0)*exp(-1.*(pp0/pat/dlamda)**dksi)
          end if

        end select
c*************************************************************************
     ! calculate stress invariants and lode angle effect
c       Call Lode_TS (Sig_temp,cm,dMratio)

        dMratio=1.0
       select case(psi_selection)
```

```
          case(1)

          psi=eec/e0
          if (psi .lt. 1.d-7) psi=1.d-7
          fi_p=atan(tan(fi)*psi**dnp)
          fi_d=atan(tan(fi)*(1./psi)**dnd)
c
          dMp    = 6.*sin(fi_p)/(3.-sin(fi_p))*dMratio
          dMpt   = 6.*sin(fi_d)/(3.-sin(fi_d))*dMratio


          case (2)

          psi=e0-eec
          dM = 6.*sin(fi)/(3.-sin(fi))*dMratio
          dMp = dM*exp(-dnp*psi)
          dMpt = dM*exp(dnd*psi)

          case default

          psi=eec/e0
          if (psi .lt. 1.d-7) psi=1.d-7
          fi_p=atan(tan(fi)*psi**dnp)
          fi_d=atan(tan(fi)*(1./psi)**dnd)
c
          dMp    = 6.*sin(fi_p)/(3.-sin(fi_p))*dMratio
          dMpt   = 6.*sin(fi_d)/(3.-sin(fi_d))*dMratio

        end select
c***************************************************************************
       epsd_p = qq0*akp/(-qq0+dMp*pp0+dMp*bond0)+0.0001
       evol_p = ad*(dMpt-qq0/(pp0+bond0))*epsd_p
       epsp1 = epsd_p+(1./3.)*evol_p
       epsp3 = (1./3.)*evol_p-(1./2.)*epsd_p
       hsv(2) = epsp3*1.1
       hsv(3) = epsp1*1.1 ! to make sure the first step is elastic, "1" is vertical
       hsv(4) = epsp3*1.1

      hsv(1) = e0                      ! void ratio
      hsv(8) = pm0
      hsv(9) = bond0                   ! bonding parameters
      hsv(18)= 0.                      ! equivalent plastic strain
      hsv(10)= 0.                      ! plastic work for breakage
      hsv(19)= eec                     ! void ratio in critical state line eec
      hsv(20)= psi                     ! the state parameter
      hsv(21)= pp0                     ! effective mean stress
      END IF
! -----------------------------------------------------------------------------------
! Strain Load
! -----------------------------------------------------------------------------------
      deps_temp  = deps*(-1.)
      Eps0       = -Eps0

     Call CSsand (cm,deps,sig0,hsv,sig)
! -----------------------------------------------------------------------------------
! UPDATE STRESS STRAIN AND STATE VARIABLES
! -----------------------------------------------------------------------------------
      STRESS(1:4) = (-1.0)*SIG(1:4)
      if( NSHR .EQ. 3 ) then
          STRESS(5:6) = (-1.0)*SIG(5:6)
      end if
      STATEV = hsv
c     write(*,21) hsv(1)
! -----------------------------------------------------------------------------------
! Give the DDSDDE BY NUMERICAL DISTURBED METHOD
! -----------------------------------------------------------------------------------
```

```
      Call MATRIXDE_SS(sig,cm,hsv,D)
      Do i=1,NTENS
         Do j=1,NTENS
            DDSDDE(i,j)=D(i,j)
         End Do
      End Do
      DDSDDE(4,4)=D(4,4)*0.5      ! change to engineering strain for abaqus-umat
      if( NSHR .GT. 1 ) then
          DDSDDE(5,5)=D(5,5)*0.5
          DDSDDE(6,6)=D(6,6)*0.5
      end if
c-----------------------------------------------------------------------------------
      Return
      END SUBROUTINE UMAT


!-----------------------------------------------------------------------------------
      Subroutine CSsand (cm,deps,sig,hsv,sig_new)
      INCLUDE 'ABA_PARAM.INC'
c     Implicit none
      Character*80 cmname
      Double Precision cm(*),hsv(*),deps(6),deps0(6),epsp(6),depsp(6)
      Double Precision sig0(6),sig(6),CC(6,6),dsig(6),dpdsig(6)
      Double Precision sig_new(6),dqdsig(6),dum(6),Eps0(6)
      Double Precision dF1dsig(6),dG1dsig(6),dF2dsig(6),dG2dsig(6)
      Double Precision dumG1(6),dumG2(6)
      INTEGER   CSL, psi_selection
      Logical   converged              ! convergence for iteration

      Parameter( zero = 0.D0, one = 1.D0, two = 2.D0, three = 3.D0,
     1 third = one/three, half = 0.5D0, twoThirds = two/three,
     2 threeHalfs = 1.5D0 )

21    format(6(1x,1p,e12.5))
      !parameters
c------- Relating to Elasticity
      e0     = cm(1)   ! Initial void ratio
      dK0    = cm(2)   ! dK=dK0*pat*(2.97-ee)**2/(1.+ee)*((p_module+bond)/pat)**dnn
      dG0    = cm(3)   ! dG=bG0*pat*(2.97-ee)**2/(1.+ee)*((p_module+bond)/pat)**dnn
      dnn    = cm(4)   ! defaut value: 0.6667

c------- Relating to Critical State Lines: eec=ecu+(ec0-ecu)*dexp(-1.*(pp0/pat/dlamda)**dksi)
      ec0    = cm(5)    ! Initial critical void ratio for initial grading when p=0
      ecu0   = cm(6)    ! =emin, Initial critical void ratio for initial grading when p=inifinity
      dlamda = cm(7)
      dksi   = cm(8)
      fi     = cm(9)   ! critical friction angle to computer M
      fi     = fi*3.14159265358979323239/180.   ! change the unity
      pat    = 101.325
c------- Relating to shear mechanism f1=q/(p+bond)-Mp*epspd/(akp+epspd), d=ad*(Mpt-q/(p+bond))
      akp    = cm(10)   ! f=q/(p+bond)-Mp*epspd/(akp+epspd)
      ad     = cm(11)   ! d = ad*(Mpt-q/(p+bond))
      dnp    = cm(12)   ! fi_b=atan(tan(fi)*(ec/e)**dnp) defaut value=1
      dnd    = cm(13)   ! fi_d=atan(tan(fi)*(e/ec)**dnd) defaut value=2
c------- Relating to normal mechanism f2= 1/dn*(eta/dM)^(dn+1)*p+p-pm, associate flow rule
      dn     = cm(14)   ! can be calculated by K0=1-sin(fi)
      pm0  = cm(15)   !
      dr0    = cm(16)   ! compression coefficient in loge-logp
c------- Relating to grain breakage
     dBr     = cm(17)   ! initial grading index =0.
     dParBr  = cm(18)   ! Br=wp/(dParBr+wp)
      ecuf   = cm(19)   ! eminf, fractal critical void ratio for fractal grading when p=inifinity
     Parecu  = cm(20)   ! obtained from emin-Br, ecu = ecuf+(ecu0-ecuf)*dexp(-1.*Parecu*dBr)
c------- Relating to bondong and debonding of cemented sand
      bond0  = cm(21)   ! for cemented sand
      dw     = cm(22)   ! degradation: dbond=-dw*bond*dEpsDP
c-----------------------------------------------------------------------------------
      psi_selection=cm(23)  !psi_selection=1, e/ec; psi_selection=2, e-ec
      eps70     = cm(24)
```

```
c-----------------------------------------------------------------------------------
c         WRITE(*,21) cm(22)
c ---------------- select the critical state line ---------------------------------------
      if (dlamda .lt. 1. .AND. dksi .eq. 1) then
         CSL=1      ! eec=ec0-dlamda*(pp0/pat)**dksi
      else if (dlamda .lt. 1. .AND. dksi .lt. 1) then
         CSL=2      ! eec=ec0-dlamda*(pp0/pat)**dksi
      else
         CSL=3      ! eec=ecu0+(ec0-ecu0)*exp(-1.*(pp0/pat/dlamda)**dksi)
      end if
! ---------------------------------------------------------------------------------------
! ---------------------------------------------------------------------------------------
! Change sign due to SCLAY formulation (compression positive)
! ---------------------------------------------------------------------------------------
      Sig    = -1.*Sig

   dEps    = -1.*dEps

c         dMratio= 1.0
c         WRITE(*,21) Sig
c**************************small strain *****************************************************
        Call MATRIXDE_SS(Sig,cm,hsv,CC)
c*******************************************************************************************
c         stop
c---------initialization-----------------
      depspv = 0.
      depspd = 0.
      depsp  = 0.
      deps0  = deps ! to reqive the eps back in the end
      sig0   = sig ! to reqive the eps back in the end for the case p<0.1
c
      ee          = hsv(1)
      epsp(1:6) = hsv(2:7)
      pm          = hsv(8)
      bond        = hsv(9)
      workp       = hsv(10)
      dBr         = hsv(11)
c
      call stress_invariant(Sig, dI1, dI2, dI3, p, q)
      Call strain_invariant(deps,depsv,depsd)
      Call strain_invariant(epsp,epspv,epspd)
c
      ee=ee-depsv*(1.+e0)
      if (ee .lt. 1.d-7) ee=1.d-7
         hsv(1) = ee
      if (dParBr .lt. 1.d-7) then
         dBr =0.                   ! not use breakage
```

```
          else
             dBr=workp/(dParBr+workp)  ! Br=wp/(par+wp)
          end if
          hsv(11) = dBr
c
c           calculate critical void ratio due to breakage
          ecu   = ecuf+(ec0-ecuf)*exp(-1.*Parecu*dBr)
c
          select case(CSL)

          case(1)

          if (p .le. 1.d-2) then
             ecrit= ecu-dlamda*log(0.01/pat)
          else
             ecrit= ecu-dlamda*log(p/pat)
          end if

          case(2)

          if (p .le. 1.d-2) then
             ecrit=ecu-dlamda*(0.01/pat)**dksi
          else
             ecrit=ecu-dlamda*(p/pat)**dksi
          end if

          case(3)

          ecu   = ecuf+(ecu0-ecuf)*exp(-1.*Parecu*dBr)
          if (p .le. 1.d-2) then
             ecrit=ecu+(ec0-ecu0)*exp(-1.*(0.01/pat/dlamda)**dksi)
          else
             ecrit=ecu+(ec0-ecu0)*exp(-1.*(p/pat/dlamda)**dksi)
          end if

          end select
c*******************************************************************************

       ! calculate stress invariants and lode angle effect
        Call Lode_TS (Sig,cm,dMratio)
c*******************************************************************************
c       save the critical void ration into state varibales Hsv
        hsv(19) = ecrit
        if (ecrit .lt. 1.d-7) ecrit=1.d-7

        select case(psi_selection)

        case(1)

        psi=ecrit/ee
        if (psi .lt. 1.d-7) psi=1.d-7
        fi_p = atan(tan(fi)*psi**dnp)
        fi_d = atan(tan(fi)*(1./psi)**dnd)
        dM   = 6.*sin(fi)/(3.-sin(fi))*dMratio
        dMp  = 6.*sin(fi_p)/(3.-sin(fi_p))*dMratio
        dMpt = 6.*sin(fi_d)/(3.-sin(fi_d))*dMratio

        case(2)

        psi=ee-ecrit
        dM = 6.*sin(fi)/(3.-sin(fi))*dMratio
        dMp = dM*exp(-dnp*psi)
        dMpt = dM*exp(dnd*psi)
```

```
        case default

        psi=ecrit/ee
        if (psi .lt. 1.d-7) psi=1.d-7
        fi_p = atan(tan(fi)*psi**dnp)
        fi_d = atan(tan(fi)*(1./psi)**dnd)
        dM   = 6.*sin(fi)/(3.-sin(fi))*dMratio
        dMp  = 6.*sin(fi_p)/(3.-sin(fi_p))*dMratio
        dMpt = 6.*sin(fi_d)/(3.-sin(fi_d))*dMratio

        end select
c
        hsv(20)=psi

        converged =.false.          ! do at least one iteration for each increment including
        inewt     = 0

!------------------------------------------------------------------------------------
!       New ton-Raphson iteration for plasticity start
        ! loop until convergence is reached
        Do While (.not.converged)
        inewt = inewt + 1
        converged=.true. ! gets false again during iteration when plasticity occurs

        ! solve plasticity
        epsp(1:6) = hsv(2:7)
        pm        = hsv(8)
        bond      = hsv(9)
        Call strain_invariant(epsp,epspv,epspd)


          ! set stress strain increment
           deps = deps - depsp
c          write(*,21) depsp
c          write(*,21) deps
           Call MatVec( CC, 6, deps, 6, dsig)
                sig = sig0+dsig
           call stress_invariant(Sig, dI1, dI2, dI3, p, q)
c         if (abs(p+bond) .le. 1.d-7) bond=bond+0.001
          if ((p+bond) .le. 1.d-7) bond=bond+0.001
          f1= q/(p+bond)-dMp*epspd/(akp+epspd)
          f2= 1./dn*(q/(p+bond)/dM)**(dn+1.)*(p+bond)+p-pm
c         write(*,*) f1
c         write(*,*) f2
        !-------------------------------------------------------------------------------

 !-------------------------------------------------------------------------------------
         If (f1 .gt. 1.d-7) Then    !======== for the case of shear, compression is included

         df1dp       = -1.*q/(p+bond)**2
         df1dq       = 1./(p+bond)
         df1dbond    = -1.*q/(p+bond)**2
         df1depspd = -1.*dMp/(akp+epspd)+dMp*epspd/(akp+epspd)**2
         dbonddepspd=-dw*bond
         depspdddepspd=1.

         dG1dp = ad*(dMpt-q/(p+bond))
         dG1dq = 1.

         call dpqdsig(Sig, dpdsig, dqdsig)
         dF1dsig = df1dp*dpdsig + df1dq*dqdsig
         dG1dsig = dG1dp*dpdsig + dG1dq*dqdsig

         Call MatVec( CC, 6, dG1dsig, 6, dum)
         dd = -dF1dsig(1)*dum(1)-dF1dsig(2)*dum(2)-dF1dsig(3)*dum(3)
        &  -2.*(dF1dsig(4)*dum(4)+dF1dsig(5)*dum(5)+dF1dsig(6)*dum(6))
         hh = df1dbond*dbonddepspd*dG1dq + df1depspd*depspdddepspd*dG1dq

         dF1dLam = dd + hh
         XLam1   = -f1 / dF1dLam
```

```
c***********************************************************************
        if(isnan(XLam1))then
          write(*,*) 'XLam1',XLam1
          XLam1=0.
        end if
c**********************************************************************c
       depsp    = XLam1 * dG1dsig
c
        if(isnan(depsp(1)))then
          depsp=0.
        end if
c
       epsp     = epsp + depsp
       Call strain_invariant(depsp,depspv,depspd)

       hsv(2:7) = epsp(1:6)
       dkappa=(1.+e0)*(p+bond)/(dK0*pat*(2.97-ee)**2/(1.+ee)
     &       *(abs(p+bond)/pat)**dnn)
c***********************************************************************

      if (depspv .lt. 0.) depspv=0.
         dr1=dr0*ee
      if (dr1 .le. dkappa) dr1 = 1.01*dkappa
      hsv(8)    = pm + (1.+e0)*pm*depspv/(dr1-dkappa)
      hsv(9)    = bond - dw*bond*depspd

        !set convergence criterion
        converged = .False.

      End If                                  !======== end for the case of shear

!------------------------------------------------------------------------------------
      If ((f1 .lt. 1.d-7) .and. (f2 .gt. 1.d-7)) Then  !======== for the case of normal compression
!   f2= 1./dn*(q/(p+bond)/dM)**(dn+1.)*(p+bond)+p-pm
c
       df2dp = 1.-(q/((p+bond)*dM))**(dn+1.)
       df2dq = (q/((p+bond)*dM))**dn*(dn+1.)/dn/dM
       df2dbond = df2dp-1.
       df2dpm = -1.
       dbonddepspd=-dw*bond
       dkappa=(1.+e0)*(p+bond)/(dK0*pat*(2.97-ee)**2/(1.+ee)
     &       *(abs(p+bond)/pat)**dnn)
       dr1=dr0*ee
      if (dr1 .le. dkappa) dr1 = 1.01*dkappa
      dpmdepspv = (1.+e0)*pm/(dr1-dkappa)

!  associated flow rule
       Call dpqdsig(Sig, dpdsig, dqdsig)
       dF2dsig = df2dp*dpdsig + df2dq*dqdsig
       dG2dsig = dF2dsig

       Call MatVec( CC, 6, dG2dsig, 6, dum)
        dd = -dF2dsig(1)*dum(1) - dF2dsig(2)*dum(2) - dF2dsig(3)*dum(3)
     &  -2.*(dF2dsig(4)*dum(4) + dF2dsig(5)*dum(5) + dF2dsig(6)*dum(6))
        hh = df2dbond*dbonddepspd*df2dq + df2dpm*dpmdepspv*df2dp
        dF2dLam = dd + hh
        XLam2 = -f2 / dF2dLam
        depsp = XLam2 * dG2dsig
c
        if(isnan(depsp(1)))then
          depsp=0.
        end if
c
       epsp  = epsp + depsp
       Call strain_invariant(depsp,depspv,depspd)

       hsv(2:7) = epsp(1:6)
       hsv(8)    = pm + (1.+e0)*pm*depspv/(dr1-dkappa)
       hsv(9) = bond-dw*bond*depspd

        !set convergence criterion
        converged = .False.
       End If                                 !======== end for the case of normal compression

!------------------------------------------------------------------------------------

       End Do ! While not converged end of plasticity

!============================================================ end of plastic iteration
```

```
101    depsp = deps0-deps
          if (inewt .gt. 1000) then
             write(*,*) 'goto101'
             write(*,21) depsp
          end if

       Call strain_invariant(depsp,depspv,depspd)
       dpeeq = depspd
       hsv(18)= hsv(18)+dpeeq


       hsv(12:17) = depsp(1:6)
       dworkp=sig(1)*depsp(1)+sig(2)*depsp(2)+sig(3)*depsp(3)
      &    +2.*(sig(4)*depsp(4)+sig(5)*depsp(5)+sig(6)*depsp(6))
       hsv(10) = workp + dworkp
       p=(sig(1)+sig(2)+sig(3))/3.0
c************************************************************************
        if (p .lt. -1.*bond) then
            sij11=sig(1)-p
            sij22=sig(2)-p
            sij33=sig(3)-p
            p = -1.*bond+0.001
            sig(1)=sij11+p
            sig(2)=sij22+p
            sig(3)=sij33+p
       end if
c
       hsv(21)=p
c-------------------------------------------------------------------------
       sig_new=sig

       Return
       End

!***********************************************************************************
      Subroutine stress_invariant(Sig, dI1, dI2, dI3, p, q)
      INCLUDE 'ABA_PARAM.INC'
c     Implicit none
      character*80 cmname
      Double Precision Sig(6)

      dI1 = sig(1)+sig(2)+sig(3)
      dI2 = sig(1)*sig(2)+sig(2)*sig(3)+sig(1)*sig(3)
     &    -sig(4)**2-sig(5)**2-sig(6)**2
      dI3 = sig(1)*sig(2)*sig(3)+2.*sig(4)*sig(5)*sig(6)
     &      -sig(1)*sig(4)**2-sig(2)*sig(5)**2-sig(3)*sig(6)**2

      p  = dI1/3.
      s1 = Sig(1) - p
      s2 = Sig(2) - p
      s3 = Sig(3) - p
      s4 = Sig(4)
      s5 = Sig(5)
      s6 = Sig(6)
      dJ2= 0.5*(s1*s1+s2*s2+s3*s3+2.*s4*s4+2.*s5*s5+2.*s6*s6)
      q  = sqrt(3.*dJ2)

      Return
      End
!-----------------------------------------------------------------------------------

      Subroutine strain_invariant(eps, eps_v, eps_d)
      INCLUDE 'ABA_PARAM.INC'
c     Implicit none
      character*80 cmname
      Double Precision eps(6)

      eps_v = eps(1) + eps(2) + eps(3)
      e1 = eps(1) - eps_v / 3.
      e2 = eps(2) - eps_v / 3.
      e3 = eps(3) - eps_v / 3.
      e4 = eps(4)
      e5 = eps(5)
      e6 = eps(6)
      eps_d =Sqrt(2./3.*(e1*e1+e2*e2+e3*e3+2.*e4*e4+2.*e5*e5+2.*e6*e6))

      Return
      End

!-----------------------------------------------------------------------
```

```
      Subroutine multip_vec(C, m, n, A, B)
      INCLUDE 'ABA_PARAM.INC'
      Character*80 cmname
      Double precision A(m,n),B(n),C(m)
c
      do 10 i = 1, m
         C(i) = 0.
         do 20 j = 1, n
          C(i) = C(i) + A(i, j) * B(j)
20       continue    !  Next j
10     continue    !  Next i

       Return
      End
!-----------------------------------------------------------------------------------

C***********************************************************************
      Subroutine MatVec(xMat,IM,Vec,N,VecR)
C***********************************************************************
C
C     Calculate VecR = xMat*Vec
C
C I   xMat  : (Square) Matrix (IM,*)
C I   Vec   : Vector
C I   N     : Number of rows/colums
C O   VecR  : Resulting vector
C
C***********************************************************************

C***********************************************************************************
      INCLUDE 'ABA_PARAM.INC'
c     Implicit none
      Character*80 cmname
      Double Precision xMat(IM,*),Vec(*),VecR(*)
C***********************************************************************************
      Do I=1,N
        X=0
        Do J=1,N
          X=X+xMat(I,J)*Vec(J)
        End Do
        VecR(I)=X
      End Do
      Return
      End
!-----------------------------------------------------------------------------
      Subroutine ZeroM(m, n, A)
      ! make a matrix zero
       INCLUDE 'ABA_PARAM.INC'
       Character*80 cmname
       Double Precision A(m,n)

      do 10 i = 1, m
      do 20 j = 1, n
        A(i, j) = 0.
```

```
20     continue    !  Next j
10    continue    !  Next i

      Return
      End
!-----------------------------------------------------------------------------------------
      Subroutine dpqdsig(Sig, dpdsig, dqdsig)
      INCLUDE 'ABA_PARAM.INC'
c     Implicit none
      Character*80 cmname
      Double Precision Sig(6), dpdsig(6), dqdsig(6)

      dpdsig(1) = 1. / 3.
      dpdsig(2) = 1. / 3.
      dpdsig(3) = 1. / 3.
      dpdsig(4) = 0.
      dpdsig(5) = 0.
      dpdsig(6) = 0.

     p = (Sig(1) + Sig(2) + Sig(3)) / 3.
     s1 = Sig(1) - p
     s2 = Sig(2) - p
     s3 = Sig(3) - p
     s4 = Sig(4)
     s5 = Sig(5)
     s6 = Sig(6)
     dJ2 = 0.5*(s1*s1+s2*s2+s3*s3+2.*s4*s4+2.*s5*s5+2.*s6*s6)
     q = Sqrt(3. * dJ2)

     If (q .gt. 1.d-7) Then
       dqdsig(1) = 3. * s1 / (2. * q)
       dqdsig(2) = 3. * s2 / (2. * q)
       dqdsig(3) = 3. * s3 / (2. * q)
       dqdsig(4) = 3. * s4 / (2. * q)
       dqdsig(5) = 3. * s5 / (2. * q)
       dqdsig(6) = 3. * s6 / (2. * q)
      Else
       dqdsig = 1.5
     End If

     Return
     End
```

```fortran
c-----------------------------------------------------------------------------
!      SUBROUTINE MATRIXDE(Sig,cm,hsv,CC)
c=============================================================================
c                      Calculation Elastic Matrix D for SIG=D*EPS
c=============================================================================
      IMPLICIT DOUBLE PRECISION(A-H,O-Z)
      DIMENSION SIG(6),cm(*),Hsv(*),CC(6,6)

        dK0     = cm(2)    ! dK=dK0*pat*(2.97-ee)**2/(1.+ee)*((p_module+bond)/pat)**dnn
        dG0     = cm(3)    ! dG=bG0*pat*(2.97-ee)**2/(1.+ee)*((p_module+bond)/pat)**dnn
        dnn     = cm(4)    ! defaut value: 0.6667
        p_module = (Sig(1) + Sig(2) + Sig(3)) / 3.
        If (p_module .lt. 0.1)  p_module = 0.1
        pat = 101.325
        ee  = hsv(1)
        bond= hsv(9)
!        calculate bulk and shear modulus
        if (ee .lt. 1.d-7) ee=1.d-7
        gg=dG0*pat*(2.97-ee)**2/(1.+ee)*(abs(p_module+bond)/pat)**dnn
        bkk=dK0*pat*(2.97-ee)**2/(1.+ee)*(abs(p_module+bond)/pat)**dnn
        emodule = 9.*bkk*gg/(3.*bkk+gg)    ! need to be modified for anisotropic elasticity
        pois = (3.*bkk-2.*gg)/(6.*bkk+2.*gg)
        if (pois .gt. 0.49) pois = 0.49
        Call ZeroM(6, 6, CC)          ! initial Matrix

        F1 = emodule/(1.+pois)/(1.-2.*pois)
        CC(1,1) = F1*(1.-pois)
        CC(2,2) = F1*(1.-pois)
        CC(3,3) = F1*(1.-pois)
        CC(4,4) = F1*(1.-2.*pois)
        CC(5,5) = F1*(1.-2.*pois)
        CC(6,6) = F1*(1.-2.*pois)
        CC(1,2) = F1*pois    ! Engineering shear strain is used, if use normal shear strain keep constant
        CC(1,3) = F1*pois
        CC(2,3) = F1*pois
        CC(2,1) = CC(1,2)
        CC(3,1) = CC(1,3)
        CC(3,2) = CC(2,3)

        Return
        end
!-----------------------------------------------------------------------------

c-----------------------------------------------------------------------------
      SUBROUTINE MATRIXDE_SS(Sig,cm,hsv,CC)
c=============================================================================
c
c                      Calculation Elastic Matrix D for SIG=D*EPS
c
c=============================================================================
      IMPLICIT DOUBLE PRECISION(A-H,O-Z)
      DIMENSION SIG(6),cm(*),Hsv(*),CC(6,6)

        dK0     = cm(2)    ! dK=dK0*pat*(2.97-ee)**2/(1.+ee)*((p_module+bond)/pat)**dnn
        dG0     = cm(3)    ! dG=bG0*pat*(2.97-ee)**2/(1.+ee)*((p_module+bond)/pat)**dnn
        dnn     = cm(4)    ! defaut value: 0.6667
        eps70   = cm(24)   ! parameter of small strain stiffness
        if(eps70 .eq. 0.) eps70=0.1

        p_module = (Sig(1) + Sig(2) + Sig(3)) / 3.
        If (p_module .lt. 0.1)  p_module = 0.1
        pat = 101.325
        ee  = hsv(1)
        bond= hsv(9)

c-------For small strain stiffness
         D_reverse = hsv(49)
         eps_d_m   = hsv(22)
      if (eps_d_m .le. 1.d-3) then
          dLL = 1.
```

```
      if(D_reverse .gt. 0.) dLL = 2.
         dRatio_SS = dLL*((1.+0.003/eps70/7.)/(1.+eps_d_m*3./eps70/7.))**2
      else
         dRatio_SS = 1.
      end if
      dG0_SS = dRatio_SS*dG0
      dK0_SS = dRatio_SS*dK0
c===================================end if small strain stiffness
c           calculate bulk and shear modulus
        if (ee .lt. 1.d-7) ee=1.d-7
       gg=dG0_SS*pat*(2.97-ee)**2/(1.+ee)*(abs(p_module+bond)/pat)**dnn
       bkk=dK0_SS*pat*(2.97-ee)**2/(1.+ee)*(abs(p_module+bond)/pat)**dnn
c
        emodule = 9.*bkk*gg/(3.*bkk+gg)    ! need to be modified for anisotropic elasticity
        pois = (3.*bkk-2.*gg)/(6.*bkk+2.*gg)
        if (pois .gt. 0.49) pois = 0.49
        Call ZeroM(6, 6, CC)          ! initial Matrix

        F1 = emodule/(1.+pois)/(1.-2.*pois)
        CC(1,1) = F1*(1.-pois)
        CC(2,2) = F1*(1.-pois)
        CC(3,3) = F1*(1.-pois)
        CC(4,4) = F1*(1.-2.*pois)
        CC(5,5) = F1*(1.-2.*pois)
        CC(6,6) = F1*(1.-2.*pois)
        CC(1,2) = F1*pois   ! Engineering shear strain is used, if use normal shear strain keep constant
        CC(1,3) = F1*pois
        CC(2,3) = F1*pois
        CC(2,1) = CC(1,2)
        CC(3,1) = CC(1,3)
        CC(3,2) = CC(2,3)

         Return
         end
!------------------------------------------------------------------------------------------

!------------------------------------------------------------------------------------------
      SUBROUTINE smallstrain_sim (hsv,Eps0)
!C=========================================================================
!c
!c    reverse loading judgement based on previous and current dstrain (deps0 and deps)
!c    current version can only be used for strain or displacement control
!c
!C=========================================================================
      INCLUDE 'ABA_PARAM.INC'
      Character*80 cmname
      Double Precision hsv(*)
      Double Precision Eps0(6),Eps_temp(6)

      Eps_temp=-1.*Eps0
      call strain_invariant(Eps_temp, eps_v, eps_d)
      eps_d_m = sqrt(eps_d*eps_d+eps_v*eps_v)
      hsv(22) = eps_d_m
      hsv(49) = 0.

      Return
      END  SUBROUTINE smallstrain_sim
!------------------------------------------------------------------------------------------

         Subroutine Lode_TS (Sig,cm,dMratio)
   !=========================================================================
   !                      calculate the crrection of dM by  TS of Yao
   !                             [Yao, Hou & Zhou (2009)]
   !                      - Geotechnique 59, No.5, 451-469
   !=========================================================================
         INCLUDE 'ABA_PARAM.INC'
         Character*80 cmname
         double precision Sig(6),cm(50)

          fi0  = cm(9)
          fi   = fi0*3.141592653589793239/180.   ! change the unity
          dMc = 6.*sin(fi)/(3.-sin(fi))
          ttt = 3./(3.+dMc)   ! based on MC or SMP
         dI1 = sig(1)+sig(2)+sig(3)

         dI2 = sig(1)*sig(2)+sig(2)*sig(3)+sig(1)*sig(3)
        &        -sig(4)**2-sig(5)**2-sig(6)**2
```

```
      dI2 = sig(1)*sig(2)+sig(2)*sig(3)+sig(1)*sig(3)
     &       -sig(4)**2-sig(5)**2-sig(6)**2

      dI3 = sig(1)*sig(2)*sig(3)+2.*sig(4)*sig(5)*sig(6)
     &        -sig(1)*sig(4)**2-sig(2)*sig(5)**2-sig(3)*sig(6)**2

      p  = dI1/3.
      s1 = Sig(1) - p
      s2 = Sig(2) - p
      s3 = Sig(3) - p
      s4 = Sig(4)
      s5 = Sig(5)
      s6 = Sig(6)
      dJ2= 0.5*(s1*s1+s2*s2+s3*s3+2.*s4*s4+2.*s5*s5+2.*s6*s6)
      q  = sqrt(3.*dJ2)

      term0 = abs(dI1*dI2-9.*dI3)
      if (term0 .lt. 1.) then
          term1 =0.
      else
         term1 = abs(dI1*dI2-dI3)/term0
      end if
      term2 = abs(3.*sqrt(term1)-1.)
      if (term2 .lt. 1.) then     ! revised on 19/07/2016
         yqc=0.
      else
         yqc=2.*abs(dI1)/term2                 ! SMP criteria
      end if
        If (yqc .lt. 1.)  then
             dMratio=1.
         else
              dMratio = q/yqc
         if (q .lt. 1.) dMratio=1. ! for numerical calculation
      end if
!--------------------modified the SMP criteria for modifying the lode angle--------------------------
      if (dMratio .ge. 1.)  dMratio = 1.
      if (dMratio .lt. ttt)  dMratio = ttt
!=---------------------------------------------------------------------------------------------------
      Return
      End Subroutine Lode_TS
!======================== end of subroutines for the soil model Iso sand ================
```

Appendix G
Matlab Code for Using Bayesian Nonparametric General Regression Method

References:

Zhao LS, Zhou WH, Su LJ, Garg A, Yuen KV (2019) Selection of physical and chemical properties of natural fibers for predicting soil reinforcement. J Mater Civ Eng 31(10):04019212.

Yuen KV, Ortiz GA (2016) Bayesian nonparametric general regression. Int J Uncertain Quan 6(3):195–213.

```
%% Close open windows, clear memory and screen
close all; clear; clc;

%% Data preparation
% Input data (design matrix)
y1  = xlsread('data','sheet1','column 1');
y2  = xlsread('data','sheet1','column 2');
y3  = xlsread('data','sheet1','column 3');
%... ...

N   = size(y1,1);                    % Size dataset
X   = [y1 y2 y3 ... ...];
d   = size(X,2);                     % Number of design variables

% Output data (Real data)
y_n = xlsread('data','sheet1','d2:d102');

%% Normalize data (mean=0, std=1)
% Normalize design matrix
mean_x = mean(X);
std_x  = std(X);
X_norm = zeros(N,d);
    for i = 1:d
      X_norm(:,i) = (X(:,i) - mean_x(i))./std_x(i);
    end
```


W. Zhou et al., *Practice of Bayesian Probability Theory in Geotechnical Engineering*,
https://doi.org/10.1007/978-981-15-9105-1

```
% Normalize measurements
mean_y    = mean(y_n);
std_y     = std(y_n);
Y_n_norm = (y_n - mean_y)./std_y;

%% Define training and testing datasets
N_tr = floor(0.7*N);             % Size training dataset (70% of the data)
N_te = N - N_tr;                 % Size testing dataset (30% of the data)

idx_te           = (1:N)';
idx_tr           = (sort(randperm(N, N_tr)))';     % Indices training set
idx_te(idx_tr) = [];                               % Indices testing set

% Training set
X_tr          = X(idx_tr,:);
X_norm_tr     = X_norm(idx_tr,:);
y_tr          = y_n(idx_tr);
y_n_tr        = y_n(idx_tr);
Y_n_norm_tr = Y_n_norm(idx_tr);

% Testing set
X_te          = X(idx_te,:);
X_norm_te     = X_norm(idx_te,:);
y_te          = y_n(idx_te);

y_n_te       = y_n(idx_te);
Y_n_norm_te = Y_n_norm(idx_te);

%% Model matrix
model       = dec2bin(0:2^(d)-1)-'0';
model(1,:) = [];
N_mod       = size(model, 1);

%% Training stage %%

%% Parameters prior PDF
l_b = 0;                              % Lower boundary
u_b = 50;                             % Upper boundary

%% Pre-allocate space in memory
theta_hat    = zeros(2,N_mod);        % Optimal parameters for each model class
Y_n_norm_gr = zeros(N_tr,N_mod);      % Regression of Y computed via GR for each model
class (Normalized values)
Y_hat        = zeros(N_tr,N_mod);     % Regression of Y computed via GR for each model
class
MSE          = zeros(N_mod, 1);       % Mean-square error
like_k       = zeros(N_mod, 1);       % Likelihood function value for each model class
P_Y_X_C      = zeros(N_mod, 1);       % Evidence of each model class
```

```
%% Options minimization routine
options = optimset('Display','off','TolX',1e-4,'TolFun',1e-4);

%% Print information on screen
fprintf('Model    Theta_1     Theta_2     P(Y|theta,X,C)    MSE         p(Y|X,C)\n');
fprintf('-------------------------------------------------------------------------------\n')
;
tic;

%% Main code
for k = 1:N_mod
  % Define handle function used to compute optimal smoothing parameter
  % (General regression)
  neg_log_pos = @(theta) min_neg_log_pos(l_b, u_b, ...
                                          X_norm_tr(:,model(k,:)==1), ...
                                                         Y_n_norm_tr, theta);

  % Compute optimal value of variance of the kernel used in GR using
  % minimization routine
  while theta_hat(1,k) <= 0
      theta_hat(1,k)=fminsearch(neg_log_pos, 0.1*rand, options);
  end

  % Compute optimal value of prediction-error variance
  [Y_tmp, exp_X] = gr_cond(X_norm_tr(:,model(k,:)==1), Y_n_norm_tr, theta_hat(1,k));
  MSE_tmp        = (Y_n_norm_tr - Y_tmp)'*diag(exp_X)*(Y_n_norm_tr - Y_tmp);
  theta_hat(2,k) = MSE_tmp / N_tr;

    % Compute likelihood function value
    like_k(k) = exp(log_like_fun(Y_n_norm_tr, Y_tmp, theta_hat(2,k), exp_X));
  %   like_k(k) = (1/((2*pi*theta_hat(2,k))^(0.5*N_tr))) * ...
  %                                          exp(-MSE_tmp/(2*theta_hat(2,k)));

    % Compute evidence of model class M_k
    P_Y_X_C(k) = evid_class(l_b, u_b, X_norm_tr(:,model(k,:)==1), ...
                                                 Y_n_norm_tr, theta_hat(:,k));

    % Compute regression of Y in X using the whole dataset
    Y_n_norm_gr(:,k) = gr_equation(X_norm_tr(:,model(k,:)==1), Y_n_norm_tr, ...
                                   X_norm_tr(:,model(k,:)==1), Y_n_norm_tr, ...
                                                        theta_hat(1,k), 0);

    % Re-scale regression of Y in X
    Y_hat(:,k) = (Y_n_norm_gr(:,k).*std_y) + mean_y;

    % Compute Mean-squared error
    MSE(k) = (y_tr - Y_hat(:,k))'*(y_tr - Y_hat(:,k))/N_tr;

    % Update information on screen

  fprintf('%2d       %2.4f       %2.4f      %2.4e        %2.4f      %2.4e\n', ...
              k, theta_hat(1,k), theta_hat(2,k), like_k(k), MSE(k), P_Y_X_C(k));
end
fprintf('-------------------------------------------------------------------------------\n')
;
fprintf('\n');
total_time = toc;

%% Compute plausilibty of each model class
P_C_Y_X = (P_Y_X_C ./ N_mod) ./ sum(P_Y_X_C ./ N_mod);

%% Identify best model
best_model = find(P_C_Y_X == max(P_C_Y_X));
sigma2_gr  = theta_hat(1,best_model);

%% Testing stage %%

%% Identify design variables best model
X_te_best = X_norm_te(:,model(best_model,:)==1);
X_tr_best = X_norm_tr(:,model(best_model,:)==1);

%% Compute regression of Y_te using optimal smoothing parameter computed in training
stage
Y_n_norm_te_gr = gr_equation(X_te_best, Y_n_norm_te, ...
                                      X_tr_best, Y_n_norm_tr, sigma2_gr, 1);
```

```
%% Re-scale output data
Y_te_gr = (Y_n_norm_te_gr .* std_y) + mean_y;

%% sub-function: min_neg_log_pos
function neg_log_pos = min_neg_log_pos(l_b, u_b, X, Y, theta)
%% neg_log_pos = MIN_NEG_LOG_POS(l_b, u_b, X, Y, theta)


%% Compute size dataset
N = size(X,1);

    if (theta < l_b) || (theta > u_b)
      %% Compute negative log-posterior PDF
      neg_log_pos = realmax;
    else
      %% Compute regression of Y using conditional PDFs and General Regression
      [Y_hat, exp_X] = gr_cond(X, Y, theta);

      %% Compute optimal value of the prediction-error variance
      R2       = (Y - Y_hat)'*diag(exp_X)*(Y - Y_hat);
      sigma2_2 = R2 / N;

      %% Compute negative log-posterior PDF
      neg_log_pos = 2*log(u_b-l_b) - log_like_fun(Y, Y_hat, sigma2_2, exp_X);
    end

%% In case of divergence, assign large positive value
    if isnan(neg_log_pos) || isinf(neg_log_pos)
      neg_log_pos = realmax;
    end

  end

  %% sub-function: gr_cond

  function [Y_gr, exp_X] = gr_cond(X_j, Y_j, theta_j)
  %% [Y_gr, exp_X] = GR_COND(X_j, Y_j, theta_j)
  %
  % This function computes the regression of Y using conditional PDFs.

  %% Beginning

  %% Compute size dataset
  N_j = size(X_j, 1);

  %% Pre-allocate space in memory
  Y_gr     = zeros(N_j, 1);
```

```
exp_X    = zeros(N_j,1);
exp_X(1) = 1;

%% Assign first value
Y_gr(1) = Y_j(1);

%% Compute Y_hat using 'bisection' method
    for jj = 2:N_j
      %% Compute mean distance between point being estimated and points in the
      conditional part
      tmppp     = gsubtract(X_j(1:jj-1,:),X_j(jj,:));
      dis_X     = mean(sum(tmppp.^2,2));
      exp_X(jj) = sum(exp(-2*sum(tmppp.^2,2)));

      %% Use GR to compute value of Y
      a = 0;
      b = 0;
          for m = 1:jj-1
            tmp = exp(-((X_j(jj,:) - X_j(m,:))*(X_j(jj,:) -
            X_j(m,:))')/(2*theta_j*dis_X+realmin));
            a   = a + Y_j(m)*tmp;
            b   = b + tmp;
          end

      Y_gr(jj) = a/(b+realmin);
    end

end
  %% sub-function: gr_equation

  function Y_hat = gr_equation(X, Y, X_k, Y_k, theta, type)
  %% Y_hat = GR_EQUATION(X, Y, X_k, Y_k, theta, type)
  % This function computes the regression of Y in X using the sample values
  % (X_k, Y_k).

  %% Beginning

  %% Compute size datasets
  N   = size(Y,1);
  N_k = size(X_k,1);

  %% Compute distances
  dis = zeros(N, 1);

      for i = 1:N
            for j = 1:N_k
              dis(i) = dis(i) + (X(i,:) - X_k(j,:)) * (X(i,:) - X_k(j,:))';
            end
      end

  %% Pre-allocate space in memory
  Y_hat = zeros(N,1);
```

```
        %% Use GR equation to compute Y_hat
            for j = 1:N
              a = 0;
              b = 0;
              sigma_1 = theta*dis(j)/N_k;
              for k = 1:N_k
                if type == 0
                  if k ~= j
                    D_2 = (X(j,:) - X_k(k,:)) * (X(j,:) - X_k(k,:))';
                    a = a + Y_k(k)*exp(-D_2/(2*sigma_1));
                    b = b + exp(-D_2/(2*sigma_1));
                  end
                else
                  D_2 = (X(j,:) - X_k(k,:)) * (X(j,:) - X_k(k,:))';
                  a = a + Y_k(k)*exp(-D_2./(2*sigma_1));
                  b = b + exp(-D_2./(2*sigma_1));
                end
              end
              Y_hat(j) = a/(b+realmin);

    end
end

%% sub-function:  log_like_fun

function log_like = log_like_fun(y, y_hat, v2, dis_X)
%% log_like = log_like_fun(y, y_hat, v2, dis_X)
%
% This function is used to compute the log-likelihood function.


%% Beginning

%% Compute size dataset
N = length(y);

%% Initialize value of log_like. Take into account that p(y_1 | y_0, v1, v2, X, C) = 1
log_like = 0;

%% Compute log_like for each conditional PDF and add the values
    for i = 1:N
      %% Acummulate value of logarithm of likelihood function
      log_like = log_like - 0.5*log(2*pi*(v2/dis_X(i))) - ...
                                           (((y(i) - y_hat(i))^2)/(2*(v2/dis_X(i))));

    end

end
 %% sub-function: evid_class

 function P_D_M = evid_class(l_b, u_b, X, Y, Theta)
 %% P_D_M = EVID_CLASS(l_b, u_b, X, Y, Theta)
 % This function computes the evidence of a model class defined by X using
 % asymptotic expansion.

 %% Handle function for negative log-posterior PDF
 target = @(theta) -log(sigma2_mar_pdf(l_b, u_b, X, Y, theta));

 %% Compute Hessian matrix
 H = hessian(target, Theta(1));

 %% Compute evidence of model class M_k
 P_D_M = exp(-1*target(Theta(1)))*sqrt(2*pi/det(H));

 end

        %% Marginal PDF of variance of the kernel used in the General Regression approach
        function p_sigma = sigma2_mar_pdf(l, u, X_j, Y_j, t)
        %% p_sigma = SIGMA2_MAR_PDF(l, u, X_j, Y_j, t)
        % This function computes the marginal PDF of the variance of the kernel used in
        % the General Regression approach.
```

```
    %% Beginning
    N_j = size(X_j,1);                % Size dataset

    %% Compute Y_hat
    [Y_hat, exp_X] = gr_cond(X_j, Y_j, t);

    %% Compute marginal PDF
    C       = (2/((u-1)^2))*(1/(pi^(N_j/2)))*gamma(0.5*N_j + 1);
    l_p_s   = log(C) + sum(0.5*log(exp_X)) - ...
                     (0.5*N_j + 1)*log(((Y_j-Y_hat)'*diag(exp_X)*(Y_j-Y_hat)));
    p_sigma = exp(l_p_s);
    end

  %% Hessian matrix using finite difference method
    function Hess = hessian(fun,x)
    % This function computes the numeric Hessian matrix of a multivariate
    % function. The Hessian matrix is computed by finite difference method.

    %% Beginning
%% Size of Hessian matrix
x   = x(:);
dim = length(x);

%% Create perturbation vectors:
delta_Theta = diag(0.01.*abs(x));

%% Evaluate the function at x
fx = feval(fun,x);diag(0.01.*abs(x))

%% Compute Hessian by central difference
Hess = zeros(dim);
for m = 1:dim
  %% Diagonal elements
  Hess(m,m) = (feval(fun, x + delta_Theta(:,m)) - ...
                            2*fx + feval(fun, x - delta_Theta(:,m))) ./ ...
                                         (delta_Theta(:,m)'*delta_Theta(:,m));

  if Hess(m,m) <= 0
    Hess(m,m) = eps;
  end

  %% Off-diagonal elements, use central difference method
  for n = m+1:dim
    delta_1   = delta_Theta(:,m);
    delta_2   = delta_Theta(:,n);
    Hess(m,n) = (feval(fun, x + delta_1 + delta_2) - ...
                     feval(fun, x + delta_1 - delta_2) - ...
                           feval(fun, x - delta_1 + delta_2) + ...
                                 feval(fun, x - delta_1 - delta_2)) ./ ...
                                     (4*(delta_Theta(m,m)*delta_Theta(n,n)));

    % Hess is a symmetric matrix
    Hess(n,m) = Hess(m,n);
  end
end
end
```

Appendix H
Matlab Code for Markov Chain Monte Carlo Sampling

References:

Qi XH, Zhou WH (2017) An efficient probabilistic back-analysis method for braced excavations using wall deflection data at multiple points. Comput Geotech 85:186–198.

```
%% MCMC_sampling
% meanPD: Mean of prior distribtion (normal)
% eg:
meanPD = [11 8.6 8.1 13 25.25...
          32 32.2 32.7 17.6 16 35 ...
          25 20 15 10 150 300];
COVPD  = [0.15,0.15,0.15,0.15,0.15,...
          0.1,0.1,0.1,0.1,0.1,0.1,...
          0.15,0.15,0.15,0.15,0.15,0.15];
nv        = length(meanPD); %number of variable to be updated
corrMatPD = diag(ones(nv,1)); %correlation matrix
covarPD   = corr2cov(meanPD.*COVPD,corrMatPD);%covariance
%transform lognormal paramter to normal paramter
[muPDNorm,sigmaPDNorm,COVarPDNorm]=lognorm2norm(meanPD,COVPD,corrMatPD);
%input statistics
meanMB    = 1.01; %mean of model bias factor,lognormal
COVMB     = 0.21; %standard deviation of model bias factor
%tansform to normal distribution
muMB      = log(meanMB./sqrt(1+COVMB.^2));
sigmaMB   = sqrt(log(1+COVMB.^2));
numS=2000;  % Number of sampling times
    for i = 1:500+numS   %the first 500 samples are discarded
        mu    = meanPD;
        covar = SF*covarPD;                         %covariance
        thetaStar = mvnrnd ( mu, covar );   %candidate
        xhq = thetaStar<0;
        if sum(xhq)>=1
            disp('error, negative value is generated')
        end
```


W. Zhou et al., *Practice of Bayesian Probability Theory in Geotechnical Engineering*,
https://doi.org/10.1007/978-981-15-9105-1

```
        OmaxDe = f(p(iSta,:),meanPD);
        NmaxDe = f(p(iSta,:),thetaStar);
        q1     =
        lognpdf(DefObs(iSta)/NmaxDe,muMB,sigmaMB)*mvnpdf(log(thetaStar),muPDNorm,COVar
        PDNorm);
        q0     =
        lognpdf(DefObs(iSta)/OmaxDe,muMB,sigmaMB)*mvnpdf(log(meanPD),muPDNorm,COVarPDN
        orm);
        r      = min(1,q1/q0);
        u      = rand;
        if u<r
            theta1 = thetaStar;
        else
            theta1 = meanPD;
        end
        if i>500
            chain(i-500,:) = theta1;
        end
        meanPD = theta1;
    end
%plot the MCMC samples
plot(1:1:size(chain,1),chain(:,1))
% Sub-function: lognorm2norm
    function [mu,sigma,COVar]=lognorm2norm(meanY,COVY,corrMatYY)
        %mean, COV, corrMat is parameter of lognormal distributed variable, y
        %corrMat=correlation matrix of lognormal vector, y
        %mu, COVar is mean and covariance of normal distributed variable, log(x)
        mu=log(meanY./sqrt(1+COVY.^2));
        sigma=sqrt(log(1+COVY.^2));
        %insure sigma is a column vector
            if size(sigma,2)>1
                sigma=sigma';
            end
        xhq=(exp(sigma.^2)-1) * (exp(sigma.^2)-1)';
        corrMatXX=log( 1+ corrMatYY.*sqrt(xhq) )./(sigma*sigma');
        COVar= corr2cov(sigma, corrMatXX);
    end
```

The manufacturer's authorised representative in the EU is Springer Nature Customer Service Centre GmbH, Europaplatz 3, 69115 Heidelberg, Germany. If you have any concerns regarding our products, please contact ProductSafety@springernature.com

Printed and bound by CPI Group (UK) Ltd, Croydon, CR0 4YY

30/09/2026

02225287-0012